More-than-Moore Devices and Integration for Semiconductors

Francesca Iacopi • Francis Balestra
Editors

More-than-Moore Devices and Integration for Semiconductors

Editors
Francesca Iacopi
Australian Research Council Centre of Excellence in Transformative Meta-Optical Systems (TMOS)
University of Technology Sydney
Ultimo, Australia

Francis Balestra
Univ. Grenoble Alpes, CNRS, Grenoble INP, IMEP-LAHC
Grenoble, France

ISBN 978-3-031-21612-1 ISBN 978-3-031-21610-7 (eBook)
https://doi.org/10.1007/978-3-031-21610-7

This Springer imprint is published by the registered company Springer Nature Switzerland AG
The registered company address is: Gewerbestrasse 11, 6330 Cham, Switzerland

Preface

Introduction

The term "More-than-Moore" appeared and was readily adopted by the semiconductor community since the early 2000s, when, in addition to the decades-long focused effort to scale down the footprint of logic and memory devices according to the well-known Moore's Law, an orthogonal trend in electronics miniaturisation had started to gain momentum.

In contrast to the aggressive pursuit of increasingly more powerful computing led by Moore's Law (More Moore), the More-than-Moore trend focuses on the combination of an increasing number of functionalities within a miniaturised system [1]. One of the main application drives for More-than-Moore had clearly been the pursuit of smart portable systems, with smartphones being one of the consumer applications spearheading this new trend. While smart portable devices in the late 1990s and early 2000s were still at a very early stage of development, mainly focused on mobile computing functionalities such as palm-sized PCs and using rudimentary connectivity and awkward user interfaces, the launch of the first Apple iPhone model – among the first wave of truly multifunctional portable smartphones, precursor to today's ubiquitous technologies – took place not too long afterwards, in 2007. The opportunity for such an extraordinary leap of smart systems with increasing complexity, number of functionalities and autonomy, all within an increasingly small form factor (More-than-Moore trend), has originated out of the simultaneous convergence of several key technologies, including the evolution of the following:

1. Mobile communications, particularly digital cellular networks, also thanks to the development of ICs for wireless communications, including power MOSFET and RF ICs.
2. Integrated power sources and energy-harvesting systems, key to ensuring autonomy.
3. Low-power ICs, specifically developed for mobile applications.
4. User interfaces such as advanced touchscreen technologies.

5. The availability of an increasing number of miniaturised functionalities, starting from the historically more advanced ones, such as digital CMOS cameras, MEMS technologies for sensors and actuators (loudspeakers, microphones, gyroscopes, etc.) and optoelectronics (LEDs, etc.)

More-than-Moore technologies have been taken into account in recent years in several International Roadmaps, in particular the NanoElectronics Roadmap for Europe (NEREID, [2]) and the IEEE International Roadmap for Devices and Systems (IRDS, [3]). While silicon technologies still play an absolutely central role, with many of the complementary functionalities to logic and memory still often using silicon as key material (MEMS, digital cameras, photonics, power, etc.), the emphasis of More-than-Moore is not so much on the miniaturisation of the single components as in the More Moore digital technologies, but rather on the miniaturisation and the improvement of the multifunctional system performance as a whole.

In addition, where alternative semiconductors to silicon are undoubtedly performing better than silicon, such as in optoelectronics, power and flexible electronics, new semiconductor technologies – from III–V to organic – are also being introduced and brought up fast to fabrication capabilities at scale either in combination with or complementary to silicon technologies [4–7]. This also means that we are seeing an increasingly more complex array of new semiconductors being brought onto a silicon fabrication platform, or different substrates to silicon being combined into the same system through advanced packaging.

An additional application drive for More-than-Moore technologies has undoubtedly been the rise of the Internet of Things, or Internet of Everything [8]. IoT usually refers to interconnected "smart" sensing nodes which could be serving any aspect from traffic to air and water quality, to healthcare, energy and manufacturing automation, as well as to a plethora of consumer applications.

Although the concept of IoT had already been discussed as early as the 1990s [9], it is not until approximately a decade later that this concept developed practically over a large scale, supported by the progress in the 1–5 technologies above, but also importantly supported by the advances in AI and Big Data analytics [10].

The schematic in Fig. 1 depicts a typical IoT node and its functionalities. The note would be generally built around a generic sensing unit, which represents a very broad range of potential sensing technologies (MEMS, optical, etc.) and specific applications. It would typically include an analog-to-digital converter (ADC) and potentially some embedded logic and memory, performing more or less extensive pre-processing operations on the acquired data. The processed data are then transmitted wirelessly to another node through the RF unit, which will include an appropriate antenna. Finally, another key component of an IoT node is the power unit and integrated power sources.

Depending on the projected consumption of the IoT node and the desired level of autonomy, the hardware needs for power can vary extensively, particularly in terms of power sources. Power considerations are in fact a key aspect for More-than-Moore technologies and an important theme of this book discussed in Chaps.

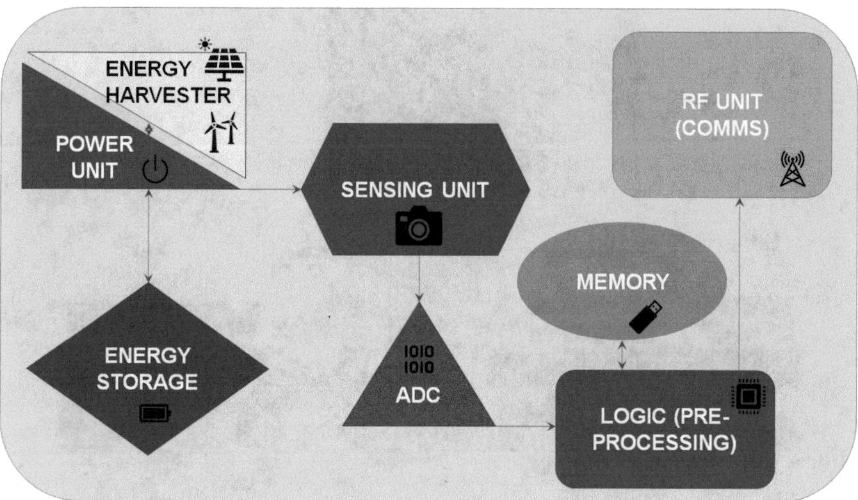

Fig. 1 Simplified diagram representing the typical units and components required in a smart node of the Internet of Things

1 and 2. Additional specific units may be required beyond what depicted in Fig. 1, depending on the sensing purpose of the specific IoT node.

IoT has generated a strong push for a further level of required miniaturisation and autonomy of multifunctional systems. Extreme miniaturisation allows for cutting-edge applications for example in healthcare – see ingestible sensors able to monitor the health of the gut with wireless transmission capabilities [11], only one of the endless possibilities offered by miniaturised, autonomous systems. Smart sensor nodes have also strongly driven the necessity for uninterrupted autonomous power, which in turn has also led to the development of efficient in-situ miniature storage such as microbatteries [12] and miniaturised supercapacitors [13], but also to ambient energy harvesting systems, as explained by Yeatman in Chap. 1. The importance of the further development and convergence of power harvesting, storage and management technologies for Moore-than-Moore systems cannot be overstated. The extraordinary recent advances made in power electronics using wide band-gap materials are therefore extremely welcome news, as explained by Zekentes et al. in Chap. 2.

Additional key enablers of the further progress of More than Moore are heterogeneous integration and advanced packaging. Over the last decades, as the complexity of integrating different semiconductors, functionalities and technology nodes within the same chip has reached higher complexity, the system-on-chip (SoC [14], Fig. 2) trend has been slowing down as heterogeneous integration through advanced packaging has been expanding and greatly diversifying to cater for the numerous different technology combinations (see also the IEEE Heterogeneous Integration Roadmap, HIR [15]). While it would be difficult to provide an exhausting description of the plethora of relevant advanced packaging technologies available, it is

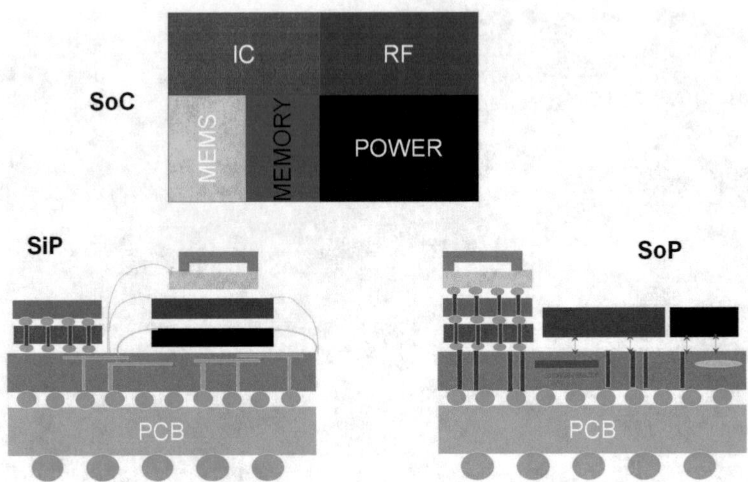

Fig. 2 Schematic depicting different system integration strategies, such as system-on-chip (SoC), and advanced packaging approaches such as system-in-package (SiP) and system-on-package (SoP)

useful to consider those enabled by the through-silicon-vias (TSV) technology, in particular the system-in-package (SiP) and system-on-package (SoP). They both aim at greatly reducing the footprint of a heterogeneous system by making use of silicon as either as a mostly passive interposer, connecting chips or "tiles" with different functionalities with much reduced pitch as compared to a PCB board (2.5D integration, [14]), or using active silicon chips to connect chips stacked in the vertical direction, also called 3D integration [14], respectively. These approaches are depicted in Fig. 2. The main aim would be to contain the whole system in a single package hence dramatically reducing the system footprint, although this may not always be possible because of specific packaging requirements for example for sensing chips, which may need to have access to the ambient to perform their functions [16] or other restrictions.

Further, there are several other trends in advanced packaging that are expected to deliver major contributions for More-than-Moore technologies. One is the "chiplets" or "dielets" route, which allows to combine two chips with an extreme pitch and is also currently providing a boost to logic chips [17]. Another important example, which represents a departure from a fully rigid silicon-based package, is the flexible hybrid electronics [18]. This approach is going to be particularly favourable when considering system integration for wearables, allowing for the most advanced logic to be combined in the same system with flexible technologies based on organic and or printed electronics, including 3D-printed components such as antennas and filters [19]. The rise of flexible electronics and the advanced combination of rigid and flexible are key aspects of More-than-Moore for smart sensing, as explained by Iniguez in Chap. 3.

The rapid evolution and versatility of advanced packaging enables electronics, and particularly More-than-Moore integration, to keep bringing more and more complementary technologies and functionalities, as they become available, under a single package. In particular, the advancement of miniaturised photonics is primed to advance particular areas of computing, including quantum, as well as interconnects and sensing [20, 21]. The advent of metamaterials and metasurfaces [22] has also opened the door to a long awaited, dramatic miniaturisation of optical components by using semiconductor materials and processing, which could lead to unprecedented functionalities that include the vastly underexploited THz gap in electromagnetic radiation, as explained by Atakaramians et al. in Chap. 4.

Undoubtedly, heterogeneous integration has grown increasingly complex in recent times, opening new fabrication and reliability frontiers which still need to be completely settled and addressed.

In particular, the appearance of silicon effectively as a packaging material with the TSV technology has created divergent views of ownership and handshake between silicon fabs and outsourced semiconductor assembly and test (OSAT) companies, which have been debated for almost a decade. This, together with the additional reliability challenges caused by the combination of vastly different technologies – very different materials, with mismatched properties like CTE, elastic modulus, lattice constants and different thermal stabilities, and not as extensively known as silicon, different technology nodes, different packaging needs – serving the broad range of required functionalities, has made More-than-Moore integration an area potentially as challenging as that of Moore's scaling.

Reliability is hence a key issue in More-than-Moore integration [23]. This book puts specific emphasis on mechanical reliability, more specifically fracture mechanisms, failure modes and their inspection and mitigation strategies (Chap. 5 by Zschech and Elizalde). Mechanical reliability is one very challenging aspect of such integrated systems, in addition to electrical reliability as well as thermal reliability, which still strongly limits the deployment of 3D integration to this date, due to the lack of an efficient heat removal technology [24, 25].

Finally, in this book we wanted to provide some tangible examples where More-than-Moore technologies are going to play an increasingly important role in enabling advanced integrated capabilities able to take full advantage of the sustained advances in artificial intelligence (AI). Chapter 6 by Delic and Afshar explains how advanced 3D packaging of advanced photonics, optoelectronics and CMOS-based neuromorphic computing can enable a portable navigation system based on event-driven imaging (LiDAR). The future development of truly neuromorphic hardware is expected to further enable autonomous miniaturised systems for low-latency, highly accurate event-driven AI operations, achieving powerful intelligent systems with minimal energy consumption.

Chapter 7 by Do, Duong and Lin, provides a snapshot into how our interaction with such a smart system could look like, thanks to brain-computer interfaces (BCIs) and AI. This aspect has a dual meaning, as it opens the door to a different way for humans to interact with electronic machines and smart systems, while also explaining how miniaturised technologies and the integration offered by More-than-Moore could advance non-invasive sensing for BCIs.

As neural interfaces, neuromorphic computing, wearable technologies, integrated power sources and integrated photonics advance further, it is clear to see what the next paradigm shift enabled by More-than-Moore integration is likely going to be. Human-computer integration [26] is going to change completely the way we interact with electronic systems, and perhaps it is going to make the demarcation line between the biological parts and electronic extensions of a human being somewhat ambiguous. In other words, More-than-Moore integration is going to be at the core of the latest generation of the Internet of the Bodies [27], where human bodies and their technological extensions, both wearable or internal/implanted, will appear seamlessly integrated – be it different types of sensors, of communication interfaces, smart prosthetic devices [28], advanced pacemakers, artificial organs [29] and other forms of technological replacement or augmentation of the biological functions of the human body.

We hope this book will make a useful and inspirational reading for academics, professionals, as well as for students in a wide range of technical disciplines.

Acknowledgements

We would like to thank the following colleagues for their help in peer-reviewing this book's material: Dr. Yang Yang and Dr. Diep Nguyen (University of Technology Sydney, Australia); Prof. Xuan-Tu Tran (Vietnam National University Hanoi), Prof. Gustavo Ardila and Prof. Pascal Xavier (University Grenoble Alpes, France); and Prof. Edwige Bano (Grenoble INP, France). FI would also like to acknowledge support from the Australian Research Council Centre of Excellence in Transformative MetaOptical Systems (TMOS, CE200100010).

Ultimo, NSW, Australia Francesca Iacopi
Grenoble, France Francis Balestra

References

1. G.Q. Zhang, M. Graef, F.V. Roosmalen, The rationale and paradigm of "more than Moore", in *56th Electronic Components and Technology Conference 2006*, (2006), pp. 151–157. https://doi.org/10.1109/ECTC.2006.1645639
2. NEREID | NanoElectronics Roadmap for Europe: Identification and Dissemination. https://www.nereid-h2020.eu/. Accessed 25 Aug 2022
3. IEEE International Roadmap for Devices and Systems – IEEE IRDS. https://irds.ieee.org/. Accessed 25 Aug 2022

4. L. Basiricò, G. Mattana, M. Mas-Torrent, Editorial: Organic electronics: Future trends in materials, fabrication techniques and applications. Front. Phys. **10**, 888155 (2022). https://doi.org/10.3389/fphy.2022.888155
5. F. Iacopi, M. Van Hove, M. Charles, K. Endo, Power electronics with wide bandgap materials: Toward greener, more efficient technologies. MRS Bull. **40**(5), 390–395 (2015). https://doi.org/10.1557/mrs.2015.71
6. M. Tang et al., Integration of III-V lasers on Si for Si photonics. Prog. Quantum Electron. **66**, 1–18 (2019). https://doi.org/10.1016/j.pquantelec.2019.05.002
7. P.J. Wellmann, Review of SiC crystal growth technology. Semicond. Sci. Technol. **33**(10), 103001 (2018). https://doi.org/10.1088/1361-6641/aad831
8. D.J. Langley, J. van Doorn, I.C.L. Ng, S. Stieglitz, A. Lazovik, A. Boonstra, The internet of everything: Smart things and their impact on business models. J. Bus. Res. **122**, 853–863 (2021). https://doi.org/10.1016/j.jbusres.2019.12.035
9. M. Weiser, The computer for the 21st century. Sci. Am. **265**, 94–104 (1991). https://doi.org/10.1038/scientificamerican0991-94
10. E. Ahmed et al., The role of big data analytics in internet of things. Comput. Netw. **129**, 459–471 (2017). https://doi.org/10.1016/j.comnet.2017.06.013
11. K. Kalantar-zadeh, N. Ha, J.Z. Ou, K.J. Berean, Ingestible sensors. ACS Sens. **2**(4), 468–483 (2017). https://doi.org/10.1021/acssensors.7b00045
12. Z. Zhu et al., Recent advances in high-performance microbatteries: Construction, application, and perspective. Small **16**(39), 2003251 (2020). https://doi.org/10.1002/smll.202003251
13. J. Liang, A.K. Mondal, D.-W. Wang, F. Iacopi, Graphene-based planar microsupercapacitors: Recent advances and future challenges. Adv. Mater. Technol. **4**(1), 1800200 (2019). https://doi.org/10.1002/admt.201800200
14. R.R. Tummala, M. Swaminathan, *Introduction to System-On-Package (SOP) : Miniaturization of the Entire System* (McGraw-Hill, New York, 2008)
15. Heterogeneous Integration Roadmap | SEMI. https://www.semi.org/en/communities/ heterogeneous_integration_roadmap. Accessed 25 Aug 2022
16. M. Chiao, Y.-T. Cheng, L. Lin, Introduction to MEMS packaging, in *Microsystems and Nanotechnology*, ed. by Z. Zhou, Z. Wang, L. Lin, (Springer Berlin Heidelberg, Berlin, Heidelberg, 2012), pp. 415–446
17. P. Gupta, S.S. Iyer, Goodbye, motherboard. Bare chiplets bonded to silicon will make computers smaller and more powerful: Hello, silicon-interconnect fabric. IEEE Spectr. **56**(10), 28–33 (2019). https://doi.org/10.1109/MSPEC.2019.8847587
18. S.S. Iyer, A. Alam, Flexible hybrid electronics using fan-out wafer-level packaging, in *Embedded and fan-out wafer and panel level packaging technologies for advanced application spaces*, (Wiley-IEEE Press, 2022), pp. 233–260
19. M. Li, Y. Yang, F. Iacopi, M. Yamada, J. Nulman, Compact multilayer bandpass filter using low-temperature additively manufacturing solution. IEEE Trans. Electron. Devices **68**(7), 3163–3169 (2021). https://doi.org/10.1109/TED.2021.3072926
20. L.A. Coldren, P.A. Verrinder, J. Klamkin, A review of photonic systems-on-chip enabled by widely tunable lasers. IEEE J. Quantum Electron. **58**(4), 1–10 (2022). https://doi.org/10.1109/JQE.2022.3168041
21. Z. Zhou, B. Yin, J. Michel, On-chip light sources for silicon photonics. Light Sci. Appl. **4**(11), e358–e358 (2015). https://doi.org/10.1038/lsa.2015.131
22. N. Yu, F. Capasso, Flat optics with designer metasurfaces. Nat. Mater. **13**(2), 139–150 (2014). https://doi.org/10.1038/nmat3839
23. R. Radojcic, *More-than-Moore 2.5D and 3D SiP Integration*, 1st edn. (Springer, Cham, 2017)
24. H. Liu et al., Thermal-mechanical reliability assessment of TSV structure for 3D IC integration, in *2016 IEEE 18th Electronics Packaging Technology Conference (EPTC)*, (2016), pp. 758–764. https://doi.org/10.1109/EPTC.2016.7861584
25. C. Bäumler, J. Franke, J. Lutz, Reliability aspects of 3D integrated power devices, in *2021 Third International Symposium on 3D Power Electronics Integration and Manufacturing (3D-PEIM)*, (2021), pp. 1–6. https://doi.org/10.1109/3D-PEIM49630.2021.9497262

26. M. Mehrali et al., Blending electronics with the human body: A pathway toward a cybernetic future. Adv. Sci. **5**(10), 1700931 (2018). https://doi.org/10.1002/advs.201700931
27. G. Boddington, The internet of bodies—Alive, connected and collective: The virtual physical future of our bodies and our senses. AI Soc., 1–17 (2021). https://doi.org/10.1007/s00146-020-01137-1
28. L.J. Marks, J.W. Michael, Artificial limbs. BMJ **323**(7315), 732–735 (2001). https://doi.org/10.1136/bmj.323.7315.732
29. X. Wang, Bioartificial organ manufacturing technologies. Cell Transplant. **28**(1), 5–17 (2019). https://doi.org/10.1177/0963689718809918

Contents

1 Energy Harvesters and Power Management 1
Michail E. Kiziroglou and Eric M. Yeatman
1 Introduction 1
2 Motion 3
3 Heat 13
4 Electromagnetic Fields 21
5 Energy Harvesting from Waves 28
6 Power Management 31
7 Conclusion 37
References 39

2 SiC and GaN Power Devices 47
Konstantinos Zekentes, Victor Veliadis, Sei-Hyung Ryu, Konstantin Vasilevskiy, Spyridon Pavlidis, Arash Salemi, and Yuhao Zhang
1 Introduction 47
2 Silicon Carbide Diodes 50
3 SiC BJTs 57
4 SiC Junction Field-Effect Transistors 59
5 SiC MOSFETs 63
6 SiC IGBTs 72
7 III-Nitrides Power Devices 75
8 Conclusions 90
References 91

3 Flexible and Printed Electronics 105
Benjamin Iñiguez
1 Introduction 105
2 Materials and Processes for Flexible and Printed Electronics 106
3 Devices for Flexible and Printed Electronics 111
4 Conclusions 121
References 122

4 Terahertz Metasurfaces, Metawaveguides, and Applications 127
Wendy S. L. Lee, Shaghik Atakaramians, and Withawat Withayachumnankul
1 Introduction 127
2 Metasurfaces 133
3 Terahertz Metawaveguides 143
4 Conclusion and Outlook 150
References 151

5 Mechanical Robustness of Patterned Structures and Failure Mechanisms 157
Ehrenfried Zschech and Maria Reyes Elizalde
1 Reliability of Microelectronic Products and Failure Mechanisms 157
2 Risks of Microcrack Propagation and Design 161
3 Characterization of Microcrack Behavior 163
4 Understanding of Microcrack Behavior 170
5 Summary and Outlook 180
References 183

6 Neuromorphic Computing for Compact LiDAR Systems 191
Dennis Delic and Saeed Afshar
1 Introduction 191
2 Innovating SPAD Array Sensors for 3D Flash LiDAR 211
3 SPAD Neuromorphic Event-Based Sensing and Processing 223
4 Future Concepts and Summary 230
References 232

7 Integrated Sensing Devices for Brain-Computer Interfaces 241
Tien-Thong Nguyen Do, Ngoc My Hanh Duong, and Chin-Teng Lin
1 Introduction 241
2 The Principles of Brain Signal Acquisition 242
3 Applications in BCI 249
4 Future Electrode Directions 251
5 Challenges and Outlooks 254
References 254

Index 259

Chapter 1
Energy Harvesters and Power Management

Michail E. Kiziroglou and Eric M. Yeatman

1 Introduction

Advances in electronics have led to a vast proliferation of commercial devices over the last several decades. Driven by cost reduction, miniaturisation and wireless communications, a large proportion of these are untethered to mains electricity, and so in most cases are battery powered. However, in an even wider range of potential applications, battery replacement or recharging imposes an unacceptable maintenance burden. As an example, the worldwide average of Internet of Things (IoT) devices per person is expected to rise beyond 10 in the short term (Fig. 1.1). In addition, the cost of portable power in comparison to the electrical grid is at least three orders of magnitude higher (Fig. 1.2). To address these challenges, a strong incentive to develop methods for powering such devices from ambient energy has been created. This approach of generating electrical power from ambient sources, at small scale, for use in a self-powered device or system, we will refer to as energy harvesting.

Ambient energy can be present in many forms, the principal ones being light, heat and motion. Use of such sources for electronics is not new – for example, pocket calculators powered by photovoltaic cells were commercially available already in the 1970s. In this chapter, we will summarise the progress made since then, and the challenges remaining. Although most reported energy harvesting devices are too large for monolithic electronic integration, we will emphasise those with the most potential for at least hybrid integration at chip-scale. Some discussion will also be included of wireless power transfer, where the target remains untethered but the

M. E. Kiziroglou · E. M. Yeatman (✉)
Imperial College London, London, UK
e-mail: e.yeatman@imperial.ac.uk

F. Iacopi, F. Balestra (eds.), *More-than-Moore Devices and Integration for Semiconductors*, https://doi.org/10.1007/978-3-031-21610-7_1

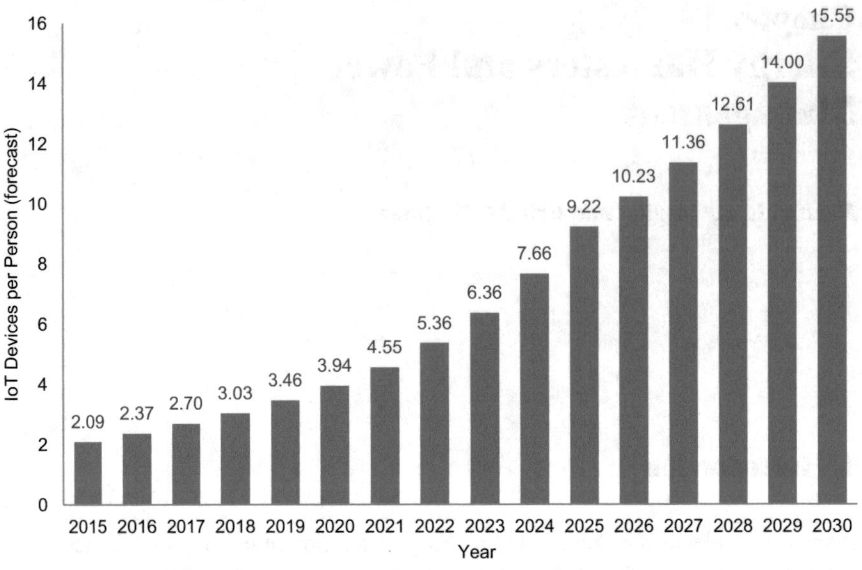

Fig. 1.1 Projected number of IoT devices per person, calculated from publicly available statistics

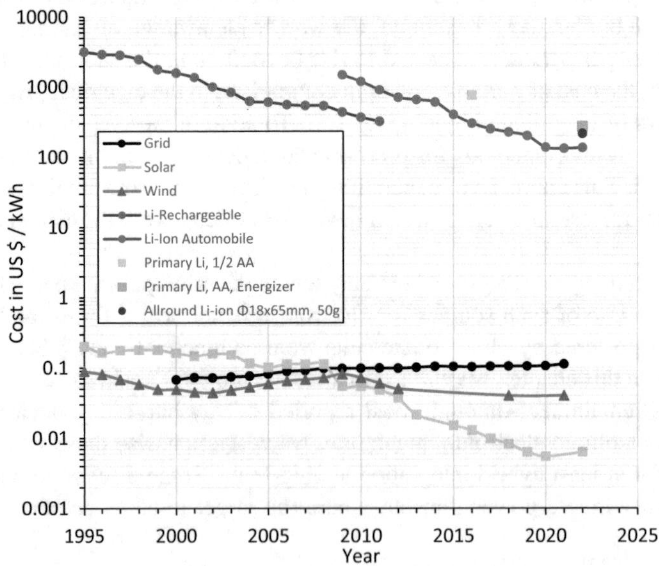

Fig. 1.2 Cost of energy from various sources and in various forms, from publicly available statistics and prices. Solar panel energy was calculated from power, for 20 years of operation at 5 hours of average exposure per day. For the battery energy cost, the commercial purchase cost is plotted. Batteries can be recharged for as high as 1000 times, but the recharging process introduces significant costs and limitations

energy source is intentionally introduced, because of the close relationship between these two technologies.

2 Motion

A great variety of mechanical energy sources exist from which energy can be harvested. Such environmental energy can come as a varying force applied directly on the microdevice such as a heel strike [1], strain on a surface [2] or a pressure [3] or as varying acceleration, such as vibrations or irregular human body motion [4]. In most cases, some force or motion translation is required from the environmental form to a form suitable for the transduction mechanism used. In the following subsections, the key features of these methods are summarised. A more comprehensive review of motion translation mechanisms for energy harvesting can be found in [5].

2.1 The Use of a Proof Mass

Harvesting of energy from a moving body without the need for connection to stationary structure is very attractive for flexibility of application, and for miniaturisation. This can be achieved by using a proof mass attached to the frame such that it can move inside the device (Fig. 1.3). When the frame experiences acceleration from the motion of the host body, the inertia of the proof mass results in relative motion between it and the frame. This relative motion is used to transduce energy. A general formulation of equations for inertial microgenerators, including piezoelectric, electrostatic and electromagnetic transduction, can be found in [6]. Inertial harvesters have effectively the same structure and operating principle as inertial sensors, although the need to maximise output power results in different design choices, particularly maximisation of the proof mass, which is a disadvantage for monolithic integration.

With a harmonic input motion but without assuming harmonic motion of the proof mass, an upper limit of the power for motion energy harvesters can be calculated as a function of device size (maximum internal displacement amplitude Z_l, mass m) and the source motion (vibration frequency ω and vibration amplitude Y_o) [7]. The maximum transduction force F_T that can be applied is the mass m times the external acceleration $\omega^2 Y_o$; otherwise, the internal motion will cease. The energy extracted is this force times the internal displacement $2Z_l$, and this can be obtained twice (once in each direction) for each period $T = 2\pi/\omega$, giving a maximum power [7]:

$$P_{\max} = \frac{2}{\pi} Y_o Z_l \omega^3 m \tag{1.1}$$

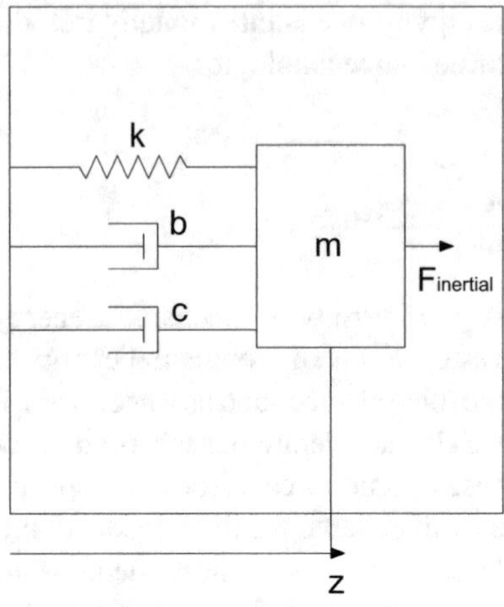

Fig. 1.3 Model of an inertial motion energy harvesting system

From this equation, one can assess the viability of particular motion harvesting applications. This is an absolute limit for inertial harvesting devices – it cannot be overcome by improved transduction methods, or by nonlinear motion structures such as frequency up-conversion methods. The only exceptions are where the input motion is rotational rather than kinematic [8]. In that case, resonant rotating devices and gyroscopic devices offer potential for large power density increases, but neither has been yet demonstrated in practice.

In Fig. 1.4, the maximum power is plotted as a function of device size for frequencies in the range expected for human motion, acceleration $\omega^2 Y_0$ of 10 m/s^2 and a proof mass density of 20 g/cm^3 occupying half of the device volume. By comparison with the power requirements and size of a typical laptop, cellphone, watch and (very low power) sensor node, one concludes that human motion harvesting is not enough for the first two applications, while there is substantial promise for the last two. Indeed, the watch application has already been commercialised in high volumes, and sensors powered by harvesting are becoming more common.

2.2 *Electrostatic*

Electrostatic transducers convert energy between kinetic and electrical form through the electrostatic force between charged bodies. This approach is especially well suited to micro-electro-mechanical systems (MEMS), because it can be implemented in many cases with standard MEMS materials and is easily adapted to a quasi-two-dimensional design. The similarity in structure to accelerometers is also an advantage. Consequently, most true MEMS energy harvesters are electrostatic in operation. This mechanism is typically described using the example of two charged parallel plates which can move with respect to each other. For a given capacitance C

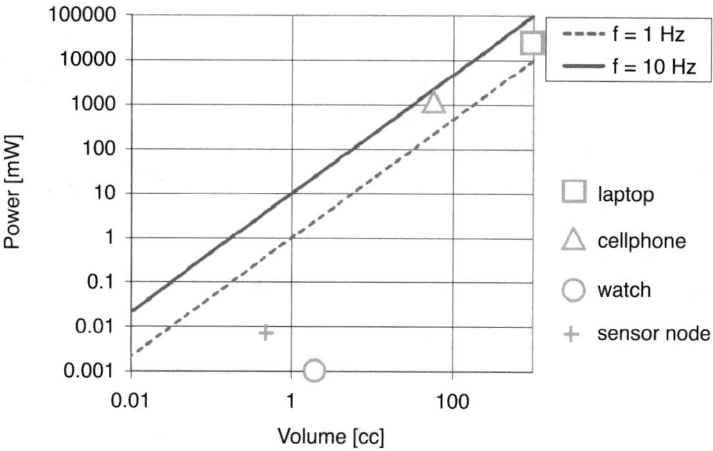

Fig. 1.4 Maximum power for motion harvesters versus size for two different excitation frequencies, with size and power requirement of various applications superimposed [7]

and voltage V, with area A and separation t, the electrostatic force F_{es} between the plates is:

$$F_{es} = \frac{1}{2}\frac{\varepsilon_0 A V^2}{t^2} \tag{1.2}$$

where ε_0 is the electrical permittivity of free space. If one of the plates is moved perpendicularly to the plate surface so that the distance t between the plates is increased, F_{es} will produce work against the motion. This work will be stored in the capacitor as electrical energy. The same will occur if the motion is parallel to the surface of the plates. This can be understood by considering that the electric field will be rotated by this motion and a component of electrostatic force arises that is parallel to the field.

Effectively, if the motion results in a change of capacitance, there will be conversion between mechanical and electrical energy, as long as there is some initial charge in the system. A common operational approach is to add a charge at a position of maximum capacitance C_{max} and subsequently extract the energy at a minimum C_{min}. The harvested energy will be given by:

$$\Delta E_{es} = \frac{1}{2}{V_{in}}^2\frac{C_{max}}{C_{min}}\left(C_{max} - C_{min}\right) \tag{1.3}$$

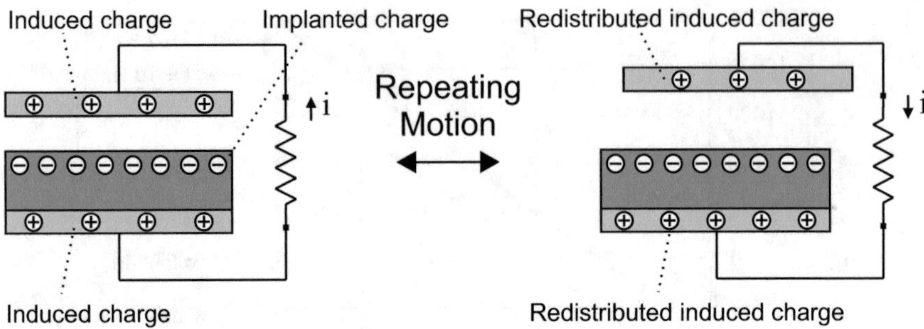

Fig. 1.5 Operation principle of electret harvester

This indicates the importance of both a large capacitance ratio and a large capacitance difference. The former will usually be limited by stray capacitance. A high priming voltage V_{in} is also essential, which creates practical challenges in the interface electronics. Another technique is to keep the voltage constant and let charge flow in or out of the plates during capacitance decrease or increase, respectively.

For the priming of electrostatic generators, inserting charge from the external circuit can be avoided by use of electrets. Electrets are dielectrics with trapped charge that allows them to have (quasi)-permanent polarisation. A typical orientation for an electret-based device is shown in Fig. 1.5 [9, 10]. Its trapped charge creates an electric field which is equivalent to charging the capacitor with a high voltage (typically hundreds of volts). Any capacitance-changing relative motion of the plates, in-plane or perpendicular-to-plane, will result in charge motion through the circuit, delivering electrical energy to a load resistance R. In this operating scheme, the electret effectively provides the initial priming of the electrostatic harvesting device. Various geometrical implementations of such devices have been proposed including in-plane shifting electrodes [9], rotating electrodes [11], patterned electrodes [12] and comb-like electrode structures [13]. A quantitative analysis of operation for such devices can be found in [12].

Beyond electret-based devices, electrostatic priming can be provided by an active circuit, although a method for device initialisation is still required in this case. Another approach is the direct use of a passive sensor with voltage output as the priming source of an electrostatic harvester [14]. An implementation using a pH sensor for priming is shown in Fig. 1.6 [14]. That device includes radio frequency (RF) transmission of the sensor information at distances up to 1 m by direct discharge of the harvested energy into a loop antenna. The main challenge of this implementation is related to the voltage range of common passive sensors which is usually lower than that required for efficient operation of electrostatic harvesters.

In electrostatic MEMS actuators, the hyperbolic increase of electrostatic field with decreasing gap distance results in an instability that causes sudden attraction to the point of minimum separation. This is called the pull-in effect and disturbs motion control for MEMS devices. In MEMS microgenerators, the induced field opposes

Fig. 1.6 Electrostatic harvester with an external proof mass, pH-sensor priming and RF transmission [14]

the motion and therefore cannot lead to such an effect. However, when using active pre-biasing techniques (outlined in the power management section of this chapter) to increase damping in electrostatic devices, the pull-in effect should be considered. A review of pull-in instability effects for MEMS actuators can be found in [15].

A variant of electrostatic motion transduction is triboelectric energy harvesting which is usually considered as a separate device concept. Triboelectricity is the effect of static charge exchange between the surfaces of two different materials when they come to contact, due to different electron energy distribution. Various device implementations have been proposed to exploit this effect, often employing nanostructures to increase contact area. A general description of the operating mechanism and a review of nanopatterning methods for triboelectric harvesters can be found in [16, 17], respectively.

2.3 *Electromagnetic*

In electromagnetic transducers, electromagnetic induction as described by Faraday's law is employed. Relative motion between a magnet and a coil results in magnetic flux variation which induces an electromotive force across the coil. The motion is damped when electrical power is taken from the output coil. While electromagnetic transduction is used in most macro-scale electric generators, it does not scale well into the micro-domain. This is due to the difficulty of designing a planar implementation suitable for MEMS fabrication, the difficulty of effectively guiding magnetic flux in such a planar or quasi-planar configuration (particularly if high permeability materials cannot be integrated) and the limited number of turns achievable in micro-scale coils, which results in low output voltages which cannot be efficiently rectified.

For sensing applications, the design of electromagnetic transducers focuses on sensitivity, signal-to-noise ratio and overall optimised measurement of the effect/quantity of interest [16]. In contrast, for energy microsystems, including energy harvesters and actuators, maximum power transduction is generally the priority, requiring a different operation point and transducer design. Coils are needed to operate such that they provide maximum power and are therefore connected to impedance matched loads, instead of high-impedance signal acquisition circuitry such as analogue to digital converters. Consequently, the coil impedance is far more important in energy microsystems than in sensing. This results in significant design differences between sensing inductive transducers and inductive transducers for motion harvesting.

On the other hand, most macro-scale generators employ rotational motion. In micro-mechanics, low-loss bearings are difficult to implement, and consequently most devices use vibrating motion and flexures rather than sliding or rolling bearings. In harvesters that have rotational input motion, the rotation can be converted to oscillation via some translation mechanism. Such methods are discussed in [16].

To achieve high flux density in the small scale, high-performance permanent magnets may be employed. Systems that usually follow this approach include microturbines for fluid flow [3] and rotational motion [18] energy harvesters. As an example, an image of a microturbine by Holmes et al. is shown in Fig. 1.7 [19]. This device is based on MEMS fabricated coils and springs, but it relies on manual placement for its 35 mg NdFeB permanent magnet. In this way, the MEMS magnet fabrication which usually involves high-temperature sintering processes is avoided. For the same reason, most electromagnetic harvesters use commercially purchased and externally assembled NdFeB as permanent magnets [20–29]. Reviews of various implementations, including performance comparison tables, can be found in [6, 30].

MEMS integration of permanent magnets by sputtering for NdFeB and SnCo [31–33], and by electrodeposition for CoPt and FePt compounds [34–40], has been demonstrated with promising potential. The deposited materials are magnetised by the application of a strong field, which can be several Tesla, during [41] or after deposition [32]. A review of magnetic material MEMS integration can be found in [42].

The challenge of employing effective magnetic materials into micromachining processes has limited the flux density availability. The use of soft magnetic materials with high magnetic permeability as flux guides has the potential of achieving high flux densities in microsystems. As this technique has been developed for inductive energy harvesting for power lines and structural currents [43], it is discussed in Sect. 1.4 of this chapter. Integrating high permeability materials into microfabrication processes may enable high-performance electromagnetic micro-transducers for energy as well as sensing and actuating microsystems in the near future. Candidate materials include conventional ferrites, nanocrystalline structures,

Fig. 1.7 A MEMS electromagnetic turbine energy harvester by Holmes et al. [19]. (Courtesy of A. S. Holmes)

ferrite films [44] as well as printable [45] and electrodeposited [46] soft magnetic materials. A promising integration method for soft cores based on atomic layer deposition and agglomeration has been reported in [47, 48].

2.4 Piezoelectric

Piezoelectricity is the electromechanical coupling between stress $\boldsymbol{T}$, strain $\boldsymbol{S}$, electric field $\boldsymbol{E}$ and surface charge density $\boldsymbol{D}$ (displacement) in a bulk material [16]. The change of interatomic distances in a strained dielectric material with certain crystal asymmetries can result in changes of dipole distribution. This causes material polarisation $\boldsymbol{P}$ and a corresponding surface charge $Q = \boldsymbol{P} \cdot \boldsymbol{A}$, where $\boldsymbol{A}$ is the surface area. Normal and shear strain in each direction can result in polarisation in all three dimensions. The relation between applied $\boldsymbol{T}$ and resulting $\boldsymbol{P}$ can be approximated as linear, with a tensor factor $\boldsymbol{d}$:

$$\boldsymbol{P} = \boldsymbol{d} \cdot \boldsymbol{T} \tag{1.4}$$

As a practical simplification, for a given selected input-output pair of stress and polarisation axes, a single effective d value can be used, although such a value may

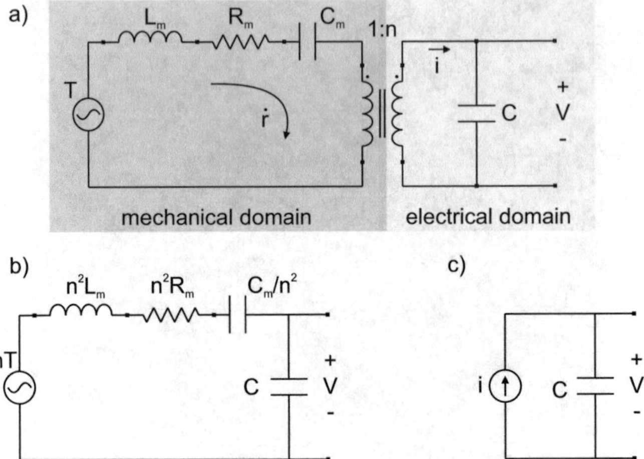

Fig. 1.8 (**a**) Lumped element model of a piezoelectric generator. (**b**) Corresponding all-electrical model. (**c**) Simplified model encompassing the excitation source and mechanical response into a current source

vary for different boundary conditions in other directions. If electrodes are deposited on two back-to-back surfaces of a piezoelectric material, a capacitor C is formed. Stress-driven polarisation moves charge Q to and from the electrodes. Hence, a piezoelectric generator can be modelled as a current source in parallel with a capacitor C. The dielectric (shunt) leakage and series resistance can also be included in such a model by adding corresponding components. The mechanical behaviour of the structure can be modelled with lumped elements. The link between the electrical and the mechanical domain can be represented by a transformer, leading to equivalent circuits such as the one shown in Fig. 1.8a [49]. In this example, force is represented as a potential quantity, analogous to voltage, while mass displacement rate $\dot{r}$ is represented as a flow quantity, analogous to current. In this way, the inductance L_{m}, resistance R_{m} and capacitance C_{m} represent the structure mass (inertia), mechanical losses and elasticity, respectively. The transformer ratio n translates force to voltage and $1/n$ translates displacement rate to current, and its units are therefore V/N or equivalently C/m.

Using circuit analysis, the mechanical components can be transferred to the electrical side as shown in Fig. 1.8b. This model can be used to analyse the electromechanical behaviour of piezoelectric transducers. The determination of L_{m}, R_{m} and C_{m} depends on the mechanical properties and geometry of the full mechanical structure. For bulk piezoelectric transducers such as in acoustic devices, the parameter values are primarily determined by the piezoelectric itself, while for devices such as accelerometers where the piezoelectric material forms just a small part of the moving structure, they are determined by the mechanical properties of the overall structure. An example is the typical unimorph structure.

Table 1.1 Summary of common piezoelectric materials

Material	d_{33} (nC/N)	d_{31} (nC/N)	Reference
PZT	0.63	−0.28	[50]
PMN-PT	1.25	−0.15	[54]
PNN-PZT	1.75	−0.44	[55]
$LiNbO_3$	0.006	−0.001	[56]
AlN	0.004	−0.002	[57]
PVDF	0.02	−0.015	[58]
KNN-BNZ-AS-Fe	0.5	–	[59]

A list of piezoelectric materials typically employed for energy applications is presented in Table 1.1 [16]. The most common is lead zirconate titanate (PZT) which exhibits a d_{33} of around 0.63 nC/N, one of the highest available in the market [50]. While such ceramic piezoelectrics are generally formed by sintering of powders, thin film processes suitable for monolithic MEMS integration have been widely investigated [51–53]. Single-crystal piezoelectrics employ the electrostrictive (relaxor) effect and can exhibit higher d values. The typical lead magnesium niobate-lead titanate (PMN-PT) exhibits constants higher than 1 nC/N in single-crystal form, e.g. 1.25 nC/N in [54]. The electrostrictive type materials are usually more challenging to use because of the high cost of single crystal and the requirement for an additional polarisation field. In microgenerators, piezoelectric materials are usually integrated in bimorph cantilever structures, exploiting their d_{31} coefficient which is smaller than d_{33}. AlN and PVDF exhibit low coupling factors but are easier to integrate. PVDF is also a flexible material allowing the implementation of high elasticity transducers.

A review of piezoelectric energy harvesting devices can be found in [60].

2.5 Comparison of Transduction Mechanisms

The broad range of differences in prototype architecture, fabrication methods, testing conditions and target application makes a fair and useful comparison very difficult to achieve, especially across the transduction mechanisms. Furthermore, in most device cases, there is little reported on performance in real environments. Instead of an exhaustive figure of merit comparison, a table with three indicative implementations for piezoelectric, electrostatic and electromagnetic transduction is presented in Table 1.2. A comparison of key features for each mechanism is presented in Table 1.3. For state-of-the-art comparative tables, the reader is referred to recent reviews [30, 60–64].

Table 1.2 Performance of three indicative MEMS energy harvesters

Mechanism	Reference	Power	Proof mass	Conditions
Piezoelectric	Elfrink et al. [65]	85 μW	0.1 g	17.5 m/s^2, 325 Hz
Electrostatic	Suzuki et al. [66]	1 μW	0.1 g	20 m/s^2, 63 Hz
Electromagnetic	Shin et al. [67]	165 μW	0.05 g	4 m/s^2, 46 Hz

Table 1.3 Comparison of features for microgenerators based on different transduction mechanisms. Green, amber and red indicate high, medium and low, respectively [16]

Mechanism	Complexity			Scalability	Power Density
	Material	Structure	Circuit		
Piezoelectric	●	●	●	●	●
Non-Electret Electrostatic	●	●	●	●	●
Electret Electrostatic	●	●	●	●	●
Electromagnetic	●	●	●	●	●
Triboelectric	●	●	●	●	●

2.6 *Opportunities*

A discussion of the main current challenges and opportunities in motion energy harvesting has been presented in [16]. Considering motion energy harvesting as a combination of a motion adaptor, an active material and a power management system with optimised interfaces may allow the combination of techniques that have been developed and are currently bound to specific device types. For example, pre-biasing has been developed for piezoelectric transducers but active driving of other materials may offer significant potential.

In practice, the main limiting factor in industrial adoption of energy harvesting microsystems is the requirement for specific device designs, tailored to a very narrow set of environmental specifications, such as the availability of vibration at a specific frequency. This leads to a demand for customised and high cost research and development for each potential application. In spite of significant research advancement in broadening these environmental requirements, microgenerator prototypes tend to operate at lower performance in real application conditions. To address this limitation, a combination of energy harvesting and wireless power transfer could potentially be adopted. This approach applies to other energy harvesting mechanisms as well and is therefore discussed in Sect. 1.7 of this chapter.

3 Heat

The environmental heat is another source that can be used to power microsystems, by the thermoelectric effect, the pyroelectric effect or gas heat engines such as the Stirling engine. This section focuses on thermoelectric energy harvesting because it is the most common mechanism used in micro energy harvesting applications. A review of pyroelectric materials including energy harvesting applications can be found in [68]. Gas engines are considered to be beyond the scope of this chapter.

The thermoelectric effect is based on the coupling between heat transfer and charge carrier transport. In a classical picture, a certain percentage of heat is carried by electron flow in metals, and either electrons or holes in semiconductors. The rest is carried through phonons, i.e. propagation of lattice vibration. In a quantum mechanical view, the net carrier transport in one direction, either along or against the heat flow, depends on the asymmetry of the density of states on either side of the Fermi level. In both interpretations, this results in a voltage difference between the hot and the cold side of a material $\Delta V = S\Delta T$, where S is a temperature-dependent factor called the Seebeck coefficient of the material and ΔT is the temperature difference. To access a ΔV on the same side of the device (either the hot or the cold one), a couple of materials with different S are required:

$$\Delta V = (S_{\mathrm{A}} - S_{\mathrm{B}}) \cdot \Delta T = S_{\mathrm{AB}} \cdot \Delta T \tag{1.5}$$

where S_{AB} is the Seebeck coefficient of the material couple. This is called a thermocouple. The efficiency of power transduction increases with the square of S and the material electrical conductivity σ and decreases with thermal conductivity k. Therefore, the following figure of merit is often used for thermoelectric materials:

$$zT = \frac{\sigma S^2}{k} T \tag{1.6}$$

In practice, because S is small (below 1 mV/K), an array of thermocouples electrically connected in series is typically employed. For a generator with total electrical resistance R_{e} and thermal conductance K, the overall figure of merit is defined in the literature as:

$$ZT = \frac{S^2}{K R_e} T \tag{1.7}$$

Note that in the theoretical case of a thermocouple with the same electrical and thermal conductivity, the device level ZT is equal to $zT/4$. This means that when a material is reported to exhibit a certain zT, its contribution to a generator device will correspond to $zT/4$. An equation for the maximum efficiency of a thermoelectric generator (TEG) as a function of ZT, hot side temperature T_{h} and cold side temperature T_{C} can be calculated to be [69]:

$$\eta_{\mathrm{TEG}} = \frac{\Delta T_{\mathrm{TEG}}}{T_{\mathrm{h}}} \cdot \frac{\sqrt{1+ZT}-1}{\sqrt{1+ZT}+\frac{T_{\mathrm{c}}}{T_{\mathrm{h}}}} = \xi\,(ZT, T_{\mathrm{h}}) \cdot \Delta T_{\mathrm{TEG}} \tag{1.8}$$

where $\xi(ZT, T_{\mathrm{h}})$ is defined in this equation. Note that this is the maximum possible efficiency of the TEG. The operating efficiency in general depends also on the electrical load connected to the TEG. The operation point of maximum efficiency is not the same with the one for maximum output power, due to the effect of current flow on the TEG heat conductivity [70, 71]. Nevertheless, the difference between the two operating points is less than 10% for typical temperature and ZT values [70], and for this reason, Eq. (1.8) is widely used in the literature to evaluate the performance of TEGs.

The TEG efficiency depends on the available ΔT, on temperature range and on ZT. The value of ZT is determined by the material properties and is independent from the device design. Research towards high ZT values has focused on developing materials with high σ and small k, e.g. by employing phonon-confining nanostructures or superlattices, or with higher Seebeck coefficients, by enhancing carrier transport asymmetry. Bismuth tellurides have been, and still are, the most commonly used thermoelectric materials. Other materials, including silicon nanostructures [72, 73], have been under investigation. A material of special interest is tin selenide (SnSe), which has been shown to exhibit zT values over 2 at high temperatures (~900 K) in single-crystal form, and over 3 in polycrystalline form [74]. Another promising technique is carrier transport energy filtering, which has been shown to improve heat and charge transport coupling [75]. This technique can be implemented by metal particle implantation which introduces thermionic emission to current transport. The energy selectivity of thermionic emission, in turn, reduces electrical conductivity σ but boosts the Seebeck coefficient S, resulting in an overall increase of the power factor σS^2 of the material. This is demonstrated in [76] by implantation of Ag nanoparticles in an antimony telluride thermoelectric material. A review of material and device level thermoelectric research can be found in [77].

Beyond improvement of zT (and the corresponding ZT), the temperature difference across a TEG has a key role in output power as heat flow is proportional to ΔT, for a given TEG thermal conductance K, and the transduction efficiency is also almost proportional to ΔT, through Eq. (1.8). Hence, the TEG output power scales with ΔT^2, and achieving the optimal ΔT is of decisive importance in practical installations. This relies on the TEG design, which defines its K, as well as on the thermal design of interfacing with the environmentally available temperature difference ΔT_{E}. The environmental ΔT_{E} can be directly available between two heat conductive bodies, with a series thermal resistance $R_{\mathrm{th,\,E}}$, or can be created artificially from fast temperature fluctuations in time. The two cases, coined static and dynamic thermoelectric energy harvesting, are discussed in the following two subsections.

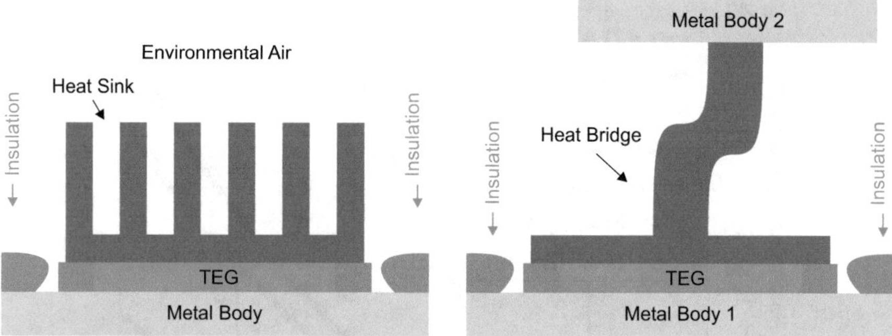

Fig. 1.9 Installation of a static thermoelectric harvester between a metal body and air using a heat sink (left) and two metal bodies using a heat bridge (right)

3.1 Static Thermoelectric Harvesting

In the case of static thermoelectric harvesting, the environmental ΔT_E is available between a heat conductive body, such as a metal structure, and the surrounding air, or between two heat conductive bodies, as illustrated in Fig. 1.9. The TEG is installed such that one of its surfaces is in direct thermal contact with one of the bodies, while the other is either in contact with air through a heat sink (Fig. 1.9, left) or with the second body through a thermal bridge (Fig. 1.9, right). The thermal resistance of the heat sink, the heat bridge or the heat conductive bodies themselves are in series with the thermal resistance of the TEG, $R_{\text{th, TEG}}$. If the total series resistance is denoted as $R_{\text{th, E}}$, then the heat flow through the TEG will be:

$$Q = \frac{\Delta T_{\text{E}}}{R_{\text{th,TEG}} + R_{\text{th,E}}} \tag{1.9}$$

The temperature difference across the TEG will be:

$$\Delta T_{\text{TEG}} = \Delta T_{\text{E}} \frac{R_{\text{th,TEG}}}{R_{\text{th,TEG}} + R_{\text{th,E}}} \tag{1.10}$$

Given that the efficiency in Eq. (1.8) varies linearly with ΔT_{TEG} ($\eta_{\text{TEG}} = \xi(ZT, T) \cdot \Delta T_{\text{TEG}}$), in good approximation (see [78] Fig. 17.14), the TEG maximum output power can be written as:

$$P_{\text{TEG}} = \eta_{\text{TEG}} \cdot Q = \xi \cdot \Delta T_{\text{TEG}} \cdot \frac{\Delta T_{\text{E}}}{R_{\text{th,TEG}} + R_{\text{th,E}}} = \xi \cdot \Delta T_{\text{E}}^2 \frac{R_{\text{th,TEG}}}{\left(R_{\text{th,TEG}} + R_{\text{th,E}}\right)^2} \tag{1.11}$$

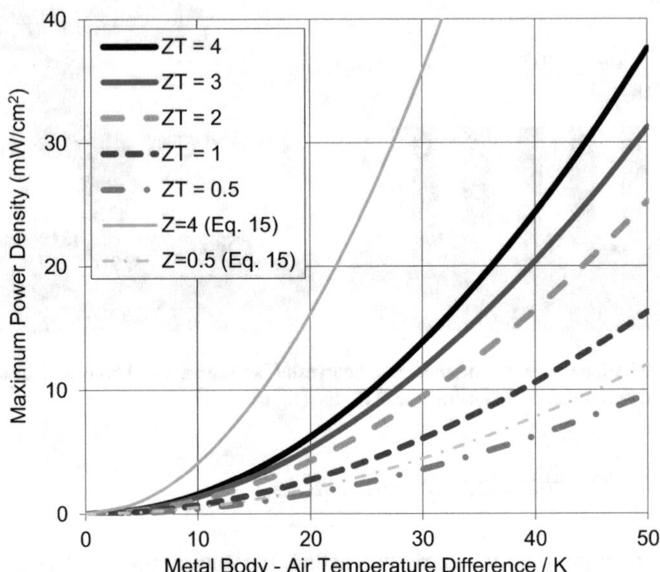

Fig. 1.10 Maximum static TEG power output as a function of environmental body-to-air ΔT, using a typical heat sink with a moderate airflow, calculated by Eq. (1.12). Calculations from the simpler Eq. (1.13) are also shown as faded lines for $ZT = 0.5$ and $ZT = 4$, for comparison. An order of magnitude agreement for $ZT = 0.5$ is observed

This expression shows that to maximise P_{TEG}, the minimum possible $R_{\mathrm{th,\,E}}$ is desirable. In practice this is limited by size, geometry and installation access restrictions. In the case of using air convection, as in the body-to-air case of Fig. 1.9 left, ventilation is a significant limiting factor. For a given application, once $R_{\mathrm{th,\,E}}$ is minimised, the $R_{\mathrm{th,\,TEG}}$ must be selected for maximum P_{TEG}. Under the approximation of a constant $R_{\mathrm{th,\,TEG}}$, which is not affected substantially by the Seebeck and the Ohmic effects, the dependence of P_{TEG} on $R_{\mathrm{th,\,TEG}}$ has a typical load-matching form. The optimal $R_{\mathrm{th,\,TEG}}$ can be calculated by zeroing the first derivative of (1.11) to be equal to $R_{\mathrm{th,\,E}}$, yielding:

$$P_{\mathrm{TEG,MAX}} = \xi \cdot \frac{\Delta T_{\mathrm{E}}^2}{4R_{\mathrm{th,E}}} = \frac{1}{T_{\mathrm{h}}} \cdot \frac{\sqrt{1+ZT}-1}{\sqrt{1+ZT}+\frac{T_{\mathrm{c}}}{T_{\mathrm{h}}}} \cdot \frac{\Delta T_{\mathrm{E}}^2}{4R_{\mathrm{th,E}}} \tag{1.12}$$

Using this equation, the energy harvesting power expected from an environment with a certain available ΔT_{E} and achievable $R_{\mathrm{th,\,E}}$ can be calculated for different thermoelectric figures of merit ZT. An indicative calculation, for $T_{\mathrm{C}} = 300$ K, and $R_{\mathrm{th,\,E}} = 20$ K/W corresponding to a typical 1 cm^2 finned heat sink at a mild 0.5 m/s airflow [79] is presented in Fig. 1.10.

As already mentioned, the calculation of maximum output power as given in Eq. (1.12) is based on three approximations. First, that the Seebeck, Peltier, Ohmic and Thomson effects do not alter the TEG thermal resistance as experienced by the heat flow, and hence the ΔT distribution across $R_{\text{th, E}}$ and $R_{\text{th, TEG}}$. If this was taken into account, the optimum thermal resistance balance would deviate from the $R_{\text{th, TEG}} = R_{\text{th, E}}$ condition. Second, that the maximum power occurs at maximum conversion efficiency. To take this into account, a different expression for TEG efficiency should be used, as given in [69, 70], corresponding to an electrical load matching, $R_L = R_e$, condition. Third, that the electrical current, as controlled by R_L, also has a negligible effect to the ΔT distribution. In summary, Eq. (1.12) does include the Peltier and Ohmic effects on the conversion efficiency (in the ξ term), but not on the optimisation of thermal and electrical resistance ratios. These approximations lead to deviations in the 10% range for small ZT and ΔT values (<1 and <20 K, respectively).

On the other hand, if the Peltier, Ohmic and Thomson effects are considered to be negligible for such small ZT and ΔT values, a much simpler calculation of $P_{\text{TEG, MAX}}$ can be made by considering the Seebeck coefficient, thermal and electrical load matching conditions:

$$P_{\text{TEG,MAX}} = \frac{\Delta V^2}{4R_e} = \frac{S^2 \Delta T_E^2}{16R_e} = \frac{Z}{4} \cdot \frac{\Delta T_E^2}{4R_{\text{th,E}}} \tag{1.13}$$

This equation can also be derived from the TEG efficiency expression at electrical load matching conditions (as given in [69, 70]), by neglecting the Peltier and Ohmic terms in the denominator. It provides an order-of-magnitude agreement with the predictions of (1.12) for small ZT and ΔT (<1 and <20 K, respectively) but deviates significantly for higher values. For comparison, the corresponding calculation curves for $ZT = 0$ and $ZT = 4$ are plotted as grey lines in Fig. 1.10.

The employment of detailed analysis, identifying the optimal TEG thermal resistance and the optimal electrical load resistance such as those in [80, 81], may offer significant improvement of device performance, especially for higher ZT and ΔT values. A photograph of an experimental setup developed for the evaluation of static energy harvesting from hot metal pipes under different Ohmic loads is shown in Fig. 1.11.

While research results demonstrating ZT values as high as 2, in practice, for room temperature applications with ΔT values around 20 K, commercially available TEGs exhibit a ZT of approximately 0.5. Overall, a power density up to 10 mW/cm^2 may be expected for direct TEG harvesting applications. As discussed in this section, the main limitation is ΔT availability and the associated practically achievable $R_{\text{th, E}}$. Dynamic thermoelectric harvesters offer an alternative method of acquiring a significant ΔT across a TEG. This concept is discussed in the following subsection.

Fig. 1.11 Static thermoelectric energy harvesting from hot pipelines in the Advanced Materials for Energy Lab, UC Berkeley in 2016

3.2 Dynamic Thermoelectric Harvesting

Dynamic thermoelectric devices comprise an insulated heat storage unit (HSU) which is in thermal contact to the environment (e.g. to a metal body) through a TEG as shown in Fig. 1.12. The HSU is filled with a phase change material (PCM) to increase thermal storage density and the time constant of its heat dynamics. When the environment undergoes temperature fluctuation in time, the HSU follows this change with a delay, achieving a substantial ΔT across the TEG, which is essential for efficient power transduction and management. This device concept was introduced in [82], studied analytically and numerically [70] and used in various implementations, including demonstrators for aircraft applications and flight tests [83] and integrated wireless sensor networks [84]. A model for phase change inhomogeneity was introduced in [85]. A practical dry fabrication method was proposed, based on 3D-printed double-wall insulation and water capsules [86].

A dynamic analysis of this device concept has shown that the maximum energy per temperature cycle, from a temperature fluctuation range Θ, using a HSU with heat capacity C and latent heat L can be written as [70]:

$$E_{\mathrm{MAX}} = 2 \cdot (\Theta \cdot C + L) \cdot \eta_{\mathrm{TEG}} \left(\frac{\Theta}{2} \right) \tag{1.14}$$

where $\eta_{\mathrm{TEG}}(\Theta/2)$ is the TEG efficiency at temperature difference $\Delta T = \Theta/2$. This means that the overall maximum possible efficiency is simply the TEG efficiency for $\Theta/2$. Indicative results using Eq. (1.8) for η_{TEG} are plotted in Fig. 1.13 as a function of ambient temperature variation. For this calculation, a latent heat density of $L/m = 334$ kJ/kg and heat capacity density of $C/m = 4.2$ kJ/(K $\cdot$ kg) were used, where m is the PCM mass. These values correspond to using water as a PCM, and

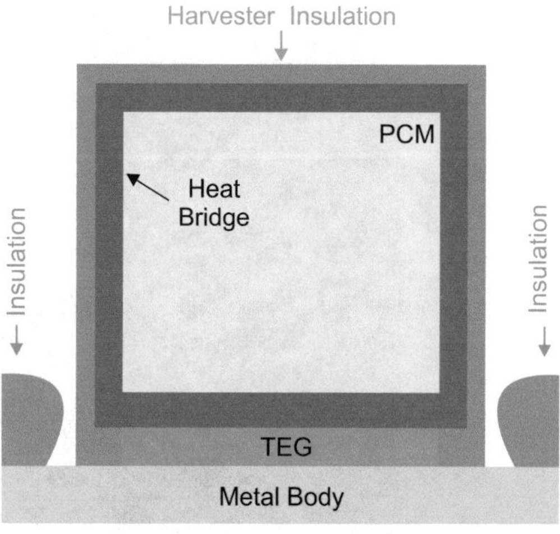

Fig. 1.12 The dynamic thermoelectric energy harvesting device concept

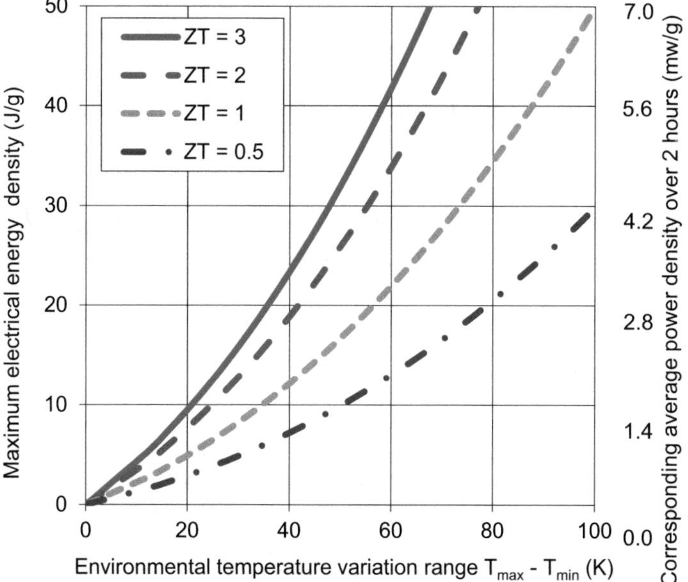

Fig. 1.13 Dynamic thermoelectric harvesting energy availability vs environment temperature fluctuation range [70]. The corresponding average power assuming a 7200 s fluctuation period (corresponding to an aircraft use case) is indicated on the right vertical axis

it is chosen because its heat storage properties are superior to other salt-based or organic solutions. The phase change temperature must lie within the environmental temperature fluctuation range, and therefore, for applications with temperatures not crossing 0 °C, other suitable PCMs may be required.

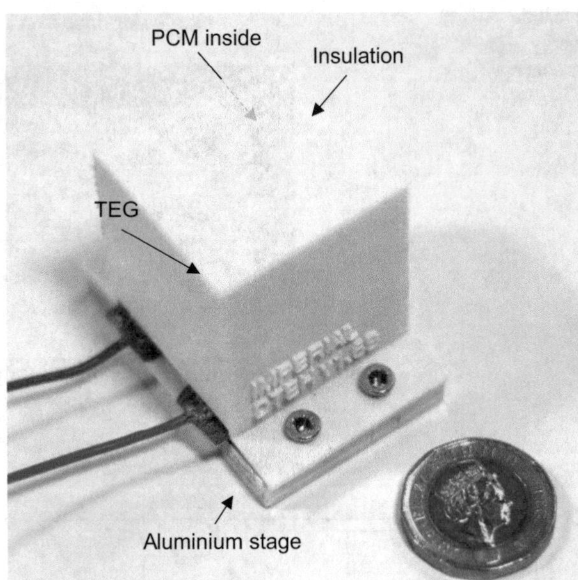

Fig. 1.14 Dynamic thermoelectric harvesting prototype developed for aircraft applications, presented in [87]

A photograph of a dynamic thermoelectric energy harvester developed for aircraft sensor power supply applications is shown in Fig. 1.14.

3.3 Opportunities

In the literature, review papers have studied the thermoelectric harvesting state of the art, offering insight in key technology aspects such as material properties [88], thermal contact improvement [89], the potential of silicides [90] and BiTe nanostructures [91], electrodeposited tellurides [92] and room temperature TEGs, including organic and carbon-based [93]. At device level, reviews focusing on TEG microfabrication [94], wearable applications [95] and thermal design prototypes [96, 97] can be found among others. Further to the key promising aspects highlighted in these works, the following are suggested as topics of particular interest for the short- and mid-term future:

- The wide temperature range of potential applications has led to a wide distribution of research effort, at the cost of slow and out-of-focus advancement. Focusing material research to temperature ranges associated with priority applications would be beneficial for the advancement of thermoelectric energy harvesters and especially energy autonomous microsystems.

- Dynamic thermoelectric harvesting has been shown to address the challenge of unreliable ΔT availability, by employing heat storage. Extension of TEG design to include dynamic response and even incorporate heat storage could lead to a more widely applicable generation of TEG devices.
- The inherent interconnection of heat flow dynamics, steady-state ΔT distribution across series thermal resistance and electrical load matching means that a more holistic dynamic design approach could potentially increase power output. Switching the thermal as well as the electrical contacts towards a fully dynamic thermoelectric operation, in combination with heat storage, could lead to significant improvement of TEG performance in real application environment.
- The impressive performance of SnSe at high temperature in single crystal as well as in polycrystalline form could offer an opportunity of further understanding the mechanisms of controlling electrical conductivity, thermal conductivity and the Seebeck coefficient. This would in turn benefit the development of new materials at lower temperatures with broader application range.
- As also pointed out in [88], energy filtering and engineering the density-of-state asymmetry may lead to materials of improved heat/charge transport coupling. Furthermore, the employment of engineered thermally and electrically asymmetric contacts, such as clean Schottky interfaces, could offer an additional means of asymmetric, energy filtered transport, especially in microfabricated thermoelectrics. An investigation of the role of contacts in general to thermoelectric performance beyond their view as parasitic resistance may lead to new understanding and opportunities in TEG fabrication.

4 Electromagnetic Fields

Another very interesting type of energy harvesting is coupling with environmental local electromagnetic fields. Such fields are usually available around power lines and other alternating current (AC) carrying structures. In this way, power can be delivered to stationary or portable microsystems that require energy autonomy, without invasion to existing electrical infrastructure. In the case of power lines, in addition to portability and installation simplicity benefits over physical Cu wire splitting, such a method also offers security, isolation and electrical decoupling, which can be highly desirable for microsystems that require separate autonomy and reliability in their functionality, such as in security or emergency systems. Relevant environments include industrial plants and machinery, the electrical power grid, electrical installations of buildings, vehicle power networks and any electrified infrastructure such as road and utility networks.

Piezoelectric [98–104], electrostatic [105–108], and inductive [109–116] coupling methods have been proposed in the literature. In most of these works, it has already been demonstrated that adequate power density can be achieved by noninvasive coupling, to support wireless sensors with continuous power. In this section, a brief overview of piezoelectric and electrostatic (capacitive) devices for energy harvesting from AC power lines is presented. Subsequently, the inductive coupling method is discussed in some more detail, because of its broader applicability to other environments and power delivery/collection applications including inductive wireless power transfer.

4.1 Piezoelectric Harvesting from AC Power Lines

The alternating magnetic field around an AC power line can be exploited by employing a permanent magnet mounted on a piezoelectric beam. If the beam is installed such that the magnetic flux density vector B lies in the direction of the beam deflection, the magnetic force can drive the beam into oscillation which is in turn transduced into electricity by the piezoelectric material. This electromechanical approach offers electrical isolation and protection of the secondary circuit, at the expense of an additional intermediate energy transduction step. For a constant given power line frequency, the beam can be designed to operate at resonance, thereby minimising losses. This method was introduced in [104], demonstrating 0.35 mW from a 13 A RMS 60 Hz current, using a 0.26 cm^3 piezoelectric beam installed on a bipolar power supply cable. A similar approach using a Halbach magnet array for field amplification was adopted in [102], demonstrating 0.52 mW from a 5 A RMS, 50 Hz current, from a 2.5 cm^3 beam.

4.2 Electrostatic Harvesting from AC Power Lines

The electric field of an AC power line can also be used for coupling to the conductor voltage. This is possible by employing a capacitor structure, located such that the field gradient creates a voltage difference between its two conductive plates. As the field alternates, charge can flow in and out of the capacitor plates, similar to the electrostatic motion energy harvesting concept, providing a current to a connected electrical load. Thereby, energy is transferred from the power line voltage to a circuit that can drive a local microsystem. A benefit of this approach is that it is functional even without current flow. On the other hand, the electric field is inversely proportional to the conductor/ground distance leading to weak coupling in certain applications. Experimental prototypes have been reported mainly for power grid applications. In [106], a 0.1-m-diameter, 0.2-m-long cylindrical device was proposed. Tests on a 60 kV/50 Hz commercial power grid line demonstrated 16 mW of harvested power. In [108] a device of similar scale was shown to provide 23 mW, in laboratory tests using a 12.7 kV/50 Hz line. In [107], a noninvasive voltage metre solely powered by capacitive electrical field harvesting was demonstrated. A comparative overview of several other implementations can be found in [108]. As mentioned, the output performance of electrostatic power line harvesters relies on the electric field strength. Therefore, applications involving high voltage, single wire non-shielded conductors and short line-to-ground distance could be of particular interest for this energy harvesting concept.

4.3 Inductive Coupling for Energy Collection

Inductive coupling is one of the main transduction methods used in energy harvesting. In addition, it is dominant in currently available wireless power transfer technologies. Therefore, research on coil design and flux engineering is of wide significance, and relevant to both environmental energy collection and wireless power distribution. In the rest of this section, a summary of indicative inductive energy harvesting devices developed for power lines is given. Subsequently, an overview of coil design and flux concentration considerations is presented, followed by a discussion of opportunities for further progress on electromagnetic energy harvesting in general.

Inductive energy harvesting has shown considerable progress as a method to power wireless sensors [109–112, 116]. Power densities in the 0.1–0.5 mW/cm^3 range for low current (~1 A RMS, 620 Hz) [110] and as high as 16 mW/cm^3 from high current (~100 A RMS, 60 Hz) power lines have already been demonstrated [111]. A power output of 0.61 mW from a 290 cm^2 device in a 7 μT RMS, 50 Hz field was demonstrated in [116], using bow-tie core structures to increase the magnetic flux density. In [117], a power density of 0.36 mW/cm^3 was obtained in the vicinity of a 140-mm-long H-shaped structural aircraft beam carrying a 25 A RMS, 360 Hz structural current. A key limiting factor in this progress is the requirement for very specific environmental and installation conditions, such as the ability to install a soft-core loop around a given power line, e.g. in a Rogowski coil geometry. Therefore, in performance comparison, the significant differences among permanent all-around, temporary wrap-around and non-wrapping installations must be taken into account.

4.4 Coil Design

When designing a coil as an energy harvesting transducer, the objective of priority is maximisation of power delivery density, rather than sensitivity, linearity or efficiency which are typically required for sensing or other power transformation applications. In addition, minimal invasion to the source cable is required, which typically results in a single primary loop current transformer approach. The coil-and-core structure must therefore be designed for maximum power density, P_{D}. An expression of P_{D} as a function of magnetic flux density amplitude and frequency, and coil geometry parameters, can be analytically derived. Assuming a cylindrical N-turn coil geometry, around a soft magnetic core cylinder with diameter D, with wire diameter d and hexagonal packing as shown in the cross section of Fig. 1.15,

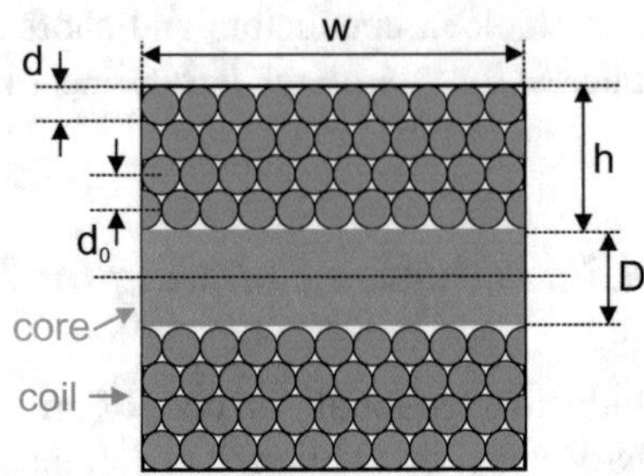

Fig. 1.15 Cross section of a cylindrical, core and coil geometry with hexagonal close coil packing [118]

the coil height h will be:

$$h = d + (I - 1)\, d_0 \tag{1.15}$$

where $I = Nd/w$ is the number of wire layers and $d_o = \sqrt{3}d/2$ is the distance between the hexagonally packed layers.

By approximating the spiral loops with circles, the total wire length can be calculated to be [118]:

$$L = \cong N \cdot \pi \cdot (D + h) \tag{1.16}$$

The coil resistance R_C and maximum output power P_o for a given coil flux amplitude Φ can then be derived, giving:

$$R_C = \rho \cdot \frac{L}{S} = \rho \cdot \frac{\pi \cdot N \cdot (D + h)}{S} = \rho \cdot \frac{\pi \cdot N^2 \cdot (D + h)}{w \cdot h \cdot \eta} \tag{1.17}$$

$$\begin{aligned} P_o = \frac{V_o^2}{8 \cdot R_c} &= \frac{N^2 \cdot \left(\frac{d\Phi}{dt}\right)^2}{8 \cdot \rho \cdot \pi \cdot N \cdot (D + h)/S} = \frac{N \cdot S \cdot \dot{\Phi}^2}{8 \cdot \rho \cdot \pi \cdot (D + h)} \\ &= \frac{w \cdot h \cdot \eta \cdot \dot{\Phi}^2}{8 \cdot \rho \cdot \pi \cdot (D + h)} \end{aligned} \tag{1.18}$$

Here S and ρ are the coil wire cross section and resistivity, respectively, and η is the coil packing filling factor (e.g. $\eta = \pi \cdot \sqrt{3}/6$ for hexagonal close packing). A detailed derivation can be found in [118]. This equation demonstrates that for a given coil size, the relative selection between number of turns N and coil wire diameter d does not affect the maximum power delivery to a matched load. The maximum power is defined by the size (in volume or mass) of the coil material used. The overall device maximum power density depends on the size balance between the core and the coil material.

In a given application with defined size restrictions, an optimal coil/core size balance can be selected. Subsequently, the number of turns can be selected to provide an adequate voltage level, from the anticipated field flux density. In this selection, the resulting coil inductance must also be taken into account. Although the coil reactance can be compensated with a series capacitor, high reactance values are difficult to match accurately. In addition, cancellation is usually achieved at a single frequency. For these reasons, a high N value may result in increased coil output impedance and therefore, moderate N values may be preferable in certain applications.

Note that inductance scales with N^2, as in the long coil approximation: $L_C = \mu N^2 A/w$, where μ and A are the core permeability and cross section area, respectively. For a fixed coil size, R_C also scales with N^2, resulting in a N – independent coil quality factor. A more detailed discussion of coil optimisation for energy collection with indicative value plots can be found in [118].

4.5 Core Structures

The presence of a soft ferromagnet in a magnetic field results in magnetic domain alignment which in turn produces a magnetisation field M. This field increases the magnetic flux density B in the material and reduces the total B around it. By shaping the soft-core structure, magnetic flux can be guided to pass through a smaller cross section, thereby amplifying B. A description of this concept, comparing a rectangular and a funnel-shaped core, is presented in Fig. 1.16. This method has been used and studied in [43, 116–118], achieving power density improvement by more than an order of magnitude. Flux concentration allows the reduction of required core and coil mass to engage with a given flux, but also reduces the required coil length and thereby its output resistance. In this way, the limited-flux disadvantage of electromagnetic coupling in small-scale devices can be moderated. By employing flux guiding structures, small devices, including microgenerators, sensors and actuators, can in turn couple more strongly to weak environmental fields. As an example, a flux funnelling inductive energy harvesting power supply under test for an aircraft application is illustrated in Fig. 1.17 [117].

4.6 Opportunities

Overall, coupling to environmental electromagnetic fields and especially to existing current distribution infrastructure provides an opportunity for delivering mW range continuous power to autonomous wireless microsystems, without invasive installation. This option is not available in all environments but could be practical for applications in industrial plants, in vehicles and along infrastructure networks

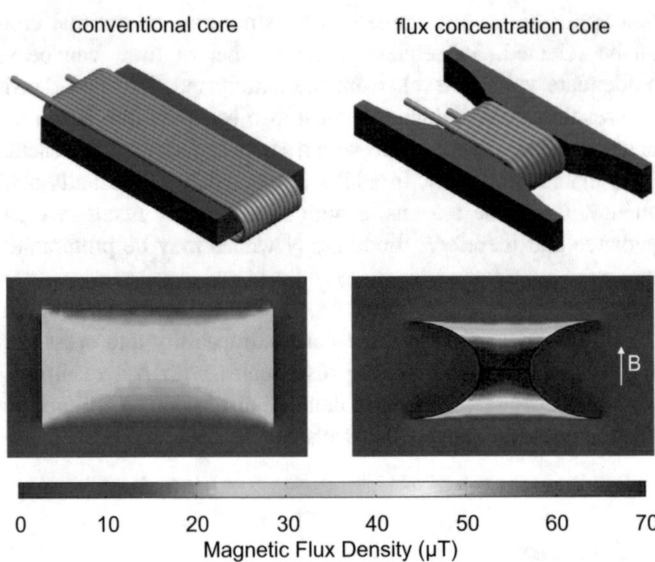

Fig. 1.16 Concept of flux concentration for higher flux density, less transducer mass and lower coil resistance. Top: Conceptual illustration of a core-coil system for a conventional and a flux concentration geometry. Bottom: Indicative COMSOL magnetic flux density simulations for a field corresponding to a current carrying aircraft structural beam (From Ref. [118])

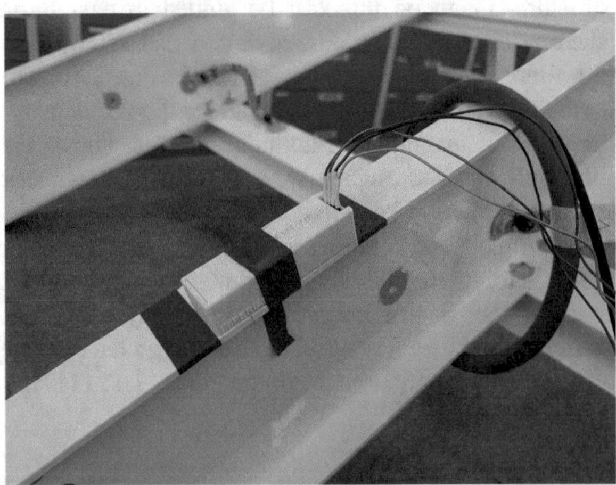

Fig. 1.17 A structural current power supply under test on an industrial aircraft beam presented in [117]

including the electrical power grid and electric railways. A summary of the performance of indicative state-of-the-art devices is given in Table 1.4.

The adoption of optimised coil-and-core design, including flux concentrating structures, is expected to improve drastically the power density of this type of energy

Table 1.4 Overview of energy harvesting devices for environmental electromagnetic fields

Paper	Method	Source line (RMS)	P mW	P_V mW/cm^3
Leland (2006) [104]	Piezo-beam and magnet	13 A, 60 Hz bipolar	0.35	1.3
Zhao (2013) [106]	Electric field, capacitive	60 kV, 50 Hz	16	0.01
Toh (2014) [110]	Tuned coil	0.9 A, 620 Hz	2.9	0.65
He (2014) [102]	Piezo-beam and Halbach	5 A, 50 Hz bipolar	0.52	0.21
Yuan (2017) [116]	Bow-tie and helical core	7 μT, 50 Hz	0.61	0.002
White (2018) [111]	Flux guidance	100 A, 60 Hz	1500	16
Kiziroglou (2021) [117]	Tuned coil, flux funnel	25 A, 360 Hz structural	0.70	0.36

Note: P and P_V denote output power and power density per device volume

harvesting. The applications of these techniques to wireless power transfer receivers may also be beneficial for certain applications, for example, in cases involving weak field coupling or in cases where the use of a soft core at the receiver is permitted.

Beyond these considerations, a concept of particular interest is the combination of energy harvesting and wireless power transfer in a single transducer. An electromagnetic energy harvesting transducer, designed for normal operation under an environmental electromagnetic field, can also be occasionally excited by an inductive wireless power transmitter, to improve power supply reliability by fully charging the storage elements, for system testing purposes, during installation or for activating power-intensive functionalities in a normally low-power microsystem network. This potential feature is not unique to inductive power harvesting, as piezoelectric or photovoltaic transducers can also be externally excited in a wireless (vibration/acoustic or optical) power transfer scheme. This hybrid approach was introduced in [118] and is illustrated in Fig. 1.18. Energy harvesting and wireless power receiver systems are usually designed separately for environmental collection and transfer, as conceptually shown in Fig. 1.18a, b. However, it may be possible to consider both operation modes for a single, hybrid transducer and a dual operation power management system as illustrated in Fig. 1.18c. For example, an inductive harvester operating in a varying magnetic field of a power line could also be occasionally charged by an inductive transmitter wand for testing purposes. In analogy, a vibration energy harvester operating at a remote location could receive power through intentionally induced vibration or acoustic wave, at its resonance frequency, thereby operating as a resonant vibration power receiver and getting a scheduled fast full charging for improved functional reliability.

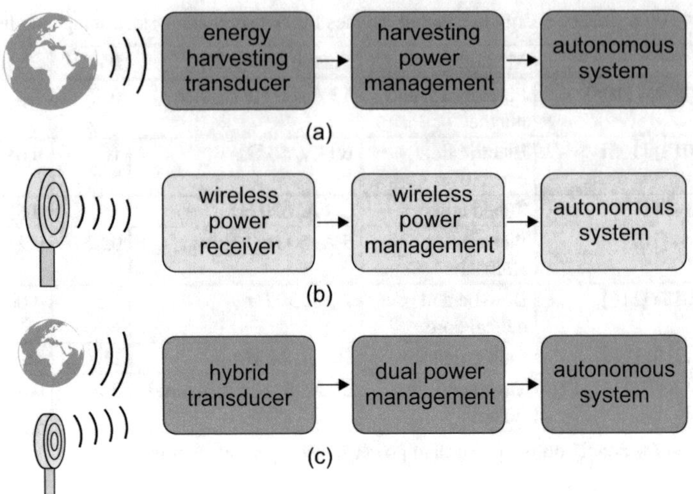

Fig. 1.18 (**a**, **b**) The concept of developing separate systems for energy harvesting and wireless power transfer. (**c**) The concept of designing for dual, harvesting/transfer functionality

5 Energy Harvesting from Waves

While this chapter is focused on motion, thermal and electromagnetic field energy harvesting techniques, here we will briefly discuss other methods.

5.1 Solar Energy Harvesting

Solar energy harvesting refers to the use of small-scale photovoltaic cells to collect energy from light in order to power a local, autonomous microsystem. It is the most mature of energy harvesting technologies and is widely commercialised at scales from millimetres to hundreds of metres. The irradiance of direct sunlight is in the range of 1 kW/m^2 [119]. The efficiency of commercially available monocrystalline and polycrystalline silicon solar cells is up to 20% (typically 17% in integrated commercial panels at a low-volume price of less than 2 $/W). The corresponding electrical power availability is 20 mW/cm^2 in direct sunlight. While silicon solar cells dominate the solar power generation market due to low cost, high-end technologies such as multi-junction and light-concentration devices may be more suitable for energy harvesting applications, in which the required device surface is much smaller and the power value is significantly higher, by at least two orders of magnitude in comparison to the electrical grid market [119]. As an example, the solar-powered, Michigan Micro Mote (M3) wireless sensor platform of the University of Michigan [120] was recently updated with a 25% efficient GaAs solar cell [121].

Beyond the efficiency increase offered by wider bandgap semiconductors and multi-junction cells (up to nearly 50% in laboratory demonstrations, e.g. including six junctions and light concentration in [122]), such high-end devices may offer adequate operation in weaker lighting conditions or under radiation with different spectra. Indeed, indirect outdoor sunlight is around one order of magnitude weaker and indoor lighting around three orders of magnitude weaker than direct sunlight. In such conditions, the conversion efficiency of monocrystalline and polycrystalline silicon is reduced to less than 10%, leading to an indoor electrical power availability in the range of 10 μW/cm^2. While this power density level may be adequate for a range of low-power sensor applications, the reliable availability of light at the desired installation location is a major limitation. The GaAs cell in [121] has been demonstrated to provide 100 μW/cm^2 at 1000 lux (~7.9 W/cm^2) and 20 μW/cm^2 at 200 lux. This allowed the M3 platform to receive around 70 nW under 200 lux, which is significantly higher than the 30 nW required by the M3 for a wireless reporting of temperature once every 30 minutes. Such implementations show promise towards energy autonomous subcutaneous biomedical sensors.

5.2 *Acoustic Energy Harvesting*

Environmental sound waves at acoustic as well as ultrasound frequencies can also be exploited to collect energy for local use by energy autonomous microsystems. Acoustic energy harvesting devices typically include an acoustic impedance matching or wave concentration interface and a piezoelectric or electromagnetic transducer.

The ambient acoustic power density availability is expressed by sound intensity $I = P \cdot U$, measured in W/m^2, where P and U are the sound pressure and velocity, respectively. In air, I is typically expressed in dB, using $I_0 = 1$ pW/m^2 as a reference, and called sound intensity level (SIL): $L_I = 10 \log (I/I_0)$. This reference value corresponds approximately to the lowest sound intensity hearable by a human ear. The corresponding sound pressure level (SPL) is also expressed in dB, using $P_0 = 20$ μPa as reference, selected such that L_I and L_P have the same value in air, given that the air characteristic-specific acoustic impedance is $Z_0 \cong 400$ Pa · s/m . The SIL (and equally SPL) available in some indicative environments are given in Table 1.5.

An important challenge in acoustic energy harvesting research is acoustic impedance matching from air to the transducers. The impedance mismatch is in the three orders of magnitude range, and for this reason, advanced matching structures have been explored, such as multilayers, 3D printed structures as well as metamaterials. In applications where both the ambient sound and the receiver are in a liquid or solid medium, such as industrial metallic infrastructure, liquid containers or maritime environments, the matching problem is greatly reduced.

Just as for ambient vibration, environmental sound is often not present at a constant and predictable frequency. This affects significantly the transducer design,

Table 1.5 Environmental sound power availability in different locations

Environment	SIP and SPL (dB)	Power density (μW/cm^2)
Jet engine (at 1 m distance) [123]	150	100,000
Rock concert [124]	110	10
Glass foundry [125]	100	1
Printing press industry [125]	90	0.1
Metal factory [125]	90	0.1
Plastic packing factory [125]	80	0.01
Aircraft interior during flight [123]	80	0.01
Busy motorway at 10 m [123]	80	0.01
Discussion in average room [124]	60	0.0001
Rainfall [124]	50	0.00001

which needs to operate at resonance for maximum power delivery, as well as the impedance matching structure. Taking in addition into account the limited ambient acoustic power density availability, the employment of acoustic energy collectors in (active) acoustic power transfer systems could be more advantageous. In this way, acoustic pressure fields of higher power density and directionality, and frequency tuned to operate the receiver at resonance and maximum output power, are possible. This also allows tuned acoustic and electrical impedance matching techniques, and the employment of phase synchronisation and beam forming, including phased array and acoustic focusing techniques [126, 127]. A review of various acoustic energy harvesting implementations can be found in [123].

5.3 RF Energy Harvesting

The environmental radio frequency (RF) electromagnetic field is also being considered as a power source for energy harvesting. This includes RF wireless communication carrier signals of G4 (up to 2.5 GHz) and G5 (3.3 GHz–4.2 GHz for the low-frequency C-band of G5) wide area networks and of IEEE 802.11 (Wi-Fi, 2.4 GHz or 5 GHz) local area networks, as well as lower frequency sources. An experimental survey in 2013 demonstrated an ambient RF power availability in the 0.1 μW/cm^2 range at the dominant wireless communication bands at the time [128]. A typical Wi-Fi router emits around 20 dBm (100 mW), which results in a 1 μW/cm^2 power density at a distance of 1 m.

The transducer typically used for RF energy harvesting is a rectifying RF antenna, called a rectenna, which is implemented by combination of an antenna loop in series with a rectifying diode. In RF transmission, the power P_r at a receiver located at distance d from the transmitter as a function of transmitter power P_t is usually calculated using Friis' formula:

$$P_r = P_d \cdot A_r = D_t \frac{P_t}{4\pi d^2} \cdot D_r \frac{\lambda^2}{4\pi}$$

Here, $P_d = D_t P_t/(4\pi d^2)$ is the power density occurring at the location of the receiver, taking into account the geometrical field spreading as well as the transmitter antenna gain D_t due to directionality. At the receiver, A_r is the effective reception area of the receiver antenna, which for a given RF wavelength λ and a typical half wave antenna size can be calculated to be $A_r = D_r\lambda^2/(4\pi)$. In this equation, D_r is the receiver antenna gain due to its own directionality. It can be seen here that antenna efficiency is a major challenge for miniaturisation of RF energy harvesting systems operating at any but the highest frequencies in common use.

The voltage drop, losses and capacitance of the rectenna diode are critical for successful energy harvesting. As an example implementation, a cold-starting RF harvesting power supply of 7.4 $\mu W/cm^3$ from ambient cellular RF carrier signals in an urban environment was presented in [128]. The low-power availability from ambient RF radiation can be addressed in certain applications by intentional transmission of an RF signal, in a far-field RF wireless power transfer approach. In [129], an energy autonomous wireless temperature sensor was reported, demonstrating an RF input power of 72 μW at 70 cm distance from a 27 dBm (500 mW) directional RF transmitter with a 3 dB antenna gain. Reviews of various approaches and implementations of ambient RF energy harvesting can be found in [130, 131].

6 Power Management

The output of energy harvesting devices is usually not in a form suitable for powering directly the intended microsystem, such as a microcontroller, sensor electronics or a wireless transmitter. An interface circuit is typically required, which includes various stages, the most common of which are active transducer driving, rectification, electrical impedance matching, voltage boosting, intermediate storage and associated charging-control circuitry, voltage bucking and regulation and overall monitoring and control. Because energy harvesting power supplies can often be completely depleted of any stored energy, dedicated circuitry for cold-starting is also typically included. This system interfacing the energy harvester output to the input of the power consumer system is called power management. A general block diagram of an energy harvesting power management system is illustrated in Fig. 1.19. In the rest of this section, some key aspects of each power management stage are outlined, including a discussion on overall power supply integration.

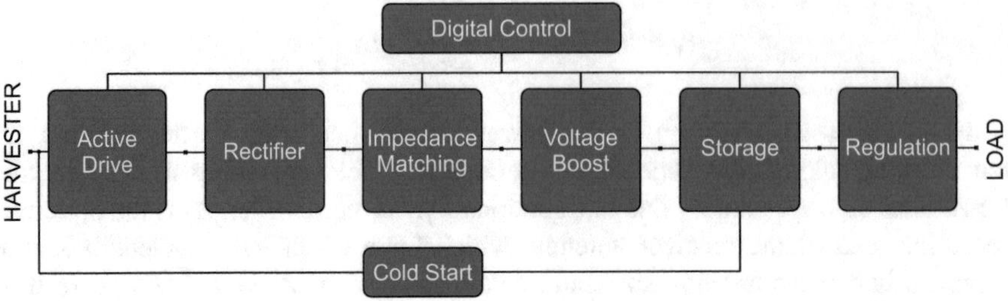

Fig. 1.19 General block diagram of an energy harvesting power management system

6.1 Active Drive

A power management system can include circuitry for applying a current or voltage waveforms to the energy harvesting transducer, in order to drive it to operation points of higher transduction. A characteristic example is the pre-biasing mechanism of piezoelectric motion harvesters. A voltage is applied to the piezoelectric material, with a polarity that is synchronised to the applied motion, such as to enforce operation at points of high strain – polarisation coupling. Specific implementations include the pre-biasing method introduced in [132] and the synchronised switch harvesting on inductor introduced in [133]. An analysis of various circuit implementation approaches can be found in [134]. As mentioned in [16], active driving may also be of interest for other transducers. For example, in inductive transducers, current flow could be switched through a capacitor to increase the current value at which a given magnetic flux variation is transduced. Active driving could also be beneficial for thermoelectric energy harvesting, for example, for exploiting dynamic heat flow and concentration effects. The employment of such methods introduces additional losses and circuit complications, which should be included in the evaluation of benefits. Nevertheless, active driving has shown potential for significant power supply performance improvement.

6.2 Rectification

The output of motion and electromagnetic field harvesters is in AC form and requires rectification before it can be used by a sensor system load or for intermediate storage in a battery or supercapacitor. In active mode operation, synchronised switched MOSFET bridges can be used, offering minimal voltage drop, losses and power consumption. The rectification synchronisation can be accurately and dynamically tuned depending on the harvester output and the electrical load demand, achieving efficiency over 95%. In cold-starting mode, diode-based rectifiers are used, in bridge or other topologies such as a voltage-doubler [117]. Because the voltage output level of energy harvesting transducers can be very low, Schottky diodes of low forward bias voltage drop are typically employed. Reverse leakage in diodes and body leakage in MOSFETs should be taken into account in the design of rectifiers for energy harvesting, because they can lead to significant drain of stored energy, especially in applications involving low duty cycles of load operation and power availability. Rectification is also required in dynamic thermoelectric harvesting, in which the ΔT and hence the output voltage is bipolar, although at a very low polarity switching speed. For such cases, special implementations allowing cold-starting only in one polarity have been considered, by employing normally closed MOSFET switches [135]. Single-polarity cold-starting and optimised bridge switching may potentially offer design and circuit simplicity and offer higher overall efficiency for energy harvesting power management systems.

6.3 Impedance Matching

In sensor applications, transducers are usually connected to electronics designed for high input resistance, so that the measured output is as close as possible to the open circuit conditions, and the influence of electronics to the transducer operation is minimised. The energy transfer in this case is very small. In the case of power transfer from a stored source, such as powering a microcontroller by a battery, the system is designed such that the current demand is small, in order to minimise the losses on the source output impedance and thereby maximise energy transfer efficiency. In contrast, in energy harvesting applications, the ambient energy is typically lost if not collected, and therefore achieving the maximum possible power extraction is priority, even at the expense of conversion efficiency. Energy harvesting devices are designed to operate at their maximum power transfer operating point, which occurs at impedance – matching conditions. The load impedance Z_L must be equal to the conjugate of the transducer output impedance Z_O: $Z_L = Z_O^*$.

The output impedance can vary substantially depending on environmental conditions. For example, the current-voltage characteristic of photovoltaic transducers depends on irradiation levels and temperature. The real and imaginary parts of capacitive and inductive transducers both vary with operation frequency. For this reason, methods for dynamically configuring the resistance presented to the transducer to maintain the maximum possible power transfer are employed. This is called maximum power point tracking (MPPT), and it is a common feature among various types of commercial power management systems designed for energy harvesting applications. The configurable input resistance is usually implemented with the combination of an input capacitor, an inductor and a switch, in combination with a sampling capacitor and a comparator. The switch duty cycle and frequency are controlled such that the voltage on the input capacitor is maintained at a value equal to a certain percentage r of the transducer output voltage, which can be sampled periodically by allowing the capacitor to reach the maximum value, and transferring that value to the sampling capacitor. The comparator allows closed-loop digital control of the voltage level. The inductor is used for efficient charge transfer from the input capacitor to the next power management stage. The value of r is determined by the expected maximum power point of a given transducer. This MPPT concept can be integrated as part of a switched voltage boosting system which is typically required for energy harvesting applications. Although in commercial power management integrated systems the r value is configurable at circuit design level, it may be beneficial to implement MPPT concepts based on monitoring and maximising the actual power income, with dynamic r reconfiguration.

Another important aspect of impedance matching is the possibility of reactance cancellation. For transducers with output impedance including a significant imaginary part, the impedance matching circuitry can include one or more inductive or capacitive components, to present an opposite reactance. In this way, the imaginary part of Z_O is counterbalanced and the overall output impedance magnitude is reduced. This technique has been employed in inductive energy harvesting [43, 136], but it may also be beneficial for certain applications of capacitive transducers, such as piezoelectric acoustic receivers [126]. Limitations in the applicability of reactance cancellation include the size and losses of the additional required components and its dependence on frequency, which may limit its effectiveness to a certain frequency range.

6.4 Voltage Boosting, Bucking and Regulation

The DC voltage output of a rectifier, or the direct output of the harvester in the case of a DC transducer, usually needs to be converted to a different level in order to be suitable for storage or for supplying power to a wireless sensor system. Various types of DC-DC converters can be used for this purpose. A typical example is a simple boost converter, which combines an inductor, a semiconductor switch and a diode. The input voltage is connected in series with the inductor, the diode and a storing capacitor. The switch periodically connects the terminal between the inductor and the diode to the ground. In this way, during this short connection, a forward current is generated in the inductor. When the switch opens, this current continues to flow, due to the inductor magnetic field current inertia, passing through the diode and into the capacitor. In this way, charge is periodically pumped into the capacitor, allowing the generation of voltage much higher than the input voltage level. The diode prevents a reverse current flow from the higher voltage output back to the input. The switching rate is controlled such that the output voltage is maintained at a certain desired, configurable level, often with the help of a comparator and a reference voltage. Voltage bucking and regulation are implemented in a similar manner, by controlled pumping charge to an output capacitor such that the voltage is maintained at a certain desirable level. Overall, this part of power management requires digital control circuitry, analogue comparators, capacitors and low-loss inductors, which are typically included as external (non-monolithic) components.

6.5 Storage

Energy storage is required as part of power management in most energy harvesting applications because of the intermittent nature of power input, but also because of the duty cycling operation of wireless sensor microsystems, which results in power demand peaks and very low-power sleep periods. These variations of input and output power are difficult and not always convenient to synchronise, and therefore some energy buffer is required to smoothen power availability. Rechargeable batteries or supercapacitors are usually employed for this purpose. Batteries offer larger energy storage density and lower leakage. On the other hand, supercapacitors offer much higher recharging cycles, higher power density and a direct measure of energy availability, through their voltage. In contrast, the status of batteries is difficult to predict due to the low and nonlinear dependence of voltage on stored energy.

Battery and supercapacitor charging and discharging is regulated through suitable circuitry, in order to ensure that the corresponding charge transfer rates are within the storage element specification. Overvoltage and under-discharge protection circuits are also employed, usually involving simple combinations of a switch, a comparator and a threshold voltage divider. Overall, the storage components physically occupy a very larger part of the overall power management system physical volume and mass, and they impose temperature, pressure and humidity limitations that are usually much stricter than the rest of the power management components.

6.6 Cold-Starting

Cold-starting refers to the ability of an energy harvesting power supply to start from a completely energy-depleted condition. It is essential because energy harvesters often experience long periods of anticipated or non-anticipated inactivity. An example is the period from fabrication to installation, which may involve environments very different from those expected at the intended installation location. In addition, installation environments may not be active all the time or may offer an energy source only occasionally, with inactivity long enough to deplete any included buffer storage. Finally, in several energy harvesting applications, devices are designed to allow complete depletion when needed. This is particularly useful in developing systems that can provide functionality even when the average incoming power is less than the leakage and sleep mode consumption.

Passive rectification and voltage boosting are the main components required for cold-starting operation. Passive rectification can be implemented using low threshold voltage diodes (such as Schottky diodes), or MOSFET bridges that allow unipolar conductivity when unbiased (such as the employment of depletion MOSFETs). Passive voltage boosting is often obtained by hysteretic switching circuits, involving positive feedback on a transistor, often through a transformer component, similar to a joule thief topology [137]. The efficiency of passive voltage boosting is very low, but it allows cold-starting by voltages as low as 30 mV [138].

Cold-starting is usually implemented in addition to active mode circuitry and accompanied by an additional small storage element which can be charged fast to supply the active mode systems, which are more efficient, as soon as possible. However, in several very-low-power energy harvesting use cases, cold-starting may in practice last for a large part of the system operation time. Therefore, the efficiency of cold-starting circuits and the employment and control of a small secondary capacitive storage can be a key part of energy harvesting power management systems.

6.7 Digital Control

Overall digital control of the different power management subsystems is usually not represented as a separate component. Powering, enabling and introducing the various subsystems to the power flow path is instead considered at each individual stage, depending on local threshold comparisons, hard-wired configurations or jumper selectors. The digital control of power supply and enabling of different subsystems could allow additional dynamic flexibility of operation reconfiguration according to power supply and demand circumstances. Such control could be implemented by a combination of ultrahigh off-state impedance MOSFET switches in combination with central surveillance logic implemented in hardware, with the support of the microcontroller that is usually available in the overall sensor node microsystem. While some commercial power management system allows significant configuration flexibility with external access to enabling various modes of operation, the adoption of a simple overall high-impedance digital control could offer a significant reduction of leakage and quiescent currents, which are very important for sleep as well as cold-starting modes of operation.

6.8 Integration

A main part of the initial conception of energy harvesting as a method to provide complete autonomy to systems with wireless communication was chip-level integration. The rapid development of energy harvesting technology has led to a wide range of meso-scale and large-scale implementations. This has been beneficial for studying and overcoming a variety of challenges including methods of coupling to the energy source (physical contact, inertial structures for motion, heat bridging, coupling to electromagnetic flux, etc.), broadband operation, fabrication and testing challenges and the power management methods of this chapter, as well as adapting to real application environments and coordinating with the priorities of sensing,

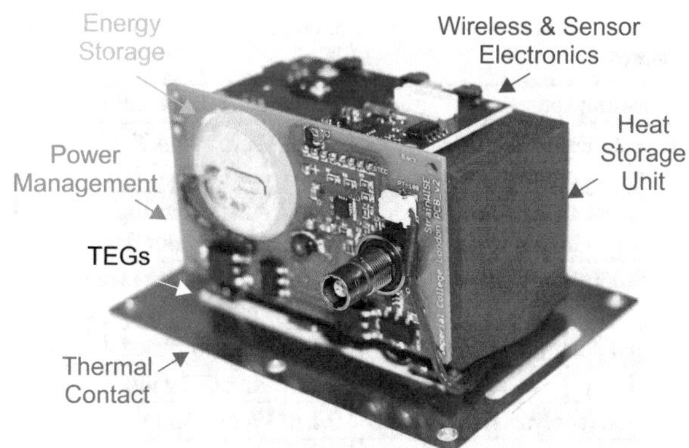

Fig. 1.20 The integrated, heat-powered STRAINWISE aircraft sensor node. (From Ref. [139]

monitoring and data exploitation technologies. Integration into fully functional energy harvesting power supplies, or further integration into energy autonomous wireless sensor nodes, is currently in the 1 cm^3–100 cm^3 range. As an example, the heat-powered STRAINWISE aircraft strain wireless sensor node is shown in Fig. 1.20.

Nevertheless, chip-level integration is still a central objective on energy harvesting technology. The integration and prevalence of MEMS accelerometers into Motion Processing Unit microchips providing integrated motion and navigation services can serve as a roadmap example towards the integration of various types of energy harvesting. In this direction, key challenges include (1) the employment of advanced, standardised and CMOS-compatible MEMS methods in the fabrication process flow of energy harvesters; (2) the introduction of compatible fabrication methods for integrating piezoelectric, thermoelectric and magnetic materials; (3) the achievement of effective power-coupling interfaces with the environment, including efficient mechanical, heat, acoustic and electromagnetic contacts; (4) the on-chip integration of low-loss inductors which are currently typically implemented as external components; and (5) the development of integrated electrolyte and large-area electrode structures for on-chip battery or supercapacitor storage. While these advancements require the addressing of significant fabrication and performance challenges, progress towards energy autonomous integrated microsystems would benefit significantly from its inclusion to microelectronics and telecommunication technological roadmaps.

7 Conclusion

In this chapter, an overview of the energy harvesting technology has been presented, focusing on the operating concepts, main benefits in particular application scenarios and key opportunities for further progress. This discussion offers a general introduction to the main energy transduction techniques for sources including motion,

Table 1.6 State-of-the-art and anticipated power density from various environmental energy sources

Powering source	State-of-the-art power density	2030 (anticipated)
Direct thermoelectric (analysis in this work)	0.1 mW/cm^2 @ $\Delta T = 5$ K 2 mW/cm^2 @ $\Delta T = 20$ K	× 10 (SeSn, nanoparticles, enhanced DOS asymmetry)
Dynamic thermoelectric (analysis in this work)	1 mW/g in 2-hour flight 1 μW/g in 24-hour day	10 mW/g in 2-hour flight 1 mW/g in 24-hour day
Outdoor solar [119]	20 mW/cm^2 @ direct sun 4 mW/cm^2 daily average	× 2 (dual bandgap stacks)
Indoor solar [121]	20 μW/cm^2, diffused 400 lm	× 2 (dual bandgap stacks)
Airflow [3]	4 mW/cm^2 @ 10 m/s	× 2 (MEMS scaling)
Motion (analysis in this work)	0.1 mW/cm^3, 100 Hz	Scaling and broadband
Inductive, power lines [43]	50 μW/g, from 25 A_{rms} 360 Hz structural current	× 10 (flux funnelling)
Inductive WPT [140]	1 mW @ range = 5 × size	Longer range (directionality, rectennas)
Acoustic WPT [141]	1 mW/cm^3 @ 1 m of metal	× 10 (phasing, Z-matching)

heat flow, varying electromagnetic fields and a brief outline of solar, acoustic and far-field RF harvesting. An overview of the main power management concepts, necessary for the development of complete power supplies, is also included, with a discussion of system-level integration. An outline of state-of-the-art and anticipated power densities for different environmental energy sources is presented in Table 1.6. The table includes some indicative power densities reported for remote inductive and acoustic power transfer for comparison and reference.

An overview of the state-of-the-art energy harvesting technology, focusing on industrial use cases, technology adoption challenges as seen by representatives from a wide range of industries and a strategic outlook, has been presented in a white paper from the Power Supply Manufacturers Association (PSMA) in 2021 [142]. In that white paper, research topics identified as key for disruptive progress towards energy autonomy include (1) spatial and time-domain concentration of environmental energy (e.g. focusing light, heat flow or electromagnetic flux), (2) broadband and multi-environment operation, (3) the combination of energy harvesting with power transfer into a single power receiver, (4) sub-μW leakage and sleep mode consumption of microelectronics and storage elements, (5) device design standardisation and (6) the concentration of research and development efforts on a few specific key applications that could be selected for importance priority and serve as industrial paradigms.

A prospect of specific interest is the combination of environmental energy harvesting with wireless power transfer. A motion, inductive or wave (solar, RF, acoustic) harvester can be designed for off-peak normal operation within a broad environmental energy source range, providing a certain relatively low duty cycle energy autonomy level to a wireless microsystem. When required or possible, the same microgenerator can be driven to its optimum power reception point, by mechanical vibration, an acoustic wave and inductive or optical wireless power transfer, to increase overall power autonomy reliability and predictability, or to allow reliable and practical testing of a wireless system network, especially at the installation phase. This combination of energy harvesting and wireless power transfer could expand the applicability of energy autonomy to a wide range of wireless microsystems and present a promising opportunity that can exploit both resonant and off-resonance operation.

As a more specific outlook, a challenge of high interest would be to exploit the multidisciplinary progress on energy harvesting of the last two decades to implement a fully wireless and autonomous, contactless monolithic or hybrid microchip, suitable for implementation in a commonly encountered environment. Such an environment could be human or animal skin, the surface of a machinery or vehicle component or a civil engineering structure.

References

1. N.S. Shenck, J.A. Paradiso, Energy scavenging with shoe-mounted piezoelectrics. IEEE Micro **21**(3), 30–42 (2001)
2. S.W. Arms et al., Energy harvesting wireless sensors and networked timing synchronization for aircraft structural health monitoring, in *Wireless Communication, Vehicular Technology, Information Theory and Aerospace & Electronic Systems Technology, 2009. Wireless VITAE 2009. 1st International Conference on*, (2009), pp. 16–20
3. D.A. Howey, A. Bansal, A.S. Holmes, Design and performance of a centimetre-scale shrouded wind turbine for energy harvesting. Smart Mater. Struct. **20**(8), 085021 (2011)
4. M.E. Kiziroglou, C. He, E.M. Yeatman, Rolling rod electrostatic microgenerator. IEEE Trans. Ind. Electron. **56**(4), 1101–1108 (Apr 2009)
5. H.-X. Zou et al., Mechanical modulations for enhancing energy harvesting: Principles, methods and applications. Appl. Energy **255**, 113871 (2019)
6. M.E. Kiziroglou, E.M. Yeatman, Chapter 17 – Materials and techniques for energy harvesting, in *Functional Materials for Sustainable Energy Applications*, ed. by J. A. Kilner, S. J. Skinner, S. J. C. Irvine, P. P. Edwards, (Woodhead Publishing, 2012), pp. 541–572
7. P.D. Mitcheson, E.M. Yeatman, G.K. Rao, A.S. Holmes, T.C. Green, Energy harvesting from human and machine motion for wireless electronic devices. Proc. IEEE **96**(9), 1457–1486 (2008)
8. E.M. Yeatman, Energy harvesting from motion using rotating and gyroscopic proof masses. Proc. Inst. Mech. Eng. C J. Mech. Eng. Sci. **222**(1), 27–36 (2008)
9. H.W. Lo, Y.C. Tai, Parylene-based electret power generators. J. Micromech. Microeng. **18**(10) (2008)
10. T. Tsutsumino, Y. Suzuki, N. Kasagi, Y. Sakane, Seismic power generator using high-performance polymer electret, in *MEMS 2006: 19th IEEE International Conference on Micro Electro Mechanical Systems, Technical Digest*, (2006), pp. 98–101
11. J. Boland, Y.H. Chao, Y. Suzuki, Y.C. Tai, Micro electret power generator, in *Mems-03: IEEE the Sixteenth Annual International Conference on Micro Electro Mechanical Systems*, (2003), pp. 538–541
12. T. Tsutsumino, Y. Suzuki, N. Kasagi, Electromechanical modeling of micro electret generator for energy harvesting, in *Transducers '07 & Eurosensors Xxi, Digest of Technical Papers, Vols 1 and 2*, (2007), pp. U436–U437

13. T. Sterken, P. Fiorini, K. Baert, R. Puers, G. Borghs, An electret-based electrostatic /spl mu/-generator, in *TRANSDUCERS, Solid-State Sensors, Actuators and Microsystems, 12th International Conference on, 2003*, vol. 2, (2003), pp. 1291–1294
14. C. He, M.E. Kiziroglou, D.C. Yates, E.M. Yeatman, A MEMS self-powered sensor and RF transmission platform for WSN nodes. Sens. J. IEEE **11**(12), 3437–3445 (2011)
15. W.M. Zhang, H. Yan, Z.K. Peng, G. Meng, Electrostatic pull-in instability in MEMS/NEMS: A review (in English). Sens. Actuator. A Phys. Rev. **214**, 187–218 (Aug 2014)
16. M.E. Kiziroglou, E.M. Yeatman, Micromechanics for energy generation. J. Micromech. Microeng. **31**(11), 114003 (2021)
17. Y. Zou, J. Xu, K. Chen, J. Chen, Advances in nanostructures for high-performance triboelectric nanogenerators. Adv. Mater. Technol. **6**(3), 2000916 (2021)
18. T. Toh, S. Wright, M. Kiziroglou, P. Mitcheson, E. Yeatman, Inductive energy harvesting for rotating sensor platforms. J. Phys. Conf. Ser. **557**(1), 012034 (2014)
19. A.S. Holmes, H. Guodong, K.R. Pullen, Axial-flux permanent magnet machines for micropower generation. J. Microelectromech. Syst. **14**(1), 54–62 (2005)
20. J.C. Park, J.Y. Park, A bulk micromachined electromagnetic micro-power generator for an ambient vibration-energy-harvesting system. J. Korean Phys. Soc. **58**(5), 1468–1473 (2011)
21. E. Bouendeu, A. Greiner, P.J. Smith, J.G. Korvink, A low-cost electromagnetic generator for vibration energy harvesting. IEEE Sensors J. **11**(1), 107–113 (2011)
22. D.A. Wang, K.H. Chang, Electromagnetic energy harvesting from flow induced vibration. Microelectron. J. **41**(6), 356–364 (2010)
23. B. Yang, C. Lee, Non-resonant electromagnetic wideband energy harvesting mechanism for low frequency vibrations. Microsyst. Technol. Micro Nanosyst. -Inf. Storage Process. Syst. **16**(6), 961–966 (2010)
24. M. El-hami et al., Design and fabrication of a new vibration-based electromechanical power generator. Sensors Actuators A Phys. **92**(1–3), 335–342 (2001). https://doi.org/10.1016/S0924-4247(01)00569-6
25. C.R. Saha, T. O'Donnell, N. Wang, R. McCloskey, Electromagnetic generator for harvesting energy from human motion. Sens. Actuator. A Phys. **147**(1), 248–253 (2008)
26. S.P. Beeby et al., A micro electromagnetic generator for vibration energy harvesting. J. Micromech. Microeng. **17**(7), 1257–1265 (2007)
27. L.-D. Liao et al., A miniaturized electromagnetic generator with planar coils and its energy harvest circuit. IEEE Trans. Magn. **45**(10), 4621–4627 (2009)
28. S.P. Beeby et al., *Macro and Micro Scale Electromagnetic Kinetic Energy Harvesting Generators* (DTIP 2006: Symposium on Design, Test, Integration and Packaging of MEMS/MOEMS 2006). pp. 286–291 (2006)
29. M. Duffy, D. Carroll, IEEE, Electromagnetic generators for power harvesting, in *Pesc 04: 2004 IEEE 35th Annual Power Electronics Specialists Conference, Vols 1–6, Conference Proceedings*, (IEEE Power Electronics Specialists Conference Records, 2004), pp. 2075–2081
30. Y.S. Tan, Y. Dong, X.H. Wang, Review of MEMS electromagnetic vibration energy harvester. J. Microelectromech. Syst. **26**(1), 1–16 (2017)
31. T. Budde, H.H. Gatzen, Thin film SmCo magnets for use in electromagnetic microactuators (in English). J. Appl. Phys. **99**(8) (2006)
32. A. Walther, C. Marcoux, B. Desloges, R. Grechishkin, D. Givord, N.M. Dempsey, Micropatterning of NdFeB and SmCo magnet films for integration into micro-electro-mechanical-systems. J. Magn. Magn. Mater. **321**(6), 590–594 (2009)
33. Y. Jiang et al., *Fabrication and Evaluation of NdFeB Microstructures for Electromagnetic Energy Harvesting Devices*, Presented at the PowerMEMS, Washington DC, December 1–4 (2009)
34. F.M.F. Rhen, E. Backen, J.M.D. Coey, Thick-film permanent magnets by membrane electrodeposition (in English). J. Appl. Phys. **97**(11) (2005)

35. I. Zana, G. Zangari, J.-W. Park, M.G. Allen, Electrodeposited co-Pt micron-size magnets with strong perpendicular magnetic anisotropy for MEMS applications. J. Magn. Magn. Mater. **272–276**(Supplement 1), E1775–E1776 (2004). https://doi.org/10.1016/j.jmmm.2003.12.990
36. L. Vieux-Rochaz et al., Electrodeposition of hard magnetic CoPtP material and integration into magnetic MEMS. J. Micromech. Microeng. **16**(2), 219 (2006)
37. F.M.F. Rhen, G. Hinds, C. O'Reilly, J.M.D. Coey, Electrodeposited FePt films. IEEE Trans. Magn. **39**(5), 2699–2701 (2003)
38. C.T. Pan, Y.M. Hwang, H.L. Hu, H.C. Liu, Fabrication and analysis of a magnetic self-power microgenerator. J. Magn. Magn. Mater. **304**(1), e394–e396 (2006). https://doi.org/10.1016/j.jmmm.2006.01.202
39. Y.Z. Wang, B.Y. Jimenez, D.P. Arnold, IEEE, 100-mu m-thick high-energy-density electroplated CoPt permanent magnets, in *2020 33rd IEEE International Conference on Micro Electro Mechanical Systems*, (Proceedings IEEE Micro Electro Mechanical Systems, New York: IEEE, 2020), pp. 558–561
40. Y.Z. Wang, J. Ewing, D.P. Arnold, Ultra-thick electroplated CoPt magnets for MEMS (in English). J. Microelectromech. Syst. **28**(2), 311–320 (2019)
41. H.J. Cho, S. Bhansali, C.H. Ahn, Electroplated thick permanent magnet arrays with controlled direction of magnetization for MEMS application. J. Appl. Phys. **87**(9), 6340–6342 (2000)
42. D.P. Arnold, N. Wang, Permanent Magnets for MEMS. J. Microelectromech. Syst. **18**(6), 1255–1266 (2009)
43. S.W. Wright, M.E. Kiziroglou, S. Spasic, N. Radosevic, E.M. Yeatman, Inductive energy harvesting from current-carrying structures. IEEE Sensors Letters **3**(6), 1–4 (2019)
44. J.V. Ahuir, *Selection and Characteristics of WE-FSFS*. Available: https://www.we-online.com/web/en/electronic_components/produkte_pb/application_notes/auswahlundeigenschaftenvonwefsfs.php (2021)
45. E. Peng, X. Wei, T.S. Herng, U. Garbe, D. Yu, J. Ding, Ferrite-based soft and hard magnetic structures by extrusion free-forming. RSC Adv. **7**(43), 27128–27138 (2017). https://doi.org/10.1039/C7RA03251J
46. B.-Y. Zong et al., Electrodeposition of granular FeCoNi films with large permeability for microwave applications. J. Mater. Chem. **21**(40), 16042–16048 (2011). https://doi.org/10.1039/C1JM13398E
47. T. Lisec et al., A novel technology for MEMS based on the agglomeration of powder by atomic layer deposition, in *2017 19th International Conference on Solid-State Sensors, Actuators and Microsystems (TRANSDUCERS)*, (2017), pp. 427–430
48. M. Paesler, T. Lisec, H. Kapels, Novel integrated BEOL compatible inductances for power converter applications, in *2020 IEEE Applied Power Electronics Conference and Exposition (APEC)*, (2020), pp. 2647–2652
49. S. Roundy, P. KWright, A piezoelectric vibration based generator for wireless electronics. Smart Mater. Struct. **13**, 1131–1142 (2004)
50. APC International Ltd, *Physical and Piezoelectric Properties of APC Materials*. Available: https://www.americanpiezo.com/apc-materials/piezoelectric-properties.html (2017)
51. E.E. Aktakka, R.L. Peterson, K. Najafi, Wafer-level integration of high-quality bulk piezoelectric ceramics on silicon. IEEE Transactions on Electron Devices **60**(6), 2022–2030 (2013)
52. C.-B. Eom, S. Trolier-McKinstry, Thin-film piezoelectric MEMS. MRS Bull. **37**(11), 1007–1017 (2012)
53. D.L. Polla, L.F. Francis, Processing and characterization of piezoelectric materials and integration into microelectromechanical systems. Annu. Rev. Mater. Sci. **28**, 563–597 (1998)
54. C. He, D. Zhou, F. Wang, H. Xu, D. Lin, H. Luo, Elastic, piezoelectric, and dielectric properties of tetragonal Pb(Mg1/3Nb2/3)O3–PbTiO3 single crystals. J. Appl. Phys. **100**(8), 086107 (2006)
55. X. Gao, J. Wu, Y. Yu, Z. Chu, H. Shi, S. Dong, Giant piezoelectric coefficients in Relaxor piezoelectric ceramic PNN-PZT for vibration energy harvesting. Adv. Funct. Mater. **28**(30), 1706895 (2018)

56. A.W. Warner, M. Onoe, G.A. Coquin, Determination of elastic and piezoelectric constants for crystals in class (3m). J. Acoust. Soc. Am. **42**(6), 1223–1231 (1967)
57. J. García, J.L. Sanchez-Rojas, A. Ababneh, U. Schmid, S. González, E. Iborra, *Piezoelectric Characterization of Ain Thin Films on Silicon Substrates*, Presented at the XXII Eurosensors 2008, Dresden, Germany, 07/09/2008–10/09/2008 (2008)
58. S.J. Rupitsch, *Piezoelectric Sensors and Actuators* (Springer, 2019)
59. M. Wu et al., High-performance piezoelectric-energy-harvester and self-powered mechanosensing using lead-free potassium–sodium niobate flexible piezoelectric composites. J. Mater. Chem. A **6**(34), 16439–16449 (2018). https://doi.org/10.1039/C8TA05887C
60. M. Safaei, H.A. Sodano, S.R. Anton, A review of energy harvesting using piezoelectric materials: State-of-the-art a decade later (2008–2018). Smart Mater. Struct. **28**(11), 113001 (2019)
61. H.C. Liu, J.W. Zhong, C. Lee, S.W. Lee, L.W. Lin, A comprehensive review on piezoelectric energy harvesting technology: Materials, mechanisms, and applications. Appl. Phys. Rev. **5**(4), 041306 (2018)
62. N. Sezer, M. Koc, A comprehensive review on the state-of-the-art of piezoelectric energy harvesting. Nano Energy **80**, 105567 (2021)
63. K. Murotani, Y. Suzuki, MEMS electret energy harvester with embedded bistable electrostatic spring for broadband response. J. Micromech. Microeng. **28**(10), 2018, 104001
64. Y. Suzuki, Electrostatic/electret-based harvesters, in *Micro Energy Harvesting*, (2015), pp. 149–174
65. R. Elfrink et al., Vacuum-packaged piezoelectric vibration energy harvesters: Damping contributions and autonomy for a wireless sensor system. J. Micromech. Microeng. **20**(10), 104001 (2010)
66. Y. Suzuki, D. Miki, M. Edamoto, M. Honzumi, A MEMS electret generator with electrostatic levitation for vibration-driven energy-harvesting applications. J. Micromech. Microeng. **20**(10), 104002 (2010)
67. A. Shin et al., A MEMS magnetic-based vibration energy harvester. J. Phys. Conf. Ser. **1052**, 012082 (2018)
68. D. Zhang, H. Wu, C.R. Bowen, Y. Yang, Recent advances in pyroelectric materials and applications. Small **17**(51), 2103960 (2021)
69. D.M. Rowe, *CRC Handbook of Thermoelectrics* (CRC Press, 1995)
70. M.E. Kiziroglou, S.W. Wright, T.T. Toh, P.D. Mitcheson, T. Becker, E.M. Yeatman, Design and fabrication of heat storage thermoelectric harvesting devices. Industrial Electronics, IEEE Transactions on **61**(1), 302–309 (2014)
71. T. Becker, A. Elefsiniotis, M.E. Kiziroglou, Thermoelectric energy harvesting in aircraft, in *Micro Energy Harvesting*, (Wiley-VCH Verlag GmbH & Co. KGaA, 2015), pp. 415–434
72. E. Krali, C. Ki, K. Fobelets, Z.A. Durrani, *Seebeck Coefficient in Silicon Nanowire arrays*, Presented at the European Conference on Thermoelectrics, Thessaloniki, Greece, 28–30 September, 2011 (2011)
73. A. Hamid, K. Fobelets, J.E. Velazquez-Perez, Optimization of Thermo-electric power generators by gating the silicon nanowires, in *European Modeling Symposium on Mathematical Modeling and Computer Simulations (EMS2017)*, (Manchester, UK, 2017)
74. C. Zhou et al., Polycrystalline SnSe with a thermoelectric figure of merit greater than the single crystal. Nat. Mater. **20**(10), 1378–1384 (2021)
75. J.-H. Bahk, A. Shakouri, Electron transport engineering by nanostructures for efficient thermoelectrics, in *Nanoscale Thermoelectrics*, ed. by X. Wang, Z. M. Wang, (Springer, Cham, 2014), pp. 41–92
76. Y. Zhang et al., Hot carrier filtering in solution processed heterostructures: A paradigm for improving thermoelectric efficiency. Advanced Materials (Deerfield Beach, FL) **26** (2014)
77. N. Jaziri, A. Boughamoura, J. Müller, B. Mezghani, F. Tounsi, M. Ismail, A comprehensive review of thermoelectric generators: Technologies and common applications. Energy Rep. **6**, 264–287 (2020)

78. M.E. Kiziroglou, E.M. Yeatman, Materials and techniques for energy harvesting, in *Functional Materials for Sustainable Energy Applications*, ed. by E. Kilner, (Woodhead Publishing, 2012)
79. *M.H.S. Thermal Resistance Calculator – Plate Fin Heat Sink*. Available: https://myheatsinks.com/calculate/thermal-resistance-plate-fin/ (2021)
80. M. Freunek, M. Müller, T. Ungan, W. Walker, L.M. Reindl, New physical model for thermoelectric generators. J. Electron. Mater. **38**(7), 1214–1220 (2009)
81. T. Becker, A. Elefsiniotis, M.E. Kiziroglou, Thermoelectric energy harvesting in aircraft, in *Micro Energy Harvesting*, ed. by D. Briand, S. Roundy, E. Yeatman, (Wiley, 2014), pp. 415–433
82. D. Samson, T. Otterpohl, M. Kluge, U. Schmid, T. Becker, Aircraft-specific thermoelectric generator module. J. Electron. Mater. **39**(9), 2092–2095 (2010)
83. A. Elefsiniotis, D. Samson, T. Becker, U. Schmid, Investigation of the performance of thermoelectric energy harvesters under real flight conditions (in English). J. Electron. Mater. **42**(7), 2301–2305 (2013)
84. L.V. Allmen et al., Aircraft strain WSN powered by heat storage harvesting. IEEE Trans. Ind. Electron. **64**(9), 7284–7292 (2017)
85. M. Kiziroglou et al., Performance of phase change materials for heat storage thermoelectric harvesting. Appl. Phys. Lett. **103**(19), 193902 (2013)
86. M.E. Kiziroglou, T. Becker, S.W. Wright, E.M. Yeatman, J.W. Evans, P.K. Wright, Three-dimensional printed insulation for dynamic thermoelectric harvesters with encapsulated phase change materials. IEEE Sensors Letters **1**(4), 1–4 (2017)
87. M.E. Kiziroglou et al., *Milliwatt Power Supply By Dynamic Thermoelectric Harvesting*, Presented at the PowerMEMS, Daytona Beach, Florida. USA, December 4–7 (2018)
88. M.N. Hasan, H. Wahid, N. Nayan, M.S.M. Ali, Inorganic thermoelectric materials: A review. Int. J. Energy Res. **44**(8), 6170–6222 (2020)
89. R. He, G. Schierning, K. Nielsch, Thermoelectric devices: A review of devices, architectures, and contact optimization. Advanced Materials Technologies **3**(4), 1700256 (2018)
90. A. Nozariasbmarz et al., Thermoelectric silicides: A review. Jpn. J. Appl. Phys. **56**(5), 05da04 (2017)
91. H. Mamur, M.R.A. Bhuiyan, F. Korkmaz, M. Nil, A review on bismuth telluride (Bi2Te3) nanostructure for thermoelectric applications. Renew. Sust. Energ. Rev. **82**, 4159–4169 (2018)
92. T.J. Wu, J. Kim, J.H. Lim, M.S. Kim, N.V. Myung, Comprehensive review on thermoelectric electrodeposits: Enhancing thermoelectric performance through nanoengineering. Front. Chem. **9**, 762896 (2021)
93. Z. Soleimani, S. Zoras, B. Ceranic, S. Shahzad, Y.L. Cui, A review on recent developments of thermoelectric materials for room-temperature applications. Sustainable Energy Technologies and Assessments **37**, 100604 (2020)
94. J.B. Yan, X.P. Liao, D.Y. Yan, Y.G. Chen, Review of micro thermoelectric generator. J. Microelectromech. Syst. **27**(1), 1–18 (2018)
95. A. Nozariasbmarz et al., Review of wearable thermoelectric energy harvesting: From body temperature to electronic systems. Appl. Energy **258**, 114069 (2020)
96. O.H. Ando, A.L.O. Maran, N.C. Henao, A review of the development and applications of thermoelectric microgenerators for energy harvesting. Renew. Sust. Energ. Rev. **91**, 376–393 (2018)
97. N. Jaziri, A. Boughamoura, J. Muller, B. Mezghani, F. Tounsi, M. Ismail, A comprehensive review of thermoelectric generators: Technologies and common applications. Energy Rep. **6**, 264–287 (2020)
98. W. He, P. Li, Y. Wen, J. Zhang, C. Lu, A. Yang, Energy harvesting from electric power lines employing the Halbach arrays. Rev. Sci. Instrum. **84**(10), 105004 (2013)
99. E.S. Leland, P.K. Wright, R.M. White, A MEMS AC current sensor for residential and commercial electricity end-use monitoring. J. Micromech. Microeng. **19**(9), 094018 (2009)

100. A. Abasian, A. Tabesh, A.Z. Nezhad, N. Rezaei-Hosseinabadi, Design optimization of an energy harvesting platform for self-powered wireless devices in monitoring of AC power lines. IEEE Trans. Power Electron. **33**(12), 10308–10316 (2018)
101. A. Abasian, A. Tabesh, N. Rezaei-Hosseinabadi, A.Z. Nezhad, M. Bongiorno, S.A. Khajehoddin, Vacuum-packaged piezoelectric energy harvester for powering smart grid monitoring devices. IEEE Trans. Ind. Electron. **66**(6), 4447–4456 (2019)
102. W. He, P. Li, Y.M. Wen, J.T. Zhang, A.C. Yang, C.J. Lu, A noncontact Magnetoelectric generator for energy harvesting from power lines. IEEE Trans. Magn. **50**(11), 8204604 (2014)
103. T. Hosseinimehr, A. Tabesh, Magnetic field energy harvesting from AC lines for powering wireless sensor nodes in smart grids. IEEE Trans. Ind. Electron. **63**(8), 4947–4954 (2016)
104. E.S. Leland, R.M. White, P. Wright, Energy scavenging power sources for household electrical monitoring, in *PowerMEMS*, (Berkeley, USA, 2006)
105. F. Guo, H. Hayat, Z. Wang, Energy harvesting devices for high voltage transmission line monitoring, in *2011 IEEE Power and Energy Society General Meeting*, pp. 1, 2011–8
106. X. Zhao, T. Keutel, M. Baldauf, O. Kanoun, Energy harvesting for a wireless-monitoring system of overhead high-voltage power lines. IET Generation, Transmission & Distribution **7**(2), 101–107 (2013)
107. S. Kang, S. Yang, H. Kim, Non-intrusive voltage measurement of ac power lines for smart grid system based on electric field energy harvesting. Electron. Lett. **53**(3) (2017)
108. J.C. Rodriguez, D.G. Holmes, B. McGrath, R.H. Wilkinson, A self-triggered pulsed-mode Flyback converter for electric-field energy harvesting. IEEE Journal of Emerging and Selected Topics in Power Electronics **6**(1), 377–386 (2018)
109. O. Thorin, *Power Line Induction Energy Harvesting Powering Small Sensor Nodes*, KTH, School of Industrial Engineering and Management (ITM), Machine Design (Dept.). KTH Royal Institute of Technology, Sweden (2016)
110. T. Toh et al., Inductive energy harvesting from variable frequency and amplitude aircraft power lines. J. Phys. Conf. Ser. **557**(1), 012095 (2014)
111. R.M. White, D.S. Nguyen, Z. Wu, P.K. Wright, Atmospheric sensors and energy harvesters on overhead power lines. Sensors (Basel) **18**(1) (2018)
112. S. Yuan, Y. Huang, J. Zhou, Q. Xu, C. Song, P. Thompson, Magnetic field energy harvesting under overhead power lines. IEEE Trans. Power Electron. **30**(11), 6191–6202 (2015)
113. T. Lim, Y. Kim, Compact self-powered wireless sensors for real-time monitoring of power lines. Journal of Electrical Engineering & Technology **14**(3), 1321–1326 (2019)
114. B. Park et al., Optimization design of toroidal core for magnetic energy harvesting near power line by considering saturation effect. AIP Adv. **8**(5), 056728 (2018)
115. W. Wang, Z.B. Zhu, Q. Wang, M.Q. Hu, Optimisation design of real-time wireless power supply system overhead high-voltage power line. IET Electr. Power Appl. **13**(2), 206–214 (2019)
116. S. Yuan, Y. Huang, J. Zhou, Q. Xu, C. Song, G. Yuan, A high-efficiency helical Core for magnetic field energy harvesting. IEEE Trans. Power Electron. **32**(7), 5365–5376 (2017)
117. M.E. Kiziroglou, S.W. Wright, E.M. Yeatman, Power supply based on inductive harvesting from structural currents. IEEE Internet Things J. **9**(10), 7166–7177 (2022)
118. M.E. Kiziroglou, S.W. Wright, E.M. Yeatman, Coil and core design for inductive energy receivers. Sensors Actuators A Phys. **313**, 112206 (2020)
119. M.E. Kiziroglou et al., Speed vs efficiency and storage type in portable energy systems. J. Phys. Conf. Ser. **1052**, 012026 (2018)
120. P. Pannuto, Y. Lee, Z. Foo, D. Blaauw, P. Dutta, *M3: A mm-Scale Wireless Energy Harvesting Sensor Platform*, Presented at the Proceedings of the 1st International Workshop on Energy Neutral Sensing Systems, Rome, Italy. Available: https://doi.org/10.1145/2534208.2534225 (2013)
121. E. Moon, I. Lee, D. Blaauw, J.D. Phillips, High-efficiency photovoltaic modules on a chip for millimeter-scale energy harvesting. Prog. Photovolt. Res. Appl. **27**(6), 540–546 (2019)

122. J.F. Geisz et al., Six-junction III–V solar cells with 47.1% conversion efficiency under 143 suns concentration. Nat. Energy **5**(4), 326–335 (2020)
123. J. Choi, I. Jung, C.Y. Kang, A brief review of sound energy harvesting. Nano Energy **56**, 169–183 (2019)
124. *Common Environmental Noise Levels*. Available: https://www.chchearing.org/common-environmental-noise-levels (2022)
125. S. Gerges, G. Sehrndt, W. Parthey, 5 NOISE SOURCES, *Occupational Exposure to Noise* (2001)
126. A.Y. Pandiyan, M.E. Kiziroglou, E.M. Yeatman, Complex impedance matching for far-field acoustic wireless power transfer, in *2021 IEEE 20th International Conference on Micro and Nanotechnology for Power Generation and Energy Conversion Applications (PowerMEMS)*, (2021), pp. 44–47
127. A. Pandiyan, *Acoustic Power Distribution Techniques for Wireless Sensor Networks*, Doctor of Philosophy, Electrical and Electronic Engineering, Imperial College London (2022)
128. M. Piñuela, P.D. Mitcheson, S. Lucyszyn, Ambient RF energy harvesting in urban and semi-urban environments. IEEE Transactions on Microwave Theory and Techniques **61**(7), 2715–2726 (2013)
129. R.L. Rosa, Remotely powered temperature sensor for monitoring food cooking, in *Energy Harvesting for a Green Internet of Things*, (PSMA, 2021)
130. S. Kim et al., Ambient RF energy-harvesting Technologies for Self-Sustainable Standalone Wireless Sensor Platforms. Proc. IEEE **102**(11), 1649–1666 (2014)
131. X. Lu, P. Wang, D. Niyato, D.I. Kim, Z. Han, Wireless networks with RF energy harvesting: A contemporary survey. IEEE Communications Surveys and Tutorials **17**(2), 757–789 (2015)
132. J. Dicken, P.D. Mitcheson, I. Stoianov, E.M. Yeatman, *Increased Power Output from Piezoelectric Energy Harvesters by Pre-Biasing*, Presented at the POWERMEMS 2009 (2009)
133. D. Guyomar, A. Badel, E. Lefeuvre, C. Richard, Toward energy harvesting using active materials and conversion improvement by nonlinear processing. IEEE Trans. Ultrason. Ferroelectr. Freq. Control **52**(4), 584–595 (2005)
134. J. Dicken, P.D. Mitcheson, I. Stoianov, E.M. Yeatman, Power-extraction circuits for piezoelectric energy harvesters in miniature and low-power applications. IEEE Trans. Power Electron. **27**(11), 4514–4529 (2012)
135. T. Toh, S. Wright, M. Kiziroglou, P. Mitcheson, E. Yeatman, A dual polarity, cold-starting interface circuit for heat storage energy harvesters. Sensors Actuators A Phys. **211**, 38–44 (2014)
136. T.T. Toh, S.W. Wright, M.E. Kiziroglou, E.M. Yeatman, P.D. Mitcheson, Harvesting energy from aircraft power lines, in *Proceedings of the 1st International Workshop on Energy Neutral Sensing Systems*, (ACM, 2013), p. 13
137. *Texas Instruments BQ25570 Nano Power Boost Charger and Buck Converter for Energy Harvester Powered Applications*, Available: www.ti.com/product/BQ25570 (2019)
138. *LTC3109 Datasheet*. Available: http://cds.linear.com/docs/en/datasheet/3109fb.pdf
139. L.V. Allmen et al., Aircraft strain WSN powered by heat storage harvesting. IEEE Trans. Ind. Electron. **PP**(99), 7284–7292 (2017)
140. D.E. Boyle, M.E. Kiziroglou, P.D. Mitcheson, E.M. Yeatman, Energy provision and storage for pervasive computing. IEEE Pervasive Computing **15**(4), 28–35 (2016)
141. M.E. Kiziroglou, D.E. Boyle, S.W. Wright, E.M. Yeatman, Acoustic power delivery to pipeline monitoring wireless sensors. Ultrasonics **77**, 54–60 (2017)
142. T. Becker et al., *Energy Harvesting for a Green Internet of Things* (PSMA White Paper). Power Supply Manufacturers Association (2021)

Chapter 2
SiC and GaN Power Devices

Konstantinos Zekentes, Victor Veliadis, Sei-Hyung Ryu, Konstantin Vasilevskiy, Spyridon Pavlidis, Arash Salemi, and Yuhao Zhang

1 Introduction

In an increasingly electrified, technology-driven world, power electronics is central to the entire clean energy manufacturing economy. Power switching semiconductor devices are key enablers in a wide range of power applications, including novel lighting technologies, automotive and rail traction, on board chargers, consumer electronics, aerospace, photovoltaic, flexible alternative current transmission systems, high-voltage DC systems, microgrids, energy storage, motor drives, UPS, and data centers. Silicon power devices have dominated power electronics due to their low-cost volume production, excellent starting material quality, ease of processing, and proven reliability and ruggedness. Although Si power devices continue to

K. Zekentes (✉)
FORTH, IESL, Crete, Greece

Grenoble INP, IMEP-LaHC, Grenoble, France
e-mail: zekentesk@iesl.forth.gr

V. Veliadis
PowerAmerica/NCSU-ECE, Raleigh, NC, USA

S.-H. Ryu
Wolfspeed, Durham, NC, USA

K. Vasilevskiy
Newcastle University, Newcastle upon Tyne, UK

S. Pavlidis
NCSU-ECE, Raleigh, NC, USA

A. Salemi
Alpha & Omega Semiconductor, Sunnyvale, CA, USA

Y. Zhang
CPES, Virginia Polytechnic Institute and State University, Blacksburg, VA, USA

F. Iacopi, F. Balestra (eds.), *More-than-Moore Devices and Integration for Semiconductors*, https://doi.org/10.1007/978-3-031-21610-7_2

make progress, they are approaching their operational limits primarily due to their poor high-temperature performance and their relatively low bandgap and critical electric field, which result in high conduction and switching losses. Wide bandgap (WBG) SiC and GaN power semiconductor devices have recently emerged as highly efficient alternatives to their venerable MOSFET and IGBT Si counterparts. With smaller form factor, reduced cooling requirements, and established reliability, WBG devices are cost-effective silicon replacements at the system level while allowing for novel circuit architectures and simplification. In particular, as environmental awareness and a worldwide push for a zero emissions economy gain prominence, the energy efficiency offered by WBG solutions is a strong driver in their wide market acceptance and mass commercialization.

The compelling material properties of WBG devices are at the core of their suitability for more efficient, lighter, smaller form-factor power electronics operating at high frequencies, and at elevated temperatures with reduced cooling. The wider energy bandgap of 4H-SiC and GaN materials compared to that of Si allows for orders of magnitude lower intrinsic carrier density, which enables high-temperature operation with simplified thermal management. With a critical electric field that is seven to ten times larger than Si's, combined with their wider energy bandgap, WBG semiconductors can be used to make practical high-voltage (10 kV) power devices with reduced conduction and switching losses. This allows for efficient high-frequency operation that minimizes the weight and volume of passive components, increases power density, and lowers the overall system cost. For instance, the drift layer of a 4H-SiC power MOSFET can have one-tenth the thickness and about hundred times higher doping concentration of the drift layer of a silicon power MOSFETs with the same blocking capability. This results in a factor of ~800 reduction in drift layer resistance and enables smaller die sizes compared to those of silicon power devices with comparable on-state resistance and blocking voltage. Therefore, it is possible to achieve low switching and conduction losses for a wide range of blocking voltages and frequencies. Lower losses simplify circuit topology and control design and reduce the complexity of gate drivers. Overall, WBG power devices enable novel power electronics systems with higher efficiency and higher gravimetric and volumetric power-conversion densities.

High-yield manufacturing at volume fabs is a prerequisite for mass WBG commercialization. Numerous well-established processes from silicon technology have been successfully transferred to SiC. In addition, several fabrication processes specific to SiC have been developed and are at a stage of maturity. Today, SiC is produced in dedicated fabs as well as alongside silicon fabrication. The latter has the potential of SiC manufacturing at the economy scale of silicon and is a particularly attractive model. Overall, a vibrant worldwide fab infrastructure produces cost-effective 650 V to 1.7 kV SiC devices having successfully duplicated the integrated device manufacturer (IDM), foundry, fabless, and design-house silicon fabrication models. Similarly, lateral GaN power devices, commercially available from several vendors in the 100–650 V range, are CMOS-compatible and are fabricated cost-competitively in volume Si fabs and foundries.

Barriers to WBG mass commercialization still exist. Primarily, they are the higher than silicon device cost, reliability and ruggedness concerns, and the need for a trained workforce to skillfully insert WBG devices into power electronics systems. In many applications, at the system level, SiC-based systems are more cost-effective than those of silicon due to passive component simplifications. And this is before energy savings over the life of the system are taken into account. Device manufacturers have accumulated extensive field data that supports reliable operation over system lifetime. Ruggedness is addressed through design trade-offs and by employing intelligent gate drives with prognostic and diagnostic functions. A plethora of educational opportunities is presently available to train students and the existing workforce in WBG power technology. Without a doubt, WBG devices are rapidly overcoming barriers to system insertion and mass commercialization, with their cost-lowering benefits. The recent insertion of SiC in automotive traction inverters, by several electric vehicle manufacturers, is a good example of a volume application where WBG brings competitive advantages like longer range and faster charging.

The present chapter reviews commercial SiC power diodes, MOSFETs, junction gate field-effect transistors (JFETs), and bipolar junction transistors (BJTs) as well as promising insulated gate bipolar transistors (IGBTs) best suited for +10 kV applications. Unipolar SiC diodes are commercially available and are significantly faster than competing Si *p-i-n* diodes as they have no minority carrier current. SiC MOSFETs, JFETs, and BJTs have been developed for power applications. SiC JFETs are simpler to fabricate and have no gate oxide reliability issues. They are native normally-on (depletion mode), which is regarded as undesirable due to safety concerns, and are made normally-off in the cascode circuit configuration. The SiC MOSFET became commercially available by Cree in 2011 and is the workhorse of the SiC power electronics industry today. SiC MOSFETs are commercially available by several vendors in the 650–1700 V range. They have been demonstrated at 3.3, 6.5, and 10 kV with those voltage nodes up for commercial release over the next few years. SiC BJTs are bipolar devices with switching speeds similar to those of MOSFETs due to the absence of sizable minority carrier storage in their drift region [1]. As with all SiC bipolar devices, their long-term performance can deteriorate due to forward-bias voltage and current gain degradations. These degradations are caused by the growth of stacking faults from basal plane dislocations within the drift epitaxial layer. BJTs are current controlled devices, which makes them less attractive for certain high current power applications. Above 10 kV, the thick drift layer of MOSFETs becomes highly resistive and bipolar conduction can lower conduction losses with acceptable switching losses. SiC IGBTs exploit this trade-off and have been demonstrated in the 15 kV node. They are briefly presented in this chapter.

Both lateral and vertical GaN power devices are reviewed in this chapter. The GaN high-electron mobility transistor (HEMT) is the most mature among these, and the only one that is commercially available with voltage ratings in the range of 15 to 650 V [2]. Adoption of these devices is rapidly increasing for a number of applications, including fast chargers, wireless charging, data centers, and electrified transportation. Various GaN HEMT configurations exist, such as the p-GaN gate

HEMT and the cascode configuration. Commercial devices with integrated drivers are also available. The lateral layout of GaN HEMTs also facilitates the development of integrated circuits based on this technology. Thus, the combination of low-loss device performance and fast switching made possible by monolithic integration makes GaN HEMTs very attractive below 1 kV. Due to recent progress made in GaN substrate technology, research in vertical GaN devices has also intensified. Vertical unipolar and bipolar diodes with breakdown ratings exceeding 1 kV have been demonstrated. Vertical transistors, including MOSFETs, JFETs, and CAVETs, have also been reported, and, in some cases, it has been experimentally confirmed that GaN offers superior performance to SiC, thus moving closer to fulfilling its potential. Among the available vertical transistor topologies, GaN JFETs are the closest to commercialization. Given recent breakthroughs in epitaxy and selective area doping, it is also expected that GaN superjunction devices could play an important role, in turn intensifying the competition with SiC technology in medium- and high-voltage applications.

2 Silicon Carbide Diodes

Silicon carbide (SiC) diodes can be and are already being used in various areas of solid-state electronics. They may demonstrate parameters superior to that one of diodes made of conventional semiconductors owing to unique SiC properties including wide bandgap, high avalanche breakdown field, and excellent thermal conductivity, chemical inertness, and thermal resistance. Some types of SiC diodes are briefly described in this section.

2.1 Silicon Carbide Power Microwave Diodes

The interest to SiC microwave diodes was based on theoretical estimations of the saturated drift velocity of electrons (υ_S) in 4H polytype SiC (4H-SiC), which was expected to be 2.5 times higher than that one in silicon. Also, it was supposed that SiC microwave diodes more powerful than their Si counterparts could be fabricated due to about ten times higher avalanche breakdown field (E_B) in SiC than that one in Si. Indeed, SiC *p-i-n* diodes capable of commutating high microwave power [3] and SiC IMPATT (IMPact ionization Avalanche Transit-Time) diodes [4–6] were demonstrated. The first SiC IMPATT oscillator generated pulsed power of 300 mW in X-band frequency range (8.0–12.0 GHz) [4]. The υ_S value was measured in SiC at electric fields close to E_B (8×10^6 cm/s at about 2 MV/cm) [7, 8] and it was found to be noticeably lower than that one in silicon. Although SiC IMPATT diodes can be more powerful than that ones made of Si, they have not received further development due to the emergence of high-power microwave transistors based on GaN. SiC *p-i-n* diodes designed to switch high microwave power still can find some niches

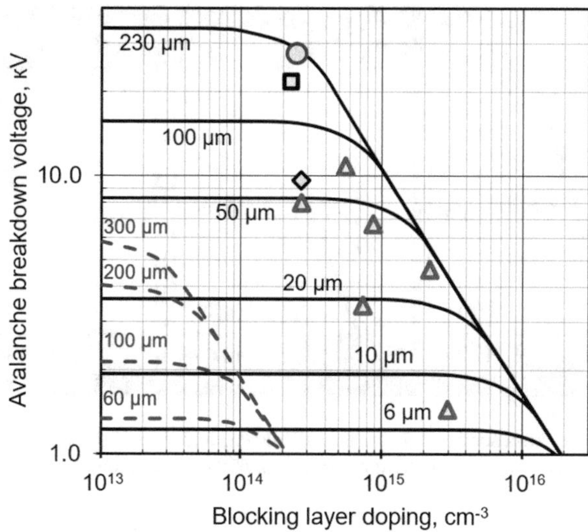

Fig. 2.1 Calculated parallel plane avalanche breakdown voltages of Si (dashed lines) and 4H-SiC (solid lines) devices as a function of n^- drift layer doping level with its thickness as a parameter. Markers denote: △, SiC SB and JBS diodes [15, 17, 18, 25]; □, SiC *p-i-n* diodes [15, 46]; ◇, SiC MPS diodes [25]; ○, SiC IGBT [122]

of application and awaiting on the appearance of sufficient commercial interest to continue their development. A comprehensive overview of SiC microwave diodes can be found elsewhere [9].

2.2 Silicon Carbide Power Diodes

High-power diodes are critical building blocks of power-conversion circuits. They are commonly used, for example, in front-end rectification bridges as well as freewheeling diodes, which are placed antiparallel to power transistors to protect them from excessive reverse voltage. They are so important for this purpose that essentially every commercial power transistor package contains such a freewheeling diode.

All rectifying diodes have a lightly doped blocking layer, which is depleted at a diode's reverse bias and does not conduct current up to designed blocking voltage (V_{BL}). As far as the E_B value in SiC is about ten times higher than in Si, SiC rectifying diodes can have about ten times thinner blocking layer at the same V_{BL}. Figure 2.1 shows calculated parallel plane avalanche breakdown voltages (V_{BR}) of Si (dashed lines) and 4H-SiC (solid lines) power devices as a function of n^- blocking layer doping level (N_D) with its thickness (L) as a parameter.

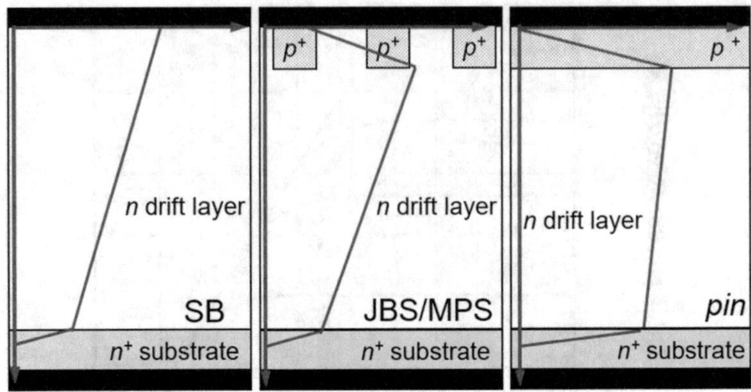

Fig. 2.2 Schematic cross sections of SiC SB, JBS/MPS, and *p-i-n* diodes (not in scale). The red graphs show schematic electrical field distributions in reverse biased diodes

Rectifying diodes may have unipolar or bipolar conductivity in on-state. Unipolar diodes, which are Schottky barrier (SB) and junction barrier Schottky (JBS) diodes, are designed to conduct current by majority charge carriers which concentration does not exceed the N_D level. *p-i-n* diodes are bipolar rectifiers, which involve a thick and lightly doped *n*-region (*i*-region), which is sandwiched between highly doped *p*- and *n*-regions. *p-i-n* diodes instead of *p-n* ones are used for power switching because the *i*-region is needed to produce a high blocking voltage. *p-i-n* and merged *p-i-n* Schottky (MPS) diodes are designed to conduct current by minority charge carriers injected in a blocking layer. In the on-state mode, the *i*-layer is conductivity modulated when the concentration of injected electrons and holes is higher than the doping concentration and thus the $R_{SP\text{-}ON}$ value is reduced as the current increases. Schematic cross sections of SiC SB, JBS/MPS, and *p-i-n* diodes are shown in Fig. 2.2. All these types of SiC high-power rectifying diodes are briefly discussed below.

2.3 Schottky Barrier Power Diodes (SBDs)

SBDs have a rectifying metal-semiconductor contact with low built-in voltages (V_{bi}) in comparison to that one in *p-n* junctions. The blocking layer conductivity in SBDs is unipolar, and hence, these diodes have a low reverse-recovery charge density ($Q_{RR\text{-}ON}$). On the other hand, the lack of the conductivity modulation in the case of SBDs results in a bend-over in the SBD characteristics at high currents due to the resistance of the lightly doped drift region. Another feature of SB diodes is the large reverse leakage current that can lead to non-negligible off-state power dissipation primarily due to thermionic field emission of carriers from the metal into the semiconductor, and exacerbated by the barrier-lowering effect. SiC SB diodes

Table 2.1 Demonstrated 4H-SiC SB diodes

V_{BL} (kV)	V_{ON} (V)	$R_{SP\text{-}ON}$(mΩ·cm^2)	References
1.2	1.35 at 200 A/cm^2	–	[11]
1.4	2 at 732 A/cm^2	1.5	[12]
1.7	2 at 126 A/cm^2	8.7	[13]
3	7.1 at 100 A/cm^2	34	[14]
4.6	2.3 at 20 A/cm^2	10.5	[15]
5	2.4 at 25 A/cm^2	17	[16]
6.7	4 at 60 A/cm^2	43	[17]
10	11.75 at 20 A/cm^2	97.5	[18]

have a higher V_{bi} value in comparison to that of Si due to higher barrier height. The higher V_{bi} value results in smaller reverse leakage currents, thus making it possible to fabricate SiC SB power diodes with very high V_{BL} voltages which are unattainable in Si SBDs as shown in Table 2.1 [10].

A recent review on SiC SB diodes with a deep description on their operation is given in [19].

High-voltage 4H-SiC SB diodes have been introduced to the market since 2001 [20], and for a long time, they have been the only SiC diodes commercially available despite the advantages of p-n diodes for very high voltages. Note that using SiC SB diodes for V_{BL} higher than 600 V is impractical due to a non-negligible off-state power dissipation, and for this reason, JBS diodes have been introduced. Practically, most of commercial SiC Schottky diodes are of JBS type.

2.4 *JBS and MPS Diodes*

The problem of high leakage current in reverse direction of SBDs was overcome when the JBS/MPS design [21] was first implemented in development of SiC high-power diodes [22]. The structure of the JBS/MPS diodes consists of interdigitated pin and Schottky diodes, electrically connected in parallel (Fig. 2.3a [23]). Under reverse bias, the diodes operate like *p-i-n* diodes minimizing thus off-state losses. Indeed, in this case, the multiple p^+ regions push the maximum of electrical field away from the Schottky contact toward the bottom of the p^+ region (see Fig. 2.2) reducing the electrical field under metal contact and, hence, the leakage current.

The difference between a MPS and a JBS rectifier is that the p^+-n junctions in JBS diodes do not turn on during on-state operation, while in the MPS rectifier, the SB regions are very narrow so that the p^+-n junctions are turned on resulting in minority carrier injection and reduction of $R_{SP\text{-}ON}$.

More precisely, the JBS and MPS represent modes of operation under forward bias (Fig. 2.3b). For low forward current values, most of the current is conducted through the Schottky areas of the diodes resulting in no minority-carrier-charge-stored, and thus the turn-off transient is fast, minimizing switching loss (JBS mode).

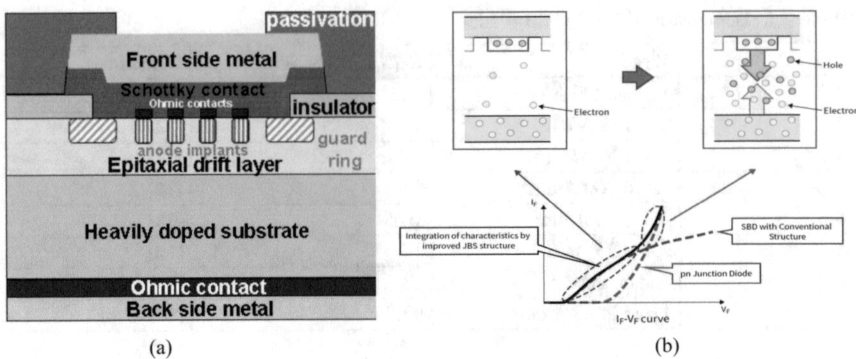

Fig. 2.3 (**a**) The structure of a 4H-SiC JBS/MPS diode. The p^+ anode regions are spaced far enough apart that their depletion regions do not touch under zero or forward bias. (From [19]). (**b**) Forward *I-V* of *p-i-n*, SB, and MPS diodes showing that the characteristic of the MPS diode is a combination of that of SB (low current values-unipolar conduction) and *p-i-n* (high current values-bipolar conduction). Note the bend-over in the SBD characteristics at high currents (1000 Acm^{-2} [24]) due to the resistance of the lightly doped drift region resulting in diode overheating. (From [23])

Table 2.2 Demonstrated 4H-SiC JBS/MPS diodes

V_{BL} (kV)	V_{ON} (V)	$R_{SP\text{-}ON}$(mΩ·cm^2)	References
0.98	3.1 at 100 A/cm^2	19	[26]
1.6	1.4 at 100 A/cm^2	7.5	[27]
1.7	1.6 at 100 A/cm^2	2.9	[28]
2.8	2 at 100 A/cm^2	7.5	[29]
5	3.5 at 108 A/cm^2	25.2	[30]
6.5	4 at 83 A/cm^2		[31]
10	3.37 at 20 A/cm^2	100	[32]

However, with no conductivity modulation, the series resistance of the drift region leads to a voltage drop that dominates the total voltage drop in Schottky areas at high currents [24]. At a certain point, the voltage drop reaches a value “turning-on” the PN areas of the diodes inducing injection of minority carriers (MPS mode). So, significant minority charge storage occurs reducing the on-state loss at high current densities, but the stored charge increases the switching loss. The crossover point between JBS and MPS mode in current density value decreases with the blocking voltage of the diode. This means that the MPS mode dominates above quite low current density values for blocking voltages above 10 kV [24, 25].

Table 2.2 summarizes some recent results on JBS diodes [10].

As mentioned above, nowadays all commercial high-power 4H-SiC diodes are of JBS type even if they often are mentioned as Schottky diodes. Samples of SiC MPS diodes with maximum rating 3300 V/50 A are available from GeneSiC Semiconductor Inc. [33].

2.5 p-i-n Power Diodes

p-i-n diodes can be more efficient at higher blocking voltages than the unipolar diodes, thanks to the conductivity modulation effect leading to the significant reduction of $R_{SP\text{-}ON}$. Since the $R_{SP\text{-}ON}$ does not depend on N_D at conductivity modulation, a punch-through design can be realized in *p-i-n* diodes for further reduction of $R_{SP\text{-}ON}$ and switching time. The reverse leakage in *p-i-n* diodes is primarily due to thermal generation and is extremely small in comparison to SB diodes. Furthermore, *p-i-n* diodes are distinguished by their inherent better reliability and thermal stability. At V_{BL} ratings exceeding 6 kV, reduction of $R_{SP\text{-}ON}$ resulting from the conductivity modulation in SiC *p-i-n* diodes compensates additional resistance of a p^+ layer and larger V_{bi} voltage (~2.8 V vs. 0.9 V) in comparison with SiC SB diodes, and using SiC *p-i-n* diodes becomes more preferable.

Injected minority carriers must have a lifetime long enough to drift through the full length of a blocking layer (about 40 μm for $V_{BL} = 6$ kV in 4H-SiC *p-i-n* diodes) for an effective conductivity modulation. That was the first pitfall on the way to SiC power bipolar devices because SiC epitaxial layers grown in the 1990s suffered from very low minority carrier lifetimes not exceeding 100 ns. Thanks to the introduction of new epitaxial methods [34], minority carrier lifetime of the order of 2 μs was measured in 4H-SiC *n*-type (10^{16} cm^{-3}) epilayers in 2001 [35]. This lifetime value corresponds to the diffusion length of about 30 μm but still remains too low for conductivity modulation of thick layers required for high-voltage SiC *p-i-n* diodes (~100 ÷ 200 μm). An effective solution of this problem was found in 2007. A two times increase of minority carrier lifetime in 4H-SiC epilayers after carbon ion implantation into the shallow surface layer and subsequent post-implantation annealing (PIA) was reported [36]. In 2009, T. Hiyoshi and T. Kimoto replaced the implantation and PIA by a single processing step of thermal oxidation [37]. Since then, the lifetime enhancement thermal oxidation has become a standard step in processing SiC power devices. Recently, S. Ryu, et al. reported carrier lifetimes ranging from 15 μs to 20 μs (corresponding to the diffusion length of 90 μm at ambipolar diffusion coefficient of 4 cm^2/s) in *n*-type 4H-SiC layers (140 μm thick, 2×10^{14} cm^{-3}) measured after the thermal oxidation at 1450 °C for 5 hours [38].

Another obstacle on the way to the SiC bipolar power devices was identified in 2000. H. Lendenmann et al. reported that the voltage drop in 4H-SiC *p-n* junction diodes anomalously increased during their operation at a forward bias [39]. It was observed that triangular planar defects interpreted as stacking faults (SF) lying in basal planes of SiC originating from basal plane dislocations (BPDs) appeared and expanded in SiC epilayers concurrently with the degradation of *I-V* characteristics. It was found that the energy of electron-hole recombination in SiC is high enough to induce a SF nucleation and expansion. Since the carrier recombination is a fundamental process in bipolar devices and cannot be avoided, the development of SiC bipolar devices was significantly hampered. Moreover, SFs can be created during device processing and special care has to be taken to avoid this

[40]. Tremendous efforts were spent to overcome this problem [41–44]. As a result, degradation-free SiC *p-i-n* diodes with active area of 0.22 cm^2 and $V_{BL} = 6.5$ kV were reported [45].

Resolving the problems of low minority carrier lifetime and forward-bias degradation paved the way to successful development of 4H-SiC *p-i-n* diodes. In 2012, H. Niwa et al. [46] reported SiC *p-i-n* diodes with $V_{BL} = 21.7$ kV. Nowadays, SiC *p-i-n* diodes with maximum rating 15 kV/1 A and 8 kV/2 A are offered by GeneSiC Semiconductor Inc. [33].

2.6 Edge Termination

Due to a very high E_B value in SiC, one of technical challenges that must be addressed in design of high-voltage SiC devices is the surface electrical field reduction at the edge of a device. This is especially important for SiC SB diodes where the maximum of electrical field is located at the metal-semiconductor interface. Numerous planar edge termination techniques have been demonstrated in SiC diodes, most of them based on similar concepts used in Si power devices [47]. Typical ones are junction termination extension (JTE) [16], floating field ring (FFR) [48], field plates [49], mesa structure [50], bevel structure [51], and hybrid solution methods [25–28]. The JTE and FFR are regarded as the most effective methods for high-voltage SiC devices. Although a single-zone JTE conceptually works, it shows a narrow window of dose optimization range to achieve the desired voltage. Therefore, multiple zone JTE is mostly used [15]. The multiple zones are formed either by performing different dose implantation steps or by creating unsymmetrical shapes and/or distances among the zones. In the case of FFR termination method, a single implant can be used, thereby reducing the processing steps. However, the optimization of the spacing between the floating zones is complex and challenging.

2.7 Main Points on SiC Power Diodes

The most important SiC diode rectifier device design trades off roughly parallel well-known silicon rectifier trade-offs, except for the fact that numbers for current densities, voltages, power densities, and switching speeds are typically much higher in SiC. Indeed, the high breakdown field of SiC allows for low R_{ON}, V_{ON}, values, and practical absence of reverse recovery and thus permitting operation of SiC diodes at much higher voltages, current densities, and switching speeds. Moreover, the higher SiC bandgap in comparison to Si allows for a higher barrier height of Schottky diodes by almost 1 eV and thus reducing the reverse current by 17 orders of magnitude at room temperature [24]. On the other hand, SiC *p-i-n* diodes have a larger built-in voltage (~2.8 V) due to its wider bandgap than the Si PiN diodes, but they have a lower forward voltage drop at high current density and higher switching speed due to the much thinner i-region. Furthermore, SiC diodes are distinguished by the inherent better reliability and thermal stability of their electrical characteristics as well as their possibility to operate at temperatures higher than 125 °C. Indeed, most commercial SiC power diodes are rated up to 175 °C [1], while diodes operating well above 200 °C have been demonstrated.

Unipolar SiC diode is the main commercially available diode on the market; its typical voltage ratings are 600 V, 650 V, 1.2, and 1.7 kV. Some 3.3 and 8 kV products also are available but their current rating is limited by the thick drift layer and the associated resistance [1]. For instance, the current rating for 8 kV SiC diode is only 50 mA. At 10–20 kV voltage ratings, 4H-SiC *p-i-n* rectifiers offer the best trade-off between on-state voltage drop, switching losses, and high-temperature performance as compared to Si *p-i-n* or SiC Schottky/JBS rectifiers.

SiC diodes have a long and enthralling story of their development and commercialization which is still ongoing and far from over. Currently, the main driving force for further development of high-power SiC diodes is a rapidly growing demand of highly efficient switches and rectifiers for automotive and industrial applications with blocking voltages and commutated power ranging from ~600 V/100 kW in invertors for electrical vehicles to ~1.1 kV/13 GW in convertors for high-voltage DC power transmission.

3 SiC BJTs

Bardeen, Brattain, and Shockley invented the bipolar junction transistor (BJT) in 1947 at Bell Laboratories [52]. BJT is a three-terminal power device, schematically shown in Fig. 2.4, available in the market for more than 50 years [53]. Muench et al. reported the first SiC BJT in 1977 [54]; however, the first SiC BJT that got attention was the first high-voltage 4H-SiC BJT reported by Ryu in 2001 [55]. 4H-SiC BJT has been extensively developed in recent years for high-voltage and high-temperature applications due to its unique properties such as low on-resistance, normally-off behavior, fast switching, and lack of gate-oxide reliability issues.

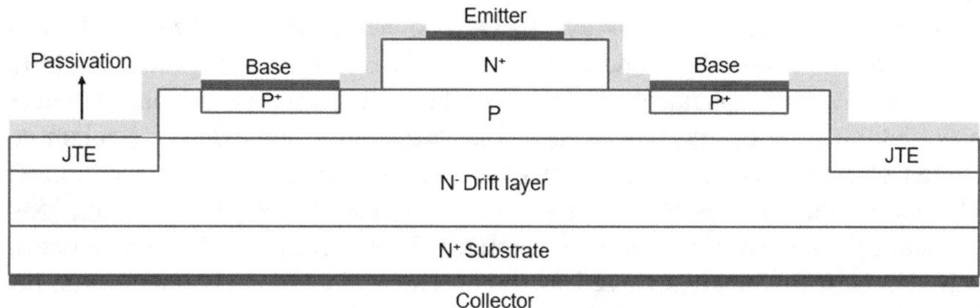

Fig. 2.4 Simplified cross sections of SiC BJT (not in scale)

Table 2.3 Recent reported high-voltage 4H-SiC BJTs

BV (kV)	Ron (mΩ•cm^2)	Current gain (β)	References
1.8	4.4	40	[62]
2.7	4	132	[63]
3.2	28	28	[64]
5.65	18.8	44	[58]
5.85	28	40	[59]
6	28	–	[65]
10.5	110	75	[66]
16.25	579	139	[60]
23.5	321	7	[67]

SiC BJTs have a large *safe operating area (SOA)* in which the second breakdown occurs at very high current densities (outside the range of possible operation) [56]. Moreover, SiC BJTs have some advantages compared to SiC MOSFETs:

1. Possibility to have conductivity modulation in the drift layer, thus lowering the on-resistance and power losses in on-state mode.
2. Lower fabrication cost.
3. The positive temperature coefficient of the on-resistance and negative temperature coefficient of the current gain (β) results in an easy device paralleling configuration.
4. Non-dependency on a gate oxide and no suffering from the oxide reliability for high-voltage, high-temperature, and harsh environment applications.

It should be noted that the on-resistance for a 4H-SiC unipolar device like MOSFETs above 15 kV increases to a point where it is impractical from a yield standpoint and cost [57]. Bipolar devices like SiC BJTs are good candidates to replace them. However, to be fully competitive with SiC MOSFETs and SiC IGBTs in the market, the SiC BJT characteristics need to be improved. In recent years, there has been a growing investigation to improve the on-resistance, current gain, current density, and breakdown voltage. Table 2.3 summarizes some of these works.

The first obstacle to the commercialization of SiC BJTs was the forward-bias degradation observed in all SiC bipolar devices originating from BPDs existing in epitaxial layers (see above part on SiC diodes). This problem has been hopefully resolved [45]. The second obstacle to the commercialization of SiC BJTs was the formation of new BPDs and lifetime killer defects during the processing [40]. For example, ion implantation followed by high-temperature annealing produces new lifetime killer defects that cause bipolar degradation and reduce the common-emitter current gain. To overcome this problem, the ion implantation for forming JTE and p^+ region base (Fig. 2.4) can be easily replaced by etched-JTE-zone [58–60] and by epitaxial regrowth of p^+ region [61], respectively. The third obstacle was surface recombination caused by the presence of interface trap density at the SiC/SiO_2-passivation-layer interface, thus lowering the current gain. Different techniques have been addressed in recent years for overcoming this issue [10].

In recent years, high-temperature SiC BJTs ICs have also been investigated in depth at device physics, circuit, and process integration [68, 69]. The short-circuit ruggedness of 10 kV SiC BJTs with a 16 μs withstand time was recently reported. It shows that the SiC BJTs can handle without failing about three times the critical short-circuit energy of the commercial SiC MOSFETs [70].

4 SiC Junction Field-Effect Transistors

The SiC junction field effect transistor (JFET) is capable of high-power and high-temperature switching as it only uses *p-n* junctions in the active device area, where the high electric fields occur, and can therefore fully exploit the high-temperature properties of SiC in a gate voltage-controlled switching device. Provided the gate-to-source junction of the JFET is biased below its built-in potential, negligible gate current is needed to drive the device and voltage controlled switching is realized. JFETs are free of MOS native oxide problems like low channel mobility, threshold voltage instability, and lack of reliability at elevated temperatures. They have demonstrated electrostatic discharge immunity to 16 kV (V Veliadis, Private communication) and as unipolar devices do not suffer from forward voltage degradation at the same degree as bipolar devices [39–42]. SiC power JFETs are almost exclusively implemented in a vertical configuration and are native depletion mode or normally-on (Non).

Several SiC JFET designs have been demonstrated over the years schematically shown in Fig. 2.5. The first power 4*H*-SiC JFETs were reported by H. Mitlehner et al. in 1999 [71]. Those were vertical JFETs (VJFETs) with lateral channel (Fig. 2.5a) and they have been further developed by D. Stephani et al. [72] and S-H. Ryu et al. [73]. This JFET is similar to a SiC double-implanted MOSFET (DMOSFET), with the oxide controlled inversion channel having been replaced with a bulk channel. This eliminates the SiC/SiO2 interface with its channel mobility and reliability drawbacks and creates a bulk channel where the SiC mobility is fully utilized.

Zhao et al. have implemented a JFET design with relatively critical dimensions (Fig. 2.5b) [74]. The design requires three implantation events. In order to implant the lower portion of the sidewalls, the wafer must be tilted against the direction of the ion beam and rotated. No epitaxial regrowth is needed.

A simplified cross-sectional schematic of a trenched-gate vertical channel p^+ ion-implanted depletion mode (normally-on) 4H-SiC JFET is shown in Fig. 2.5c [75]. This representative for all JFET designs will be used to analyze, in the following, the SiC JFETs electrical characteristics, thermal performance, and ruggedness. A series of papers reported on the detailed fabrication and the related electrical characteristics [76–78] of this SiC JFET design.

The inherent simplicity of the design shown in Fig. 2.5c, which does not require epitaxial regrowth, is the reason for the demonstration of reliable JFETs with excellent yields and parameter uniformity [75]. 1680 V SiC JFETs with an active

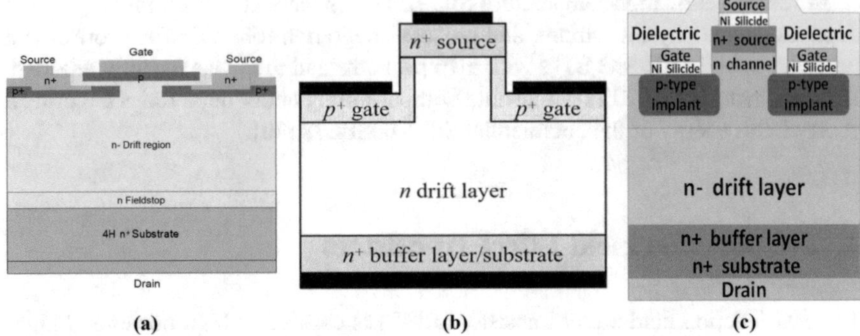

Fig. 2.5 Simplified cross sections of (**a**) a vertical JFET with a lateral channel (the n^+ source is embedded in the p-well and reaches below the p gate, in order to minimize source resistance), (**b**) a trench and implanted vertical JFET, (**c**) a depletion mode ion-implanted SiC vertical-channel JFET

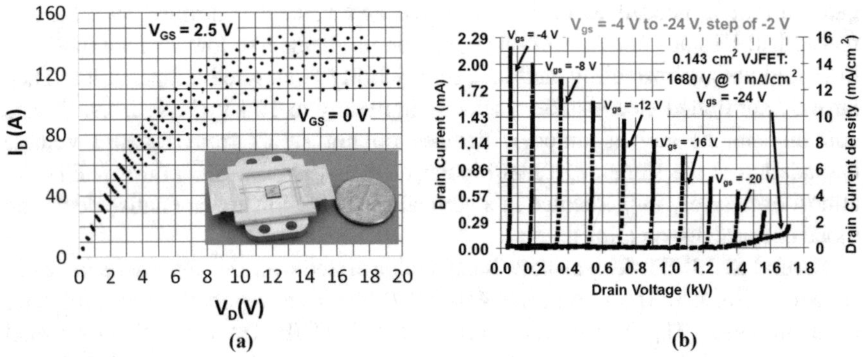

Fig. 2.6 (**a**) On-state drain current vs. drain voltage and (**b**) blocking voltage characteristics of a single 1680 V, 0.143 cm^2 packaged SiC JFET

area of 0.143 cm^2 (0.19 cm^2 total area) and on-state current capability of 50 A were fabricated by Veliadis et al. in seven photolithographic levels, with a single masked ion implantation event that simultaneously implanted the p$^+$ gates and guard rings [5]. Indeed, the simplicity of the design greatly reduces process complexity and JFETs using even only four lithography steps have been demonstrated [76–78]. Room-temperature on-state drain-current vs. voltage characteristics are shown in Fig. 2.6a at a gate bias range of 0 to 2.5 V in steps of 0.5 V. To maintain voltage-control capability (high I_D/I_G gain), the gate must be biased below its 2.7 V built-in potential value. If the gate bias increases in excess of 2.7 V, significant gate current injection occurs into the channel of the JFET, and its current gain I_D/I_G degrades. At a gate-to-source bias of 2.5 V, the JFET outputs 53.6 A at a forward drain voltage drop of 2.08 V. The specific on-state resistance is 5.5 mΩ·cm^2, and the transistor current gain is $I_D/I_G = 26{,}800$.

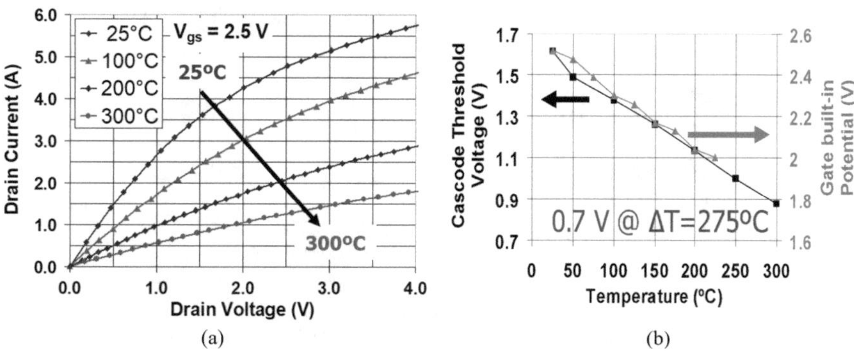

Fig. 2.7 Temperature dependence of the SiC cascode's (**a**) on-state drain current at $V_{GS} = 2.5$ V and (**b**) threshold voltage (left axis) and gate-junction built-in potential (right axis). At a temperature swing of 275°C, the threshold voltage only shifts by 0.7 V

The blocking voltage characteristics of the 0.143 cm^2 active area JFET at gate biases of −4 to −24 V, in steps of −2 V, are shown in Fig. 2.6b. At a gate-to-source bias of −24 V and a low drain current density of 1 mA/cm^2, the JFET blocks 1680 V.

An important contribution to SiC device reliability is eliminating threshold voltage instability. In SiC MOSFETs, threshold voltage instability is primarily due to the oxide traps at the SiC/gate-oxide interface (see corresponding part below).

To investigate the threshold voltage stability of JFETs, a JFET-based all-SiC normally-off switch was implemented by combining a 1200 V normally-on JFET with a low-voltage normally-off (enhancement mode) JFET in the cascode configuration [79]. To evaluate threshold voltage shift with temperature, the I_{DS}-V_{DS} characteristics of the all-SiC cascode switch were measured at junction temperatures of 25 °C, 100 °C, 200 °C, and 300 °C and are shown in Fig. 2.7a [80]. The cascode's on-state resistance is 6.2 mΩ·cm^2 and was extracted from the data of Fig. 2.7 at $V_{DS} = 0.5$ V. The increase in on-state resistance with temperature agrees well with the theoretical reduction of the electron mobility in 4H-SiC. The cascode threshold voltage was extracted and is plotted as a function of temperature on the left axis of the graph of Fig. 2.7b.

The threshold voltage decreases from 1.6 V to 0.9 V as the temperature increases from 25 °C to 300 °C; the cascode switch remains normally-off at 300 °C. The cascode's gate-junction built-in potential variation with temperature was also extracted and was plotted on the right axis of the graph of Fig. 2.7. As the temperature increases from 25 °C to 225 °C, the cascode's threshold voltage decreases by 0.54 V, while its gate-junction built-in potential decreases by 0.52 V. This excellent agreement confirms that the decrease in cascode threshold voltage with temperature stems from the reduction of its gate-junction built-in potential as expected from solid-state physics. Thus, SiC JFETs have remarkably stable threshold voltages due to the fact that it uses p-n junctions instead of gate oxides to control the current flow.

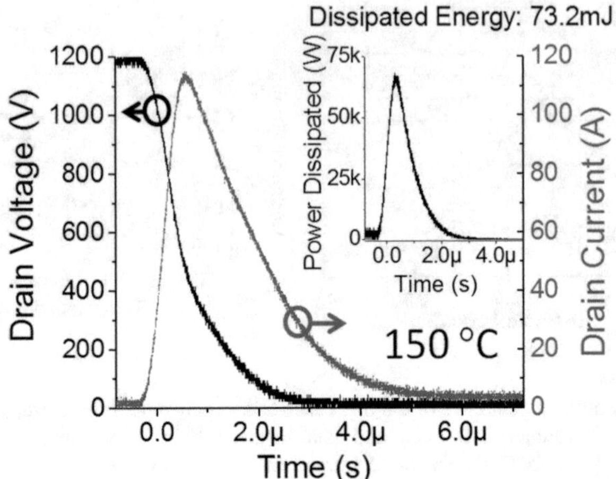

Fig. 2.8 Representative 1200 V/115 A hard switching waveforms of the SiC JFET testing at 150 °C. The energy dissipated by the JFET during each hard switching event is 73.2 mJ (inset), and the peak dissipated power is 68.2 kW

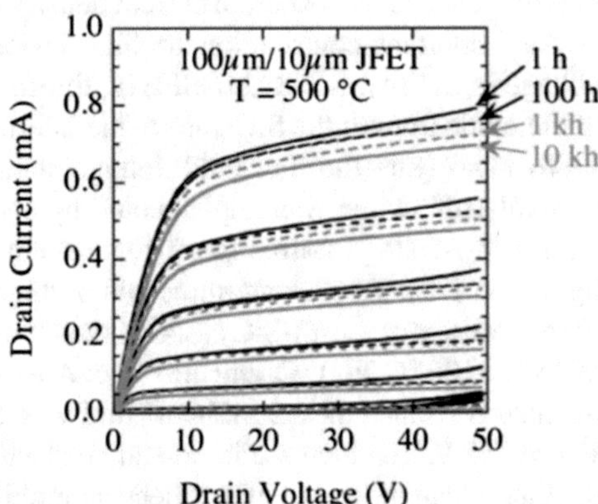

Fig. 2.9 I_D-V_D characteristics of a descrete 6H-SiC JFET. (From [84])

To evaluate ruggedness, a SiC JFET was subjected to over 2.4 million 1200 V/115 A hard switching events at 150 °C, at what is 13 times its 8.8 A 150 °C rated current (Fig. 2.8) [81]. The JFET drain voltage is plotted in black (left axis) in Fig. 2.8, while the current through the JFET is plotted in gray (right axis). By multiplying voltage by current, the power dissipated by the JFET is calculated and plotted in the inset of Fig. 2.8. The energy dissipated by the JFET during each 1200 V/115 A switching transient is 73.2 mJ, and the peak dissipated power is 68.2 kW. Finally, it has been shown [81] that the electrical characteristics (on- and off-state) do not degrade with stressing (Fig. 2.9).

The ruggedness of SiC JFETs especially at high temperatures has been validated by the group of Philip Neudeck at NASA John Glenn Research Center. They demonstrated a short-term operation of packaged 4*H*-SiC junction field effect transistor (JFET) logic integrated circuits (ICs) at ambient temperatures exceeding 800 °C in air [82]. They also demonstrated SiC lateral JFETs with operating time of 6000 hours at 500 °C which was limited by the thermal degradation of a metal stack used for the formation of ohmic contacts in these devices [83, 84].

Today, power SiC JFETs are commercially available as discrete components in the 650–1700 voltage range [85]. The MOSFET dominates SiC-based power electronics. JFETs are being inserted in systems in smaller numbers. As SiC power electronics continue to gain ground, the JFET has the potential to be the device of choice for rugged high-temperature applications.

5 SiC MOSFETs

A power MOSFET is a high-speed, easy-to-drive device, which makes it a very attractive option for power switching applications. The main advantage of the power MOSFET structure is the high impedance gate, which does not require steady-state gate current, and the gate drives are only required to provide relatively small amount of current to charge and discharge the capacitances. The current conduction in the power MOSFET structure occurs through transport of majority carriers in the drift region and does not involve minority carrier injection. Hence, there are no delays associated with storage or recombination of minority carriers in power MOSFETs. It is also easy to parallel multiple power MOSFETs because of the positive temperature coefficient of the forward voltage drop, due to the decrease in carrier mobility at elevated temperatures, which prevents subsequent thermal runaways. In addition, the power MOSFETs do not go through second breakdown like bipolar junction transistors and offer excellent safe operating area.

On-resistance of a power MOSFET increases quite rapidly with blocking voltage of the device. For silicon power MOSFETs, which is the most commonly used semiconductor material, the on-resistance of a power MOSFET can be very small if the design voltage of the device is 200 V or less [47]. Silicon power MOSFETs designed for voltages greater than 200 V have an unacceptably high on-resistance, large chip area, and significant increases in parasitic capacitances. Researchers in silicon power devices addressed this issue by placing an injecting junction at the drain side of the device, which reduced the drift layer resistance by minority carrier injection, or conductivity modulation of the drift layer [86, 87], resulting in the development of insulated gate bipolar transistors (IGBTs). The other approach used to reduce the on-resistance of silicon power MOSFETs is the use of superjunction (SJ) structure [88], which utilizes alternating n- and p-columns with relatively heavy doping concentrations. Excellent results have been achieved for devices with blocking voltages up to 950 V [89].

A silicon (Si) IGBT provides significantly lower forward voltage drops compared to a conventional Si power MOSFET in higher blocking voltage (>600 V) rated devices and at high current levels. However, this reduction in on-state forward voltage drop comes with some serious drawbacks such as (i) at lower current levels, the forward voltage drop of power MOSFETs can be lower than in IGBTs; (ii) unlike power MOSFETs, IGBTs cannot conduct currents in the reverse direction; and (iii) IGBTs exhibit longer switching times and substantially higher switching losses, when compared to a power MOSFET due to the minority carrier injection into the drift region.

Power MOSFETs in silicon carbide (SiC) can address these issues [90]. The wide bandgap properties of SiC provide high breakdown electric field, which allows thinner drift layer with significantly higher doping concentration. This makes possible designs of unipolar SiC power devices with extremely low on-resistance, which addresses most of the issues discussed above. Fabrication processes, including techniques to form high-quality gate oxide films and selective doping methods, are well established in silicon carbide, which culminated in successful development and commercialization of silicon carbide power MOSFETs.

5.1 4H-SiC DMOSFETs

Figure 2.10 shows a simplified cross section of a power double-implanted MOSFET (DMOSFET) in SiC, which was the first commercially available power MOSFET structure in SiC. The n^+ sources and MOS channel regions are built in implanted p-wells. The n^+ source regions and the p-wells are tied together using common contacts to source, to keep the potential difference between the two regions at minimum. The device turns on when a positive bias exceeding the threshold voltage of the device is applied to the gate electrode. In the on-state, electrons flow from the n^+ source regions through the MOS channel formed in the p-well into the junction field effect transistor (JFET) region. The JFET regions are defined as the n-type region formed between adjacent p-wells. The length of the MOS channel is determined by the distance from the edges of the n^+ regions and the p-wells. The electrons then spread into the drift layer and then flow into the n^+ substrate and exit the structure through the drain electrode. In the off-state, a bias less than the threshold voltage of the device is applied to the gate electrode, which removes the inversion channel in the MOS region in the p-well and isolates the n^+ regions from the JFET regions. The device turns into a *p-i-n* diode structure, which can block the voltage when a positive bias is applied to the drain electrode and allow current to flow through when a negative bias is applied to the drain electrode. It should be noticed that the depletion regions from the p-wells merge and provide shielding to the gate oxide layers. The doping concentration and the width of the JFET region should be set carefully to provide adequate shielding to the gate oxide in the off-state, as well as low resistance during the on-state operation of the device [91].

Theoretical specific on-resistance ($R_{\text{SP-ON}}$) values based on drift resistance calculations are plotted in Fig. 2.11 for silicon and 4H-SiC. Performance points of Wolfspeed power MOSFETs with blocking voltages ranging from 900 V to 15 kV are also shown on the plot [92]. For devices with blocking voltages of 6.5 kV or higher, the performance points are close to the ideal silicon carbide 1-D limit, since

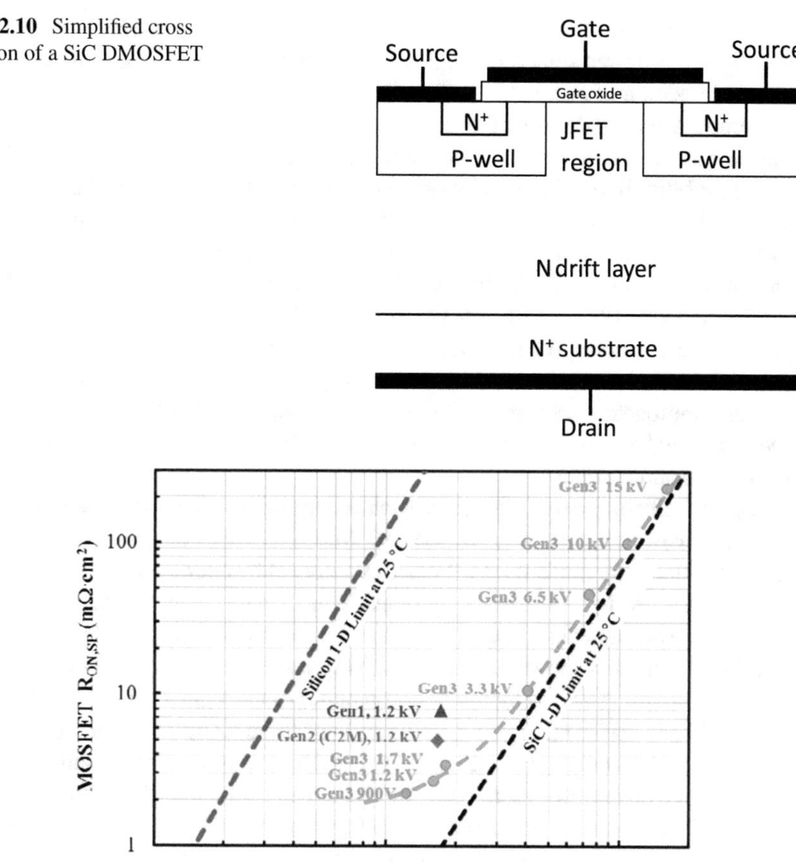

Fig. 2.10 Simplified cross section of a SiC DMOSFET

Fig. 2.11 Experimental $R_{SP\text{-}ON}$ values of SiC DMOSFET for blocking voltages ranging from 900 V to 15 kV. (From [92])

the on-resistance of the devices is dominated by the drift resistance. For devices with blocking voltages of 3.3 kV or lower, the performance points deviate from the ideal 1-D limit due to impacts of other parasitic resistance components, most significantly from the MOS channel resistance.

5.2 *MOS Channel Resistance Issue*

For optimization of power DMOSFETs in 4H-SiC, it is very important to minimize the MOS channel resistance. First successful approach for reducing the MOS channel resistance was to utilize self-aligned ion implantation to reduce the MOS channel length [93]. This approach resulted in 2 kV power MOSFETs in 4H-SiC with a specific on-resistance of 10.3 mΩ·cm^2 and provided a foundation for the commercialization of power MOSFETs in silicon carbide.

Attempts were made to further improve MOS interface properties, by incorporating impurities other than nitrogen into the gate oxide to achieve greater MOS channel mobility than what can be obtained using NO or N_2O anneals. Doping of the oxide layers with phosphorus [94] and boron [95, 96] has been investigated. A MOS channel mobility of 98 cm^2/(V·s) is achieved using phosphorus doping approach [94], and boron doping approach resulted in a MOS channel mobility of 102 cm^2/(V·s) [95], and a 4.5 kV power DMOSFET was demonstrated [96]. Approximately a factor of 3 improvement in MOS channel mobility over nitridation using NO or N_2O was observed using this approach. However, it was determined that these approaches were not suitable for commercial 4H-SiC power MOSFETs since a reasonable threshold voltage stability could not be achieved (P. Godignon, private communications).

Usage of alkaline earth elements, such as strontium (Sr) and barium (Ba), as interface passivation materials for 4H-SiC MOSFETs was also investigated [97]. The passivation was performed by placing a very thin interlayer material directly on the 4H-SiC surface, followed by deposition of gate dielectric layer, typically SiO_2, which was annealed in O_2/N_2 ambient for densification [97]. Sr passivation of the MOS interface showed a very promising result, resulting in a MOS channel mobility of 40 cm^2/(V·s), which was comparable to the values achievable using an NO anneal [97]. Passivation using Ba turned out to be significantly more efficient, resulting in a MOS channel mobility of 85 cm^2/(V·s) at room temperature, which is approximately double the value from an NO annealed sample [97]. A comparison of MOS channel mobility at temperatures ranging from 25 °C to 150 °C is shown in Fig. 2.12. A test lateral MOSFET with conventional NO anneal process and a device with barium interlayer (Ba IL) passivation process were used for this comparison [97]. The samples were fabricated on $5 \cdot 10^{15}$ cm^{-3} doped p-type epilayers on 4H-SiC substrates. The MOS channel mobility of the Ba IL-passivated sample decreases with temperature, as expected due to phonon scattering effects. This is in contrast to the NO annealed sample, which showed an increase in MOS channel mobility with temperature due to the higher interface density near the conduction band.

Recently, C-C bonds formed during thermal oxidation of SiC were identified as one of the important factors limiting MOS channel mobility [98]. The impacts of C-C bonds are also present on samples that received sacrificial oxidation, where the resulting thermal oxide layer was chemically removed [99]. Various approaches to form gate oxide layers with minimum thermal oxidation have been presented. This includes a deposition of a thin Si film, which was converted to SiO_2 by low-temperature oxidation [100] as well as a direct deposition of SiO_2 layer onto SiC surface [101]. For both approaches, H_2 treatment of SiC surface to etch away thermally oxidized region and interface nitridation to achieve a low density of interface traps (D_{it}) were performed. The elimination of thermal oxidation process resulted in approximately a factor of 2 increase in MOS channel mobility, as shown in Fig. 2.13.

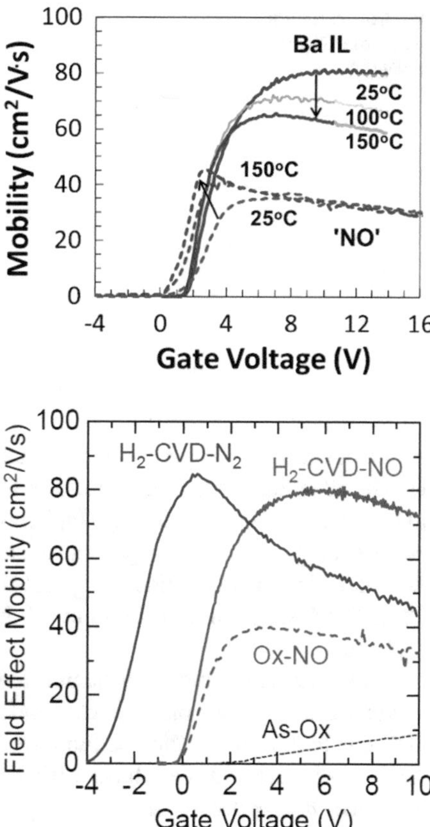

Fig. 2.12 MOS channel mobility as a function of temperature from Ba IL processed device. Results were compared to values from an NO annealed sample

Fig. 2.13 MOS channel mobility of n-channel MOSFETs, fabricated using various processes [99]

Reduction in surface nitridation temperature to avoid in situ oxidation and material decomposition was attempted using supercritical N_2O fluid (SCN_2O) [102]. The approach utilized gas-like high penetration property and liquid-like solubility of supercritical fluids. 4H-SiC MOS interface with thermally grown gate oxide was processed with supercritical N_2O fluid at a temperature of 120 °C. A MOS channel mobility of 72.3 $cm^2/(V{\cdot}s)$ was reported, showing significant improvement in MOS channel properties over devices that received thermal nitridation processes [102].

5.3 4H-SiC Trenched MOSFETs

A trench MOSFET structure, shown in Fig. 2.14, can provide devices with significantly smaller cell pitch because it places the MOS channel on the etched sidewalls. Additional real estate necessary for proper MOSFET operations, such as gate-to-source overlap and gate-to-contact metal gap, can also be placed on the sidewalls. Such design can result in a huge increase in gate packing density,

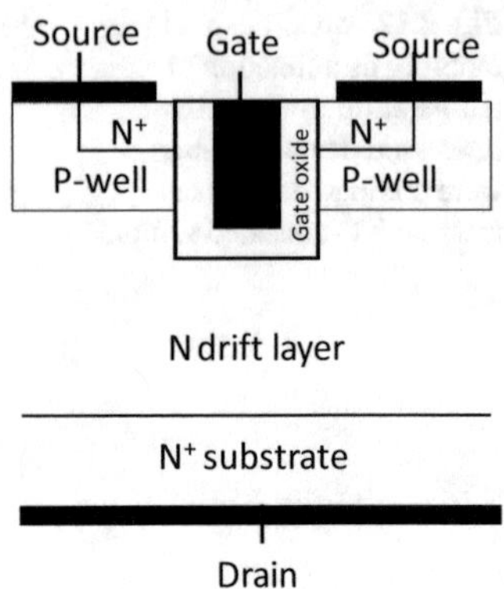

Fig. 2.14 Simplified cross section of a SiC trench MOSFET

and associated gate-to-source capacitance which is a highly desirable feature for high-speed switching power applications since it improves immunity of the power MOSFET to self-turn-on during a high *dv/dt* turn-off event. However, the associated increase in gate-to-drain capacitance has to be avoided [103]. The first trench MOSFETs in silicon carbide, using 6H-polytype, were reported by Palmour et al. [104]. The first trench MOSFETs in 4H-SiC were also demonstrated by Palmour et al. [105].

It was also experimentally demonstrated that the MOS channel mobility can be significantly higher on the etched sidewall compared to the Si-face of 4H-SiC [106] due to the anisotropic SiC mobility. The simplified trench MOSFET structure shown in Fig. 2.14 does not have a JFET region, which also helps reducing the on-resistance. It is expected that a well-optimized 4H-SiC trench MOSFET structure can offer significantly lower on-resistance compared to a 4H-SiC DMOSFET with the same voltage rating.

The simple trench MOSFET structure, shown in Fig. 2.14, works very well in silicon, since the breakdown electric field for SiO_2 is two orders of magnitude greater than that of silicon; hence, oxide breakdown is not an issue in silicon devices. However, the breakdown electrical field of 4H-SiC is only about three times lower than the theoretical breakdown electrical field of SiO_2. The electrical field increases further at the SiO_2/SiC interface by more than a factor of 2 due to the difference in dielectric constant between the two materials. This represents a huge reliability issue of 4H-SiC trench MOSFETs. For reliable operation, the gate oxide at the trench bottom must be properly shielded from the high voltage during the off-state. Figure 2.15 shows a 4H-SiC trench MOSFET structure with a p-type implanted in the bottom of the gate trench [107–109]. The p-shielding region was connected to the source region and provided excellent protection to the gate oxide at the bottom of the trench. It should be noted that this p-type protection layer and the p-base of the trench MOSFET can form a very narrow JFET region, which can add significant amount of JFET resistance, increasing the total on-resistance of the structure. This issue was addressed by placing a thin, heavier doped n-type current spreading layer (CSL) beneath the p-well layer

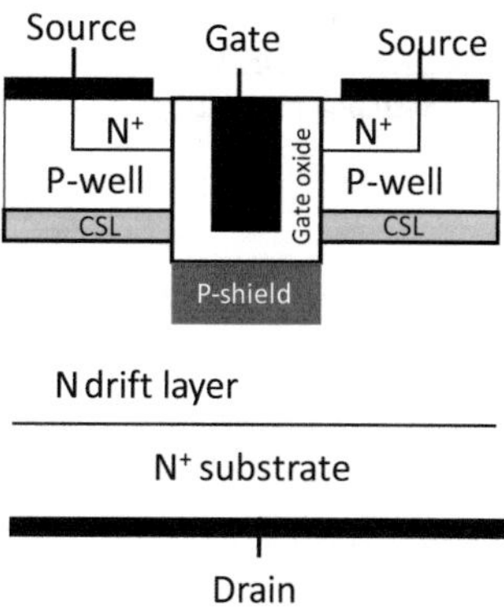

Fig. 2.15 SiC Trench MOSFET structure with trench bottom protection implants

[107]. It should be noted that the gate-to-drain capacitance is very small. However, switching losses will be very high if the resistance between the protection p-region and source is extremely high. Device layout must be optimized for this structure to achieve optimal on-state and switching performances [109]. Figure 2.16 shows a 4H-SiC trench MOSFET structure with double-trench protection approach [110]. The bottom of the gate trench was shielded by a deeper source trench. The distance between the gate trench and protection trench was approximately 2 μm in [110]. The MOS channel mobility on the etched sidewall was 11 cm^2/(V·s), which was considerably lower than that expected for a trench MOSFET in 4H-SiC. With this structure, a specific on-resistance of 0.79 mΩ·cm^2 was achieved for a 630 V 4H-SiC trench MOSFET, and an on-resistance of 1.41 mΩ·cm^2 was achieved for a 1260 V trench MOSFET, respectively. Figure 2.17 shows a 4H-SiC trench MOSFET structure with asymmetric protection implants. In this device, only one side of the trench sidewall is used as MOS channel, which is exactly aligned to the <1120> crystal plane [111]. The deep p-wells are used to limit the electric field in the gate oxide at the bottom and the corners of the trench. This cell structure has a small ratio of the Miller charge (Q_{GD}) to gate-source charge (Q_{GS}). It should be noted that this structure adds significant amount of JFET resistance. However, the added JFET regions resulted in reduced saturation currents, which improved the short-circuit withstand time (t_{scwt}).

A last point on SiC UMOSFETs related to the interface states. The etched sidewalls of 4H-SiC trench MOSFET/UMOSFETs) have lower density of interface states (D_{it}) closer to the conduction band edge than the Si-face of 4H-SiC. However, the a-face has significantly more midgap states, which may not impact the MOS channel mobility, but result in significant subthreshold hysteresis [112]. A preconditioning routine is required to measure threshold voltage from 4H-SiC trench MOSFET. This may not impact the device reliability or stability [112]. However, it is preferred to minimize the hysteresis for easier control of the devices. Further developments in MOS surface passivation techniques are needed to minimize interface trap density across the bandgap of 4H-SiC.

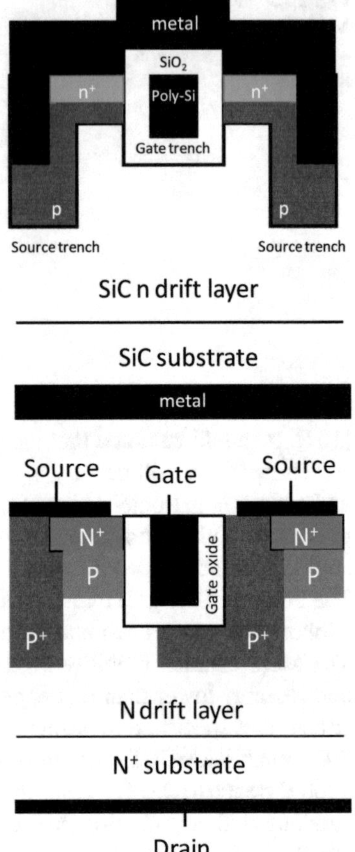

Fig. 2.16 SiC trench MOSFET structure with double-trench protection

Fig. 2.17 SiC trench MOSFET structure with asymmetric protection implants

5.4 4H-SiC Superjunction MOSFETs

Superjunction (SJ) MOSFETs in 4H-SiC were also demonstrated. The first published approach used multiple implants and epiregrowth steps to form vertical SJ structures [113–115]. Dopant diffusion cannot be utilized in the fabrication of SJ structures in 4H-SiC due to negligible diffusion coefficients in 4H-SiC [113]. Hence, for this type of approach, the SJ drift layer requires several iterations of thin epigrowth and ion implantations. Cross-sectional images of 1200 V class 4H-SiC SJ trench MOSFETs, with a pitch of 5 and 2.5 μm, are shown in Fig. 2.18 [116]. The signs of multiple epigrowths and p-type implantations are clearly visible in the image. As mentioned above, on-resistances of 4H-SiC MOSFETs, with blocking voltage less than 3.3 kV, are dominated by parasitic resistances, which include MOS channel resistance. For this reason, a 1200 V class SiC SJ MOSFET did not show any on-resistance advantage over conventional structure SiC power MOSFET at room temperature, as shown in Fig. 2.19 [114, 117]. However, the

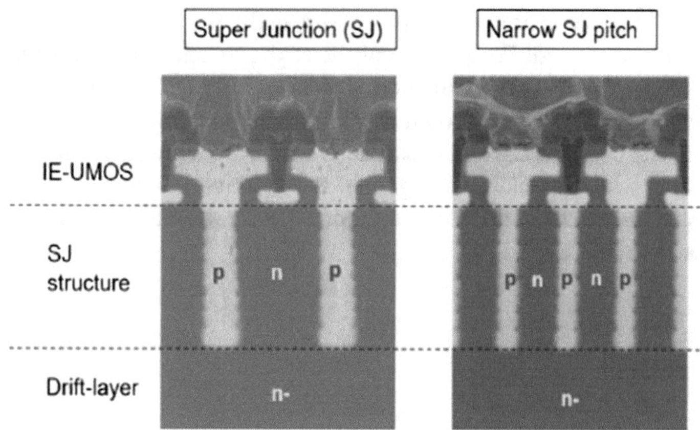

Fig. 2.18 Cross-section of 1200 V class 4H-SiC SJ trench MOSFET [116]

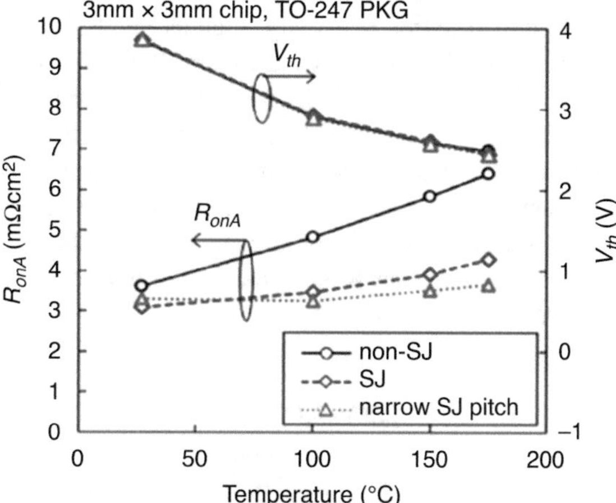

Fig. 2.19 $R_{ds,on}$ as a function of temperature for SiC SJ MOSFET, compared to a conventional SiC MOSFET [115]

SJ MOSFETs showed significantly lower on-resistance at elevated temperatures, where the MOS channel resistance reduces due to a reduction on threshold voltage, and drift resistance increases due to a decrease in bulk mobility. SJ devices showed significantly smaller rate of increase in on-resistance over temperature compared to the conventional device. The SJ device with tighter pitch and higher drift doping concentration showed smaller rate of increase compared to the SJ device with larger pitch.

The benefits of the SJ structure are greater for higher voltage (>6 kV) devices, where drift layer resistance becomes more dominant. Multiple regrowth approach used for 1200 V class SiC SJ MOSFETs is not feasible for high-voltage devices due to manufacturing costs associated with the approach. For such thick SJ structures, "trench etch and refill" approach is more reasonable. A 6.5 kV 4H-SiC MOSFET with partial SJ structure, with 23-μm-thick SJ region and 41-μm-thick,

2×10^{15} cm^{-3} doped drift region was experimentally demonstrated [118]. A trench pitch of 5 μm was used. For proper epi-fill of the trenches, the trenches must be precisely aligned to <11–20> direction. 4H-SiC power DMOSFET structure with a cell pitch of 10 μm was built on the SJ drift layer to complete the fabrication of the device. The completed 4H-SiC partial SJ MOSFET showed an on-resistance of 17.8 mΩ·cm^2 with a blocking voltage of 7.8 kV, which is significantly lower than the theoretically predicted drift resistance of a 7.8 kV 4H-SiC unipolar device with conventional structure.

6 SiC IGBTs

Ultrahigh-voltage (>10 kV) 4H-SiC MOSFETs have very high specific on-resistance, which leads to very large die size, resulting in increased manufacturing costs and gate drive requirements. Bipolar devices utilizing conductivity modulation to reduce drift region resistivity, such as SiC IGBTs, can be introduced to alleviate this issue. As shown in Figs. 2.20 and 2.21, SiC power MOSFETs have unipolar drift conduction, which is limited by the doping concentration of the drift layer. On-resistance and, consequently, forward voltage drop (V_{ON}), increases with temperature due to decreases in bulk electron mobility with temperature. On the other hand, SiC IGBTs depend on conductivity modulation achieved by injection of excess carriers, which significantly reduces drift region resistivity of the device, which reduces further at elevated temperatures due to enhanced charge injection and increased carrier lifetime.

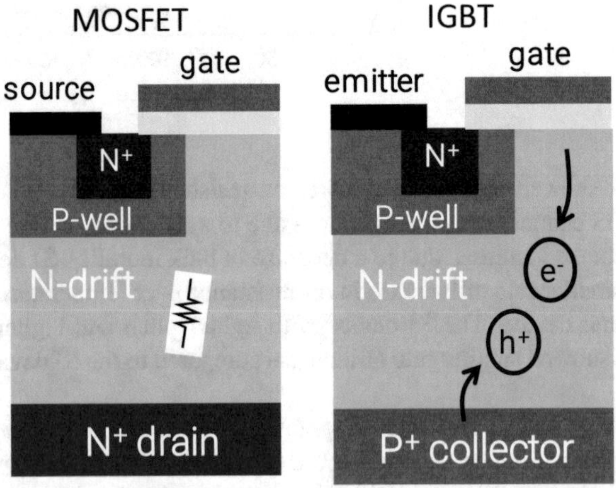

Fig. 2.20 SiC MOSFET and SiC IGBTs

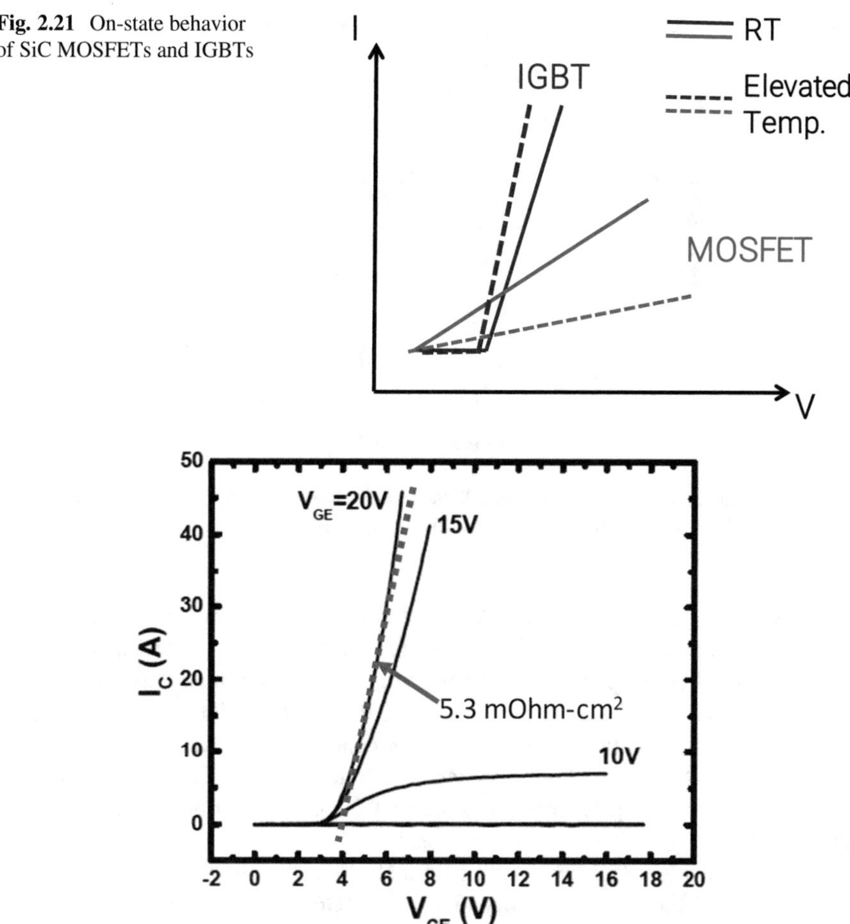

Fig. 2.21 On-state behavior of SiC MOSFETs and IGBTs

Fig. 2.22 Representative *I-V* characteristics of a 12 kV 4H-SiC IGBT. The chip size was 6.7×6.7 mm, with an active area of 0.16 cm^2

Figure 2.22 shows representative *I-V* characteristics of a 12 kV 4H-SiC n-IGBT [119]. The device used a 140-μm-thick, 2×10^{14} cm^{-3} doped drift layer, with a 6.7×6.7 mm device area (0.16 cm^2 active). A differential specific on-resistance of 5.3 mΩ·cm^2 was reported with a gate bias of 20 V at room temperature. This is significantly lower than those of 10 kV SiC power MOSFETs, which showed a specific on-resistance of around 100 mΩ·cm^2 [120].

Due to the excess carrier injection and associated increases in diffusion capacitance, switching speed of the 4H-SiC IGBTs is significantly slower, which limits the usable switching frequency of the device. The maximum controllable current in a hard switching application, with 50% duty cycle, was compared for the 15 kV 4H-SiC power MOSFETs and the 15 kV 4H-SiC n-IGBTs (see Fig. 2.23) [120,

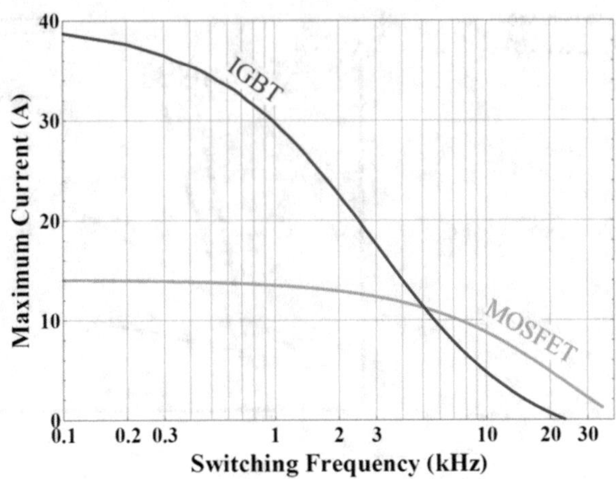

Fig. 2.23 Comparison of maximum controllable currents for the 15 kV 4H-SiC power MOSFETs and the 15 kV 4H-SiC n-IGBTs in a hard switching application

121]. A power dissipation density of 300 W/cm^2, a chip size of 8 × 8 mm (32 mm^2 active), and a supply voltage of 10 kV were assumed. A 15 kV SiC n-IGBT with a 5-μm-thick field-stop (FS) buffer layer with fast turn-off time was used for this comparison. At switching frequencies lower than 5 KHz, 4H-SiC n-IGBTs offer more advantage, showing up to 2.8 times the controllable current compared to 4H-SiC power MOSFETs. At higher switching frequencies, the 4H-SiC MOSFET has the advantage, showing 1.8 times the controllable current over the 4H-SiC n-IGBTs at 10 kHz.

The use of thick drift layer with extremely light doping concentration enables the increase of blocking capability of 4H-SiC IGBTs to beyond 20 kV [122]. A 4H-SiC IGBT, using a 230-μm-thick drift layer with a doping concentration of 2.5×10^{14} cm^{-3}, demonstrated a blocking voltage of 27.5 kV, which is the highest blocking voltage for a solid-state switching device reported to date [123]. The device had a die area of 0.81 cm^2 (0.28 cm^2 active). The drift layer received a lifetime enhancement oxidation at 1300 °C for 15 hours, and a V_{ON} of 11.8 V at a collector current of 20 A (approximately 71 A/cm^2) at 25 °C. Recently, a 4H-SiC n-IGBT using 230-μm-thick drift layer with a doping concentration of 2.0×10^{14} cm^{-3} was reported [124]. This device showed a blocking voltage of 26.8 kV. Carbon implantation and subsequent annealing processes were employed for carrier lifetime enhancement. A V_{ON} of 8.2 V at a current density of 100 A/cm^2 and a differential specific on-resistance of 36.9 mΩ·cm^2 were measured at room temperature.

Further improvements in 4H-SiC IGBTs can be achieved by enhancing the amount of carriers (both holes and electrons) in the top region of the structure [125]. Several approaches were proposed and used in silicon, including the use of cell designs with wide distance between the cells (IEGT) [126], forming micro trench and floating cells [127], and employing carrier storage layer (CSL), which

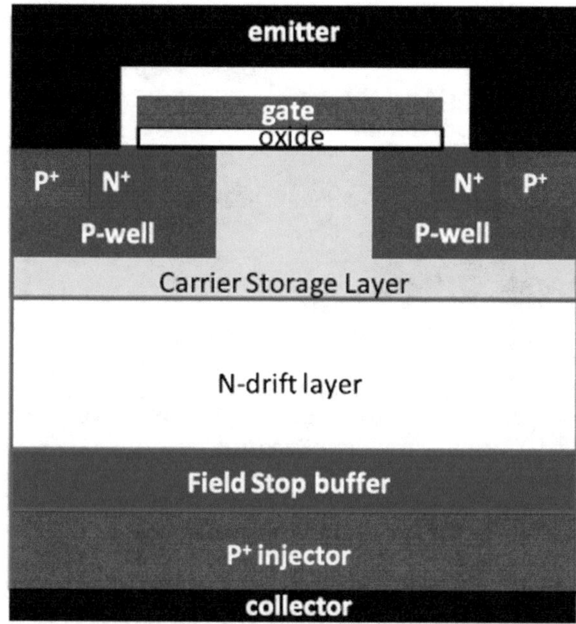

Fig. 2.24 A simplified cross section of a 4H-SiC IGBT, with carrier storage layer (CSL)

is a moderately doped n-type layer placed near the blocking junction [128]. IEGT approach is not suited for 4H-SiC IGBTs due to gate reliability issues caused by high electric field in 4H-SiC and lack of gate shielding in IEGT structure. Moreover, the use of floating cells is not favored in 4H-SiC IGBTs because of the low MOS channel mobility in 4H-SiC. However, 4H-SiC IGBT cell design can be optimized to utilize the CSL concept without compromising gate reliability (Fig. 2.24). The 4H-SiC IGBT without CSL layer showed a positive temperature coefficient in V_F due to high JFET resistance and lack of topside injection. The addition of CSL significantly reduced the JFET resistance and improved electron injection from the topside, resulting in negative temperature coefficient in V_{ON}, as shown in Fig. 2.25 [129]. It was also reported that heavier CSL doping concentration results in further reduction in V_{ON} [129].

7 III-Nitrides Power Devices

Gallium nitride (GaN) is a polar III-nitride semiconductor with a wide and direct bandgap. Due to this unique combination of highly attractive material properties, GaN has obtained significant commercial and research interest for a broad spectrum of applications, ranging from optoelectronics (e.g., displays and lighting) to high-frequency electronics (e.g., telecommunications and radar) to power electronics

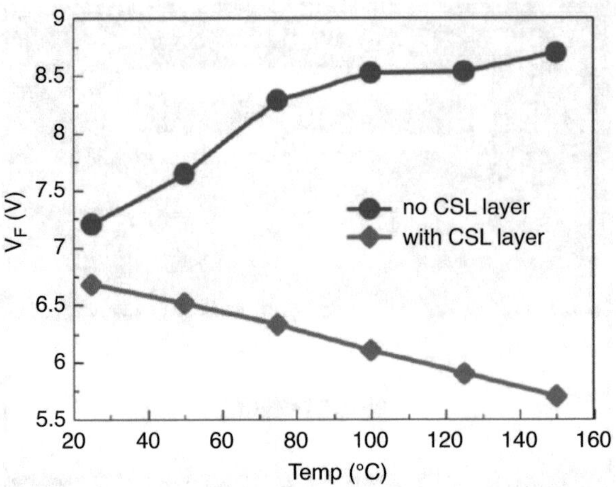

Fig. 2.25 Forward voltage drop of 15 kV 4H-SiC n-IGBTs, with and without CSL layers. V_{GE} was fixed at 20 V, and an I_C of 20 A was used for the measurements

(e.g., battery chargers and electric motor drives). This diverse ecosystem has driven rapid innovation, with advances in one application space subsequently supporting another, and vice versa. In what follows, the focus will be placed on the development of GaN devices for power electronics.

According to the Baliga Figure of Merit (BFOM), which provides a measure of a semiconductor's performance in the drift region of a vertical unipolar power device (e.g., Schottky diode), GaN can outperform conventional silicon (Si) by approximately 4000 times and silicon carbide (SiC) by approximately six times. In other words, for a given breakdown voltage and current rating, GaN devices have a significantly lower specific on-resistance (R_{ON}) and area. The latter also permits them to be driven at larger frequencies, which further reduces losses and necessitates smaller passive components, in turn shrinking the size and weight of the overall power module. The abovementioned BFOM predictions stem from the properties of bulk GaN, in particular its large critical electric field (E_C), which is linked to its wide bandgap, as well as its mobility. However, early challenges associated with manufacturing native GaN substrates steered the community away from vertical devices and spurred interest in developing lateral GaN devices on foreign substrates, such as sapphire, SiC, and Si. Consequently, it was understood that the polar nature of III-nitrides in conjunction with the use of heterojunctions (e.g., AlGaN/GaN) can be harnessed to form highly conductive channels. These heterojunctions lie at the core of GaN high-electron mobility transistors (HEMTs), which now possess record-breaking speed and power, and are the most technologically and commercially mature GaN-based power device available today. Thus, the text that follows will begin by discussing these devices. Vertical GaN devices, which have more recently been the target of resurgent attention, will then be explored, followed by an examination of future opportunities for GaN and III-nitride power devices.

7.1 Lateral GaN HEMTs

While a number of GaN-based lateral transistor topologies are technically possible, the undisputed winner is the HEMT. Very recently, the GaN HEMT has been commercialized in the 15–650 V classes [2], and its market size is projected to exceed $1.25 billion by 2027 [130]. Owing to GaN's superior physical properties over Si and SiC for power applications, GaN HEMTs allow for higher switching frequency and therefore, have already seen wide adoptions in fast chargers, wireless charging, data centers, and electrified transportation. In addition, GaN HEMTs can accommodate various substrates, e.g., Si, sapphire, SiC, and GaN. Most commercial GaN HEMTs for power electronics rely on large-diameter GaN-on-Si, and their process is CMOS-compatible, enabling a similar material and processing cost as compared to SiC power devices [131].

At the heart of the GaN HEMT is a heterojunction that provides a quasi-two-dimensional channel with high-electron density (N_S) and mobility (μ), i.e., the two-dimensional electron gas (2DEG) channel. Unlike GaAs HEMTs, the 2DEG in GaN HEMTs forms without the need for any extrinsic dopants. As shown in Fig. 2.26, when a thin aluminum gallium nitride (AlGaN) layer (typically 5–30 nm thick) is grown on top of a thicker GaN layer, a 2DEG with N_S ~10^{12}–10^{13} cm^{-2} forms below the AlGaN/GaN hetero-interface due to polarization fields and donor-like surface states [132, 133]. Electrons are vertically confined within a thin triangular potential well, allowing for a high mobility of 1500–2000 $cm^2/(V{\cdot}s)$. Note that the 2DEG can be formed in numerous combinations of group III-nitride heterostructures, and the AlGaN/GaN is the most popular heterostructure of choice. The 2DEG is a unique feature not available in Si and SiC technologies. It is the combination of high current densities obtainable via the 2DEG and the large E_C of GaN that make it such an attractive power device, despite the lateral configuration. It is also worth noting that the 2DEG forms without the application of a gate bias, meaning that GaN HEMTs are inherently depletion mode (D-mode) or normally-on devices. While this makes them directly usable for power amplifiers in telecommunications applications [134, 135], power electronics require normally-off or enhancement mode (E-mode) operation. Methods to satisfy this requirement are explored below.

7.2 Commercial and R&D Devices

Currently, four main structures are adopted in commercial power GaN HEMTs, as illustrated in Fig. 2.27. While all of commercial devices employ the 2DEG channel, main difference between them lies in the gate stack, or more specifically the techniques to enable enhancement-mode (E-mode) operation, which is highly desirable for power electronics applications [136]. The Schottky-type p-gate HEMT (SP-HEMT) (Fig. 2.27a) and gate injection transistor (GIT) (Fig. 2.27b) [137] both

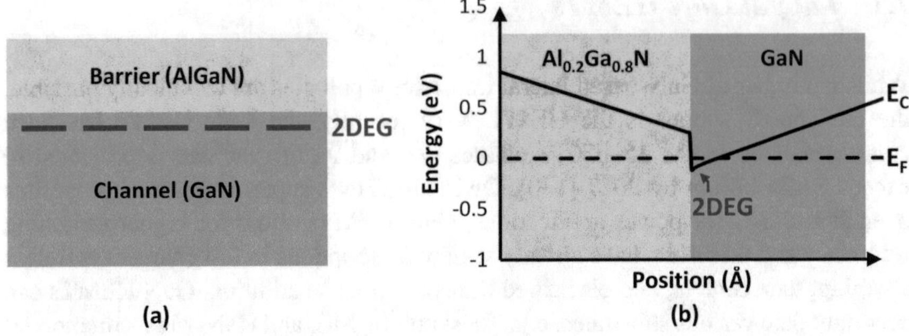

Fig. 2.26 (**a**) Schematic cross section of a heterojunction consisting of an AlGaN barrier layer and GaN channel layer. (**b**) Energy band diagram across the AlGaN/GaN heterojunction demonstrating the formation of a two-dimensional electron gas (2DEG) within the GaN layer characterized by large mobility and sheet carrier density

use p-GaN to deplete the 2DEG under the gate, but they feature different contacts between the gate metal and p-GaN. The recessed gate and ohmic gate contact in the GIT favor the hole injection and conductivity modulation, which are not present in the SP-HEMT. Given the large voltages that these devices are expected to block, field management is critical. As shown, field plates are often used to suppress field crowding. The cascode (Fig. 2.27c) and direct-drive devices usually co-package a high-voltage D-mode GaN HEMT with a low-voltage E-mode Si power MOSFET to make the composite device function like a single high-voltage E-mode transistor [138]. Direct-drive devices also co-package the gate driver and protection Si ICs with the GaN HEMTs [139].

Despite their success, the performance of commercial GaN devices has not yet reached the predicted material limit. In the last few years, two emerging GaN devices, i.e., the FinFET and trigate device as well as the multichannel device, have arguably been among the most innovative and promising lateral GaN power devices. The GaN FinFET and trigate devices have been comprehensively reviewed in [140]. As shown in Fig. 2.28a, these nonplanar GaN devices take advantage of the multi-gate fin channels to improve the gate controllability. Very different from Si FinFETs, GaN FinFETs and trigate devices have many structural innovations, such as wrapping around 2DEG channels with the MIS stack [141] or the p-n junction [142]. The superior gate controllability has not only allowed higher current on/off ratio, steeper threshold swing, and suppression of short-channel effects, but also E-mode operation, on-resistance reduction, current collapse alleviation, and enhanced thermal management [140].

Recently, the large-diameter wafer with multiple, vertically stacked 2DEG channels becomes available. This multichannel wafer allows for a sheet resistance below 120 Ω/sq, i.e., at least three times lower than that of a single 2DEG channel [143]. Despite a much lower R_{ON} of multichannel devices, it is challenging to manage the

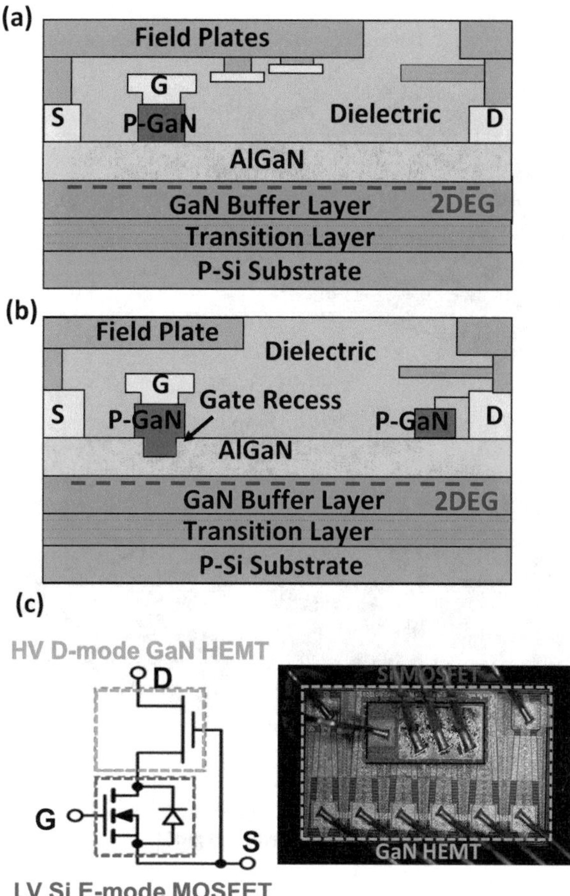

Fig. 2.27 Schematic of (**a**) SP-HEMT, (**b**) GIT, and (**c**) cascode GaN HEMT. (A de-capped device photo is shown in (**c**), which is adapted from [139])

electric field crowding at high reverse biases due to high volume charges. Various structures, e.g., trigate [144], p-GaN termination [143], 3-D junction fin [145], and reduced surface field cap layer [146], have been developed in multichannel rectifiers, which enabled their performance to exceed the 1-D SiC limit up to a voltage class of 10 kV. For multichannel HEMTs, trigate [140] and monolithic-cascode [147] designs have been innovatively applied; the monolithic-cascode device (Fig. 2.28b) demonstrates the E-mode operation with a performance surpassing SiC up to 10 kV [148].

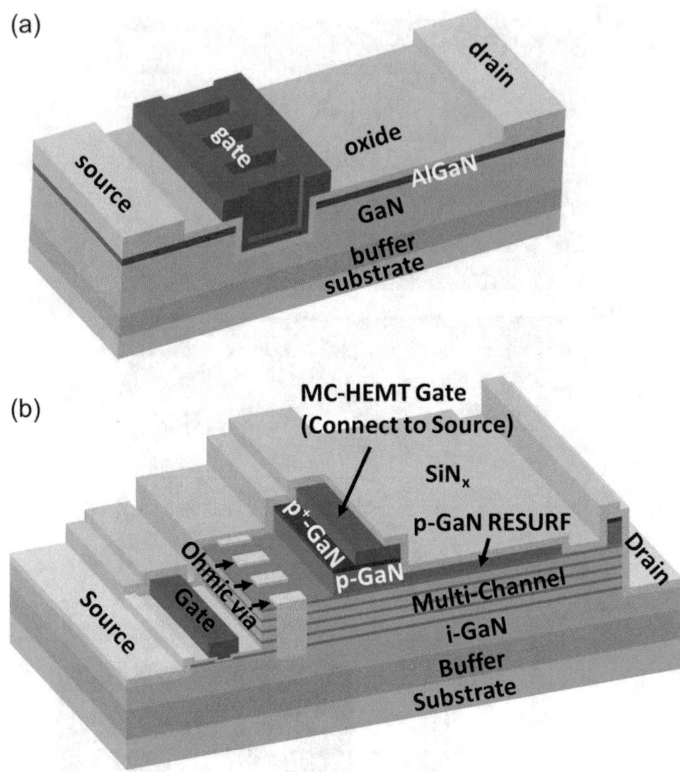

Fig. 2.28 Schematic of (**a**) trigate GaN HEMTs and (**b**) multichannel monolithic-cascode GaN HEMTs. ((**a**) is adapted from [142], and (**b**) adapted from [147])

7.3 Discrete Device Packaging

The aforementioned innovations in GaN HEMT design and chip manufacturing make it possible to push the limits of output power and switching frequency, in turn demanding careful consideration at the package level. As the junction temperature (T_J) increases, R_{ON} increases and transconductance (g_m) decreases, which contribute to increased conduction and switching losses, respectively. If left unchecked, self-heating can lead to thermal runaway and a catastrophic failure of the device. To combat these problems, thermal management and hot spot evaluation of GaN HEMTs have been extensively studied using a variety of electrical and optical techniques [148, 149]. Most commercial products call for a maximum junction temperature of approximately 150 °C for reliable operation, as defined by JEDEC standards [150].

In lower-frequency applications, well-established through-hole, lead frame packages (Fig. 2.29a), such as TO-220 and TO-247, can be used. An example of a TO-247 package is shown in Fig. 2.29b [151]. Within the package, the GaN

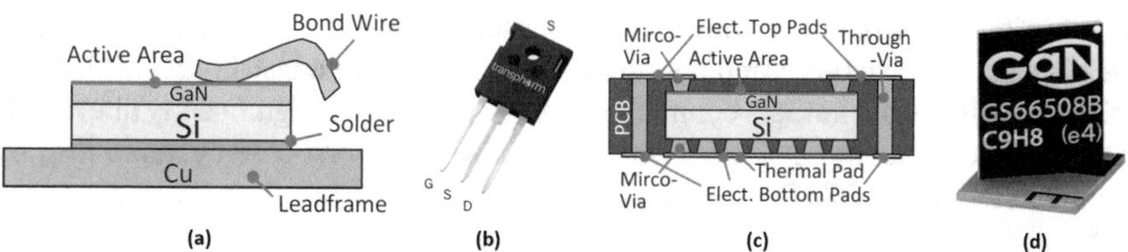

Fig. 2.29 (**a**) Schematic cross section of a lead frame package. [153] (**b**) Picture of a commercial GaN HEMT in a lead frame package. [151] (**c**) Schematic cross section of a no-lead package with an embedded GaN HEMT. [153] (**d**) Picture of a GaN HEMT in a proprietary embedded package [154]

HEMT is attached to a thermally conductive substrate, such as a direct-bonded copper (DBC) substrate. Electrical connections to the leads are made via wire- or ribbon-bonding, as shown in Fig. 2.29c. Since one of the major advantages of GaN HEMTs is their high mobility, they are extremely well-suited for high-frequency applications. As the switching frequency of the application increases, however, the parasitic inductances associated with the bond wires and leads impede performance. As a first step toward mitigating parasitics, no-lead packages, such as the Power Quad Flat No Leads (PQFN) topology have been adopted. Within these packages, however, wire-bonds are still used. For highest-frequency performance, micro-vias or flip-chip bonding [152] can instead be employed to fully embed the chip within a multilayered package, as shown in Fig. 2.29c [153] and Fig. 2.29d [154]. Thermal management is achieved by strategically introducing thermal vias and novel materials with superior thermal conductivity. The small footprint of these packages drives down the size and weight of power modules, making them highly attractive for portable applications. Chip embedding also facilitates multi-chip and heterogeneous technology integration [155].

7.4 Robustness and Reliability

The reliability and robustness challenges of GaN HEMTs come from not only the distinct device physics (e.g., the lack of p-n junctions between source and drain) but also the heterogenous GaN-on-Si epitaxy with a high dislocation density. Many issues regarding device stability, reliability, and robustness arise from the presence of traps in various regions of the epi structure. Trapping behavior is usually time-dependent. Hence, device characteristics in fast switching could significantly differ from the ones under DC conditions. A recent paper nicely overviews the relevant device physics and material issues [156]. As commercial GaN HEMTs have passed reliability qualifications, many recent studies emphasized on testing them in switching circuits to understand their robustness outside the safe operating areas.

The dynamic R_{ON} phenomenon, where R_{ON} immediately after device turn-on is higher than the DC value, is a well-known issue of GaN HEMTs that increases their conduction losses in power converters. A decade of study has revealed its origins to be associated with buffer trapping, surface trapping, and gate instability [156]. After relentless efforts by the GaN community, this issue has been significantly alleviated. Steady-state switching measurements have shown the worst-case dynamic R_{ON} of commercial GaN HEMTs to be less than two times higher than the static value [157].

Two major robustness metrics of a power transistor is the avalanche and short-circuit capabilities. The former evaluates the device's capability to pass a high avalanche current (I_{AVA}) at its avalanche breakdown voltage (BV_{AVA}) and thereby dissipate the circuit surge energy in device. GaN HEMTs have no avalanche capability and a destructive breakdown. The surge energy cannot be dissipated but induces a capacitive charging in GaN HEMTs, and they fail when the capacitive overvoltage reaches the transient BV [158]. More interestingly, this transient BV in fast switching was found to be dynamic and higher than the static BV measured in quasi-static *I-V* sweeps [159, 160]. This is attributed to the reduced buffer trapping in short pulses, where the buffer trapping intensifies the peak electric field and make it reach the critical electric field of GaN at lower drain biases [159]. Different from the standalone GaN HEMTs, the dynamic BV of cascode GaN HEMTs was found to be much lower than their static BV due to the internal Si avalanching [138].

Short-circuit robustness evaluates the device's capability of withstanding an abnormally high current in forward conduction and reverse blocking states for a certain time (usually required to be >10 μs) before the protection circuit intervenes [161]. The short-circuit robustness of GaN HEMTs is insufficient at a high blocking voltage. For example, the short-circuit withstand time of all 600/650 V rated commercial GaN HEMTs was found to be below 1 μs at a bus voltage of 400 V [162].

It is worth noting that the insufficient avalanche and short-circuit capabilities of GaN HEMTs are related to the HEMT device instead of the GaN material. In vertical GaN devices comprising p-n junctions, which will be introduced in Sect. 3, robust avalanche and short-circuit capabilities have been demonstrated. For example, a reverse I_{AVA} over 50 A [163] and forward surge current over 50 A [164] were reported in 1.2 kV vertical GaN p-n diodes; an avalanche energy comparable to Si and SiC devices has been demonstrated in vertical GaN fin-channel JFETs [165–167]. In the 650 V vertical GaN JFETs, a short-circuit withstand time over 30 μs has been reported at the 400 V bus voltage, with a short-circuit energy superior to SiC and Si MOSFETs [160].

Note that many of the above reliability and robustness challenges facing GaN HEMTs limit the further advancement of their performance. For example, due to the lack of avalanche capability, a larger voltage over-design is implemented in commercial GaN HEMTs as compared to Si and SiC MOSFETs [137, 157], which leads to larger specific on-resistance and offsets the inherent advantages of GaN devices. Addressing these challenges will not only facilitate GaN's applications but also bring considerable performance advancements to the device itself.

7.5 Approaches to GaN ICs

To take full advantage of the high switching speed of GaN HEMTs, it is necessary to minimize parasitic inductances between them and their gate driving circuits, which is desirably to be realized through monolithic integration. The E-mode/D-mode logic, which is also known as the direct coupled logic (DCL), has been employed to make the GaN HEMT-based driving circuits and integrate them with the high-voltage GaN power HEMTs [168]. Such GaN power ICs are already commercially available from companies such as Navitas Semiconductor, Power Integration, and EPC, and they have been widely used in consumer electronics systems such as fast chargers and wireless chargers.

A key limitation of the DCL-based power ICs is the high-power consumption and limited circuit design flexibility [136]. To overcome these issues, there has been extensive research on the GaN complementary technology comprising high-performance n-channel and p-channel GaN E-mode transistors [136]. For this, the key challenge is on the p-channel GaN transistor design. An early GaN CMOS demonstration employs p-channel GaN MOSFETs on a regrown epitaxial structure [169]. Later, it was proposed to utilize the p-GaN layer in standard p-gate GaN HEMTs to make the p-channel GaN transistors [170], which obviates the need for epitaxy regrowth. Subsequently, a regrowth-free GaN CMOS has been demonstrated [171].

Heterogeneous integration of GaN HEMTs with Si CMOS technology is another approach that is being pursued to enhance power and functionality. A prominent example is the use of oxide bonding at the wafer scale [172]. GaN HEMTs are first fabricated on 300 mm highly resistive Si wafers and then bonded to Si wafers. The second Si wafer is controllably removed using an etch stop layer, ultimately leaving behind single crystal Si, with which Si p-channel MOSFETs are fabricated. More recently, both Si NMOS and PMOS devices have been integrated with E- and D-mode GaN HEMTs [173] using this approach. Whereas the primary focus has been on power amplifier development thus far, the introduction of field management techniques that permit high-voltage operation would unlock several exciting opportunities in the power electronics space as well.

7.6 Vertical GaN Devices

7.6.1 Transistors

The vertical structure is often believed to favor high-voltage, high-power devices as it facilitates current spreading and thermal management [174] and allows for the realization of high voltage without enlarging the chip size. Additionally, as compared to GaN-on-Si, GaN-on-GaN homoepitaxial layers possess a much lower dislocation density, which favors the minimization of trapping effects. Over the

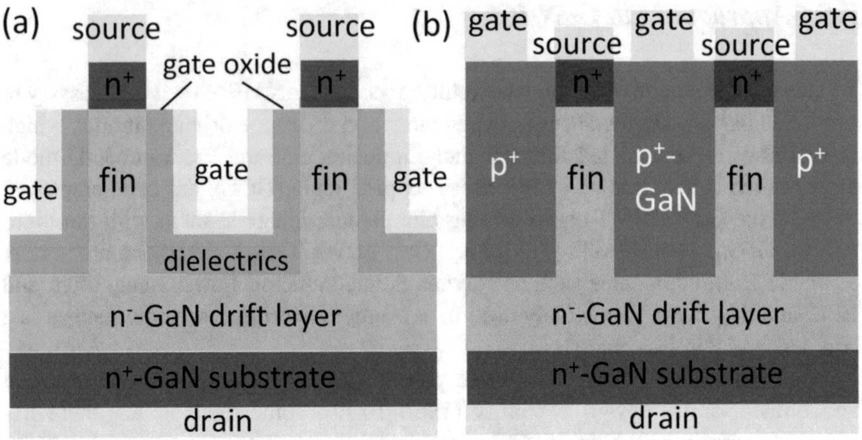

Fig. 2.30 Schematic of (**a**) the vertical GaN Fin-MOSFET and (**b**) vertical GaN Fin-JFET [175]

last several years, the cost of GaN-on-GaN wafer is dropping fast, and 4-inch freestanding wafer is widely available now [131]. Several vertical GaN transistors with a voltage class over 1.2 kV have been demonstrated recently.

The vertical GaN FinFETs leverage the digital FinFET concept and employ the submicron-meter fin-shaped channels to provide superior gate control as well as enable the E-mode operation and bidirectional conduction. The small footprint of fin channels also allows for very high channel density and thereby low channel resistance. The development of this device concept as well as its application to other materials have been reviewed in [175] and [140]. Depending on the sidewall gate stack, there are two types of power FinFETs, the Fin-MOSFET (Fig. 2.30a) and Fin-JFET (Fig. 2.30b). In each fin, the Fin-MOSFET features a bulk fin channel in parallel with two sidewall accumulation-type MOS channels [176] and the device needs only n-type GaN. In the Fin-JFET, the inter-fin region is filled with p-GaN, and the strong depletion of the lateral p-n junction allows a high doping concentration in the fin while keeping the E-mode operation [165, 166]. 1.2 kV vertical GaN Fin-MOSFETs [177, 178] and Fin-JFETs [165, 166] have both shown three- to fivefold lower specific R_{ON} and superior switching performance [165, 178] as compared to 1.2 kV SiC MOSFETs. Industrial GaN Fin-JFETs for the first time demonstrates the avalanche capability in GaN transistors [165, 166] and have a record short-circuit robustness in GaN transistors [161]. The Fin-JFET also demonstrates a unique short-circuit robustness at a bus voltage close to its $\mathrm{BV}_{\mathrm{AVA}}$ [161], which was not reported in other devices.

The AlGaN/GaN current aperture vertical electron transistor (CAVET) is a unique transistor topology that seeks to leverage the highly conductive AlGaN/GaN heterostructure's 2DEG with the voltage handling of a vertical drift region. Early demonstrations were conducted on sapphire substrates and relied on regrowth [179], but the more recent availability of native GaN substrates has improved material quality and permitted the use of Mg implantation to form the current blocking layer (CBL) and consequently define the aperture [180]. Current challenges associated with both etch and regrowth as well as ion implantation limit the quality of the CBLs, as well as their interfaces with adjacent regions. Thus, trench CAVETs were also developed, both with MIS gate [181] and p-GaN [182] gate structures to provide E-mode operation. In the latter work [182], 1.7 kV/1.0 mΩ cm^2 devices were reported, which surpassed the SiC BFOM limit at this voltage rating, and represent the current state of the art for GaN CAVETs.

While the conventional DMOSFET and UMOSFET topologies have been readily adopted in both Si and SiC technologies, they remain difficult to realize in GaN. This is primarily due to a lack of a native, high-quality oxide that can be used in the gate structure. Poor GaN/oxide interfaces lead to poor channel conduction and channel inversion. The use of ALD and MOCVD dielectrics [183, 184] as well as UID GaN interlayer/oxide (known as the "OG-FET") [185] structures has been investigated at the gate, but the performance and reliability needed to displace SiC MOSFETs have yet to be achieved. An additional challenge is selective area p-type doping, which is needed to define the channel, contacts, and edge terminations. Recently, Mg implantation has been explored for this purpose [186], but progress in this space is still needed.

7.6.2 Diodes

Early studies of vertical GaN devices started from p-n diodes, as the p-n junction is a key building block for many advanced devices. Multiple groups have reported 3.3–5 kV GaN p-n diodes with a differential R_{ON} *v.s.* V_{BL} trade-off exceeding the 1-D SiC unipolar limit [187, 188]. In addition to the aforementioned avalanche and surge capabilities, a breakdown field of 2.8–3.5 MV/cm close to intrinsic GaN limits has also been reported in vertical GaN p-n diodes. As compared to SiC p-n diodes, industrial GaN p-n diodes show a comparable robustness but smaller reverse recovery and faster switching speed [165].

However, the vertical GaN p-n diode may not be competitive as a standalone rectifier due to the large turn-on voltage due to the GaN bandgap. Advanced Schottky barrier diodes (SBDs) are highly desirable, as they combine Schottky-like forward characteristics (small turn-on voltage) and p-n-like reverse characteristics (high breakdown field and low leakage current). These advanced SBDs include the trench MIS/MOS barrier Schottky (TMBS) diode, junction barrier Schottky (JBS) diode, and merged p-n/Schottky (MPS) diode. These diodes employ either the MIS stack or the p-n junction to deplete the top part of the drift region at low reverse biases, thereby shielding the top Schottky contact from high electric field. 600–700 V GaN TMBS diodes [189] and JBS diodes [190], as well as 2 kV MPS diodes [191], have exhibited at least 100-fold lower leakage current compared to standard SBDs. Toyoda Gosei reported an industrial 10 A, 750 V GaN TMBS diode operational at over 200 °C [192].

7.6.3 Vertical Devices on Foreign Substrates

In addition to freestanding GaN substrates, vertical GaN devices can be also fabricated on low-cost foreign substrates, such as Si, sapphire, and engineering substrates. The relevant studies date back to the first demonstration of vertical GaN-on-Si diodes using a quasi-vertical structure [193]. Fully vertical GaN-on-Si devices were later realized by various approaches to handle the insulative, defective transitional layers, including the layer transfer [194], buffer doping [195], and deep backside trenches [196]. The state of the art includes 500–800 V vertical GaN-on-Si diodes [196] and MOSFETs [197] as well as 1.4 kV vertical GaN-on-sapphire SBDs [198]. Regarding the cost and performance trade-offs, the higher dislocation density in GaN-on-Si was found to induce a relatively small degradation in forward characteristics but a higher off-state leakage current at high biases [199], at the same time bringing the material and processing cost by at least tenfold [131].

7.6.4 GaN Superjunction Devices

The SJ is arguably the most conceptually innovative and commercially successful device in Si, which relies on alternative n- and p-doped pillars and can break the theoretical trade-off between R_{ON} and BV of 1-D drift regions [200]. The SJ has not reached experimental demonstrations in GaN. Instead of p-n junction, balanced polarization charges were used in lateral AlGaN/GaN devices to produce a “natural SJ” for superior E-field management [201]. However, their R_{ON} is still much higher than the 1-D GaN limit (and even the SiC limit).

The recent experimental realization of selective p-type doping in GaN devices [165, 190] has shown the promise for fabricating GaN SJ devices. Selective p-GaN/2DEG junctions with a high blocking electric field have also been reported [202]. Simulations predict that vertical GaN SJ transistors with fin channels and 2DEG channels can enable at least 20-fold smaller switching charges as compared to today’s transistors and allow multi-MHz, multi-kilovolts power switching [203].

A novel approach to realizing the GaN SJ involves the use of the lateral polar junction (LPJ), wherein neighboring n-type and p-type pillars are formed by the simultaneous growth of selectively doped N-polar and Ga-polar GaN [204]. Recently, charge balance between neighboring pillars, a critical design requirement in any SJ, was demonstrated with this approach [205]. Steps forward are also being made to reduce background doping in N-polar GaN, which is needed to achieve large blocking voltages. Given the progress being on all fronts, we believe the demonstration of a GaN SJ will arrive soon.

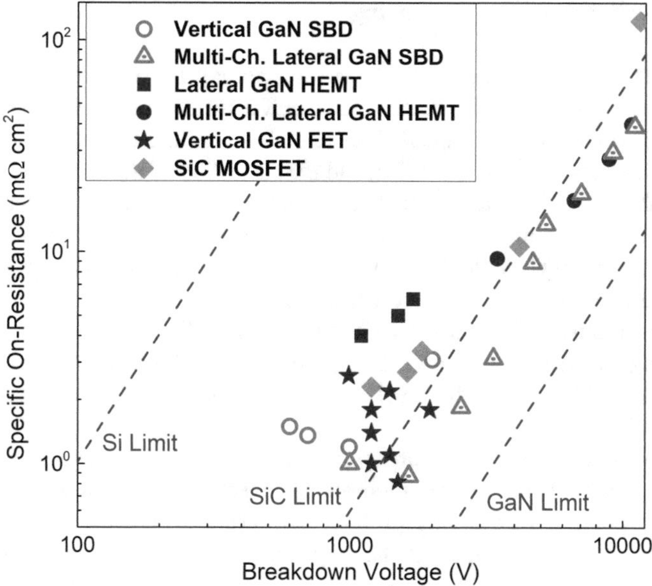

Fig. 2.31 On-resistance versus breakdown voltage trade-offs of the state-of-the-art GaN Schottky barrier diodes and power transistors as well as SiC power MOSFETs

7.7 Outlook: Research Opportunities

GaN technology has developed rapidly in the last decade, as evidenced by the significant commercial and research investments made for power electronics. Figure 2.31 shows the on-resistance versus breakdown voltage trade-offs of the state-of-the-art GaN Schottky barrier diodes and power transistors as compared to SiC power devices. The performance of commercial GaN HEMTs is between the Si and SiC theoretical limits, and many emerging GaN devices, particularly the vertical and multichannel lateral GaN devices, have shown performance exceeding the 1-D SiC theoretical limit. This suggests a good promise of expanding the GaN's application space in power electronics through device innovations. In the meanwhile, the state-of-the-art GaN device performance is still inferior to the 1-D unipolar GaN limit, suggesting room for further improvement. While many opportunities for future progress have already been presented in the preceding discussion, this section provides more detail on some of the most exciting research trends that lie ahead for GaN and III-nitrides more generally.

7.7.1 Selective Area P-Type Doping

Selective area doping is a critical process technology for both vertical and lateral power devices, including JBS diodes, MOSFETs, JFETs, and SJ devices. In Si and SiC technologies, efficient implantation of both donor and acceptor impurities has been established for this purpose. Beyond the formation of the device's active area, ion implantation is also relied on to create efficient edge terminations, such as guard rings and junction termination extensions [206]. While Si implantation is available for n-type doping of GaN, the ability to selectively form p-type GaN regions remains critical and elusive. Etch and regrowth process has been explored extensively, but is often limited by defects at the etched interface [179, 207–210]. Very recently, encouraging results have emerged using proprietary methods for etch and regrowth, which have enabled avalanche breakdown to be observed in vertical 1.2 kV GaN JFETs [166].

Mg implantation for selective area p-type doping has also been pursued for a long time, since it is expected to offer high-quality interfaces, reduced processing steps, and greater control over doping profiles. The principle challenge is suppressing decomposition of the surface at the high annealing temperatures (>~1100 °C) [211] needed to activate the Mg ions and repair the crystal damage introduced by implantation. Several annealing techniques have been proposed to activate Mg-implanted GaN while avoiding surface decomposition. Pulsed techniques such as multicycle rapid thermal annealing (MRTA) [212], gyrotron microwave annealing [213], and laser annealing [214] have successfully suppressed surface decomposition. However, the reported activation ratios are quite low (e.g., ~0.5–8%) [190, 212, 213], which limits their practical adoption. Ultrahigh-pressure annealing (UHPA) is another strategy, which calls for an N_2 ambient pressure of >300 MPa to stabilize the surface during the anneal. Most importantly, experimental results have shown that ~100% activation of the implanted Mg can be achieved with UHPA [215], though significant Mg diffusion has also been observed [216]. GaN PN junctions with kV-level blocking voltages have been recently reported with this technique [217]. How this technology develops, in particular for use in devices with even larger blocking voltages (i.e., >5 kV), could have a major impact on future generations of vertical GaN devices.

7.7.2 High-Voltage GaN Devices

As current GaN HEMTs are mainly commercialized in the low-voltage range (<650 V), there has been a popular belief that GaN devices are suitable only for low-voltage applications. Contrary to this popular belief, GaN rectifiers [146] and E-mode HEMTs [147] have been demonstrated up to 10 kV with the performance beyond the 1-D SiC unipolar limit. These results suggest that material and device landscapes for the $5 billion medium-voltage (1–35 kV) power electronics market could be reshaped, and multi-kilovolt GaN devices are very promising for applications like electric grid and renewable energy processing. From the device architecture point of view, both vertical GaN devices and multichannel lateral devices are promising candidates of the medium- and high-voltage GaN switches, as they address the challenges facing the conventional single-channel HEMTs for voltage upscaling and power upscaling (e.g., crowding current and electric field distributions, limited current capability). Looking forward, numerous research opportunities exist in physics, materials, and devices for multi-kilovolts GaN power devices, as well as reliability, robustness, and converter applications.

7.7.3 N-Polar GaN HEMTs

The band diagram for the GaN HEMT provided in Fig. 2.26 assumes the use of Ga-polar or Ga-face GaN. While it was understood early on by the community that a 2DEG could also be formed in N-polar GaN [132], virtually all GaN HEMTs reported in the literature have been based on Ga-polar technology due to its superior chemical and thermal stability of N-polar GaN [218]. Advancements in surface roughness, doping control, and in situ passivation of N-polar GaN have renewed interest in N-polar GaN. This is because the two-dimensional electron gas (2DEG) is formed with an AlGaN back-barrier because of the opposite polarization field. The consequence of this is a reduced contact resistance for the source and drain, as well as easier gate length scaling. For RF applications, N-polar GaN HEMTs have been experimentally demonstrated to outperform Ga-polar GaN HEMTs [219]. An initial report of an N-polar GaN/AlGaN HEMT on sapphire achieved a breakdown voltage of 2 kV and state-of-the-art dynamic R_{on} compared to Ga-polar GaN HEMT competitors [220]. Additional performance gains are expected after transitioning to Si or SiC substrates, as well as via improvements to the surface passivation.

7.7.4 UWBG III-Nitrides

Due to the cubic dependence of the BFOM on the critical electric field (E_C) of a material, there has been a large push to develop ultra-wide bandgap (UWBG) semiconductors for power devices [221]. This is clearly visualized in Fig. 2.32a. Among these, $Al_xGa_{1-x}N$ is a III-nitride semiconductor that shares many similarities to GaN, such as piezoelectric properties and a direct bandgap, and it can be grown on single crystal AlN substrates with a state-of-the-art dislocation density <10^4 cm^{-2} [222]. For reference, if 100% Al composition (i.e., AlN) is used, the BFOM is predicted to yield an up to 34- or 80-fold increase in the BFOM over GaN and SiC, respectively, resulting in significant loss reductions.

To study the voltage handling capability of Al-rich AlGaN, $Al_{0.85}Ga0_{.15}N/Al_{0.6}Ga_{0.4}N$ HEMTs were recently fabricated on native AlN substrates [223]. The HEMT structure consists, bottom to top, of an unintentionally doped (UID) AlN layer, a 300 nm $Al_{0.6}Ga_{0.4}N$ channel layer, and a 20 nm UID $Al_{0.85}Ga_{0.15}N$ barrier layer. The upper 150 nm of the $Al_{0.6}Ga_{0.4}N$ channel layer is doped with Si at 5×10^{17} cm^{-3}. The channel layer was doped instead of the barrier layer to avoid

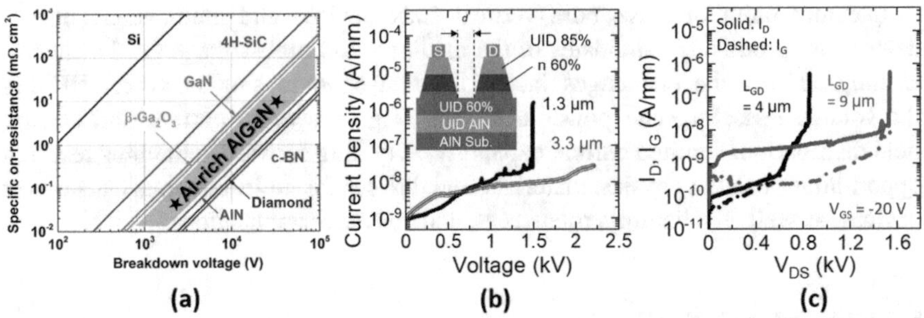

Fig. 2.32 (**a**) Trade-off of on-resistance vs. breakdown voltage for different semiconductor technologies per the BFOM. (Modified from [221]). (**b**) Two-terminal mesa breakdown characteristics for an UID $Al_{0.6}Ga_{0.4}N$ layer grown on native AlN substrate. (**c**) Three-terminal breakdown characteristics of $Al_{0.85}Ga_{0.15}N$/ $Al_{0.6}Ga_{0.4}N$ HEMT for L_{GD} of 4 and 9 μm. The L_{SG} and L_G are 1.5 μm each and V_{GS} is −20 V [223]

high gate leakage while simultaneously facilitating ohmic contact formation. The cross-sectional schematic of the resulting mesa test structure is shown in the inset of Fig. 2.32b, which shows the two-terminal mesa breakdown characteristics for two mesa separations (*d*): 1.3 and 3.3 μm. Fluorinert was used to prevent premature breakdown. At 1.3 μm separation, UID $Al_{0.6}Ga_{0.4}N$ layer breaks down at ~1500 V, which corresponds to a breakdown electric field of 11.5 MV/cm. For context, this breakdown field is ~2× and ~4× higher than that previously observed in the UID AlN buffer mesa breakdown field in AlN/$Al_{0.5}Ga_{0.5}N$/AlN HEMTs [224] and AlN MESFETs [225], respectively. The structure with 3.3 μm separation was measured up to the upper limit of the testbench used (2.2 kV) and did not break down. The three-terminal HEMT breakdown characteristics for two different gate-to-drain (L_{GD}) distances, 4 and 9 μm, are shown in Fig. 2.32c. In both cases, the source-to-gate (L_{SG}) distance and gate length (L_G) are 1.5 μm, and V_{GS} is −20 V. As seen, the 4 and 9 μm devices break down at V_{DS} 850 and 1500 V, respectively, without using any edge termination techniques. This confirms the potential for low defect density Al-rich AlGaN on AlN to be used for power devices. Following the introduction of edge terminations in the future, improvements in performance are expected. In order for AlGaN HEMT technology to truly threaten the GaN HEMT's position in the market, however, the inferior thermal conductivity of AlGaN and ohmic contacts will need to be improved significantly.

8 Conclusions

WBG devices have compelling advantages over their Si counterparts and are presently inserted in numerous power applications enabling higher efficiency, smaller form factor, higher power density, and operation at elevated temperatures with simplified circuit topologies and thermal management.

Commercial SiC power devices (diodes, MOSFETs, JFETs, and BJTs) as well as bipolar devices (IGBTs) best suited for +10 kV applications were reviewed. Unipolar SiC devices are commercially available in the 650–1700 V range with the SiC MOSFET being inserted in the vast majority of SiC power applications. SiC MOSFETs have also been demonstrated at 3.3, 6.5, and 10 kV with commercial release coming in the next few years. Above 10 kV, the thick drift layer of MOSFETs becomes highly resistive, and bipolar devices like the SiC IGBT provide a good trade-off between lower conduction and increased switching losses.

Lateral GaN power devices will most probably, with time, dominate applications in the voltage range below 600–1000 V, in particular as GaN-based ICs mature. The low-loss and fast switching speed offered by this platform promises to not only improve performance but also reduce system cost, size, and weight. Despite the many technological advancements in GaN HEMTs over the years, they have not yet fulfilled their predicted performance potential. Multichannel and trigate device configurations are closing the gap, but thermal management remains a key hurdle that must be overcome via efforts in device and materials engineering. Due to the relatively recent introduction of GaN HEMTs to the power electronics market, there remain several opportunities to better understand the mechanisms that determine reliability and robustness in these devices. As time passes and the technology matures, consumer hesitance will dissipate. Vertical GaN technology is also catching up for >1 kV applications. In order for these devices to displace their SiC counterparts, the cost of native GaN substrates must be reduced, drift region doping must be lowered, and efficient selective area doping via ion implantation must be realized.

Barriers to WBG mass commercialization include the higher than silicon device cost, reliability and ruggedness concerns, and the need for a trained workforce to skillfully insert WBG devices into power electronics systems. In many applications, at the system level, WBG-based solutions are at parity or even more cost-effective than those of silicon primarily due to passive component simplifications. WBG devices are rapidly overcoming barriers to system insertion and are entering mass production in volume fabs with its cost-lowering benefits.

References

1. X. She, A.Q. Huang, O. Lucía, B. Ozpineci, Review of silicon carbide power devices and their applications. IEEE Trans. Indus. Electron **64**(10), 8193–8204 (2017)
2. H. Amano, Y. Baines, E. Beam, M. Borga, T. Bouchet, P.R. Chalker, M. Charles, K.J. Chen, R. Nadim Chowdhury, C.D. Chu, M.M.D. Santi, S. Souza, L.D. Decoutere, B.E. Cioccio, T. Egawa, P. Fay, J.J. Freedsman, L. Guido, O. Häberlen, G. Haynes, D. Thomas Heckel, P. Hemakumara, J. Houston, M. Hu, Q. Hua, A. Huang, S.J. Huang, H. Kawai, D. Kinzer, M. Kuball, A. Kumar, K.B. Lee, X. Li, D. Marcon, M. März, R. McCarthy, G. Meneghesso, M. Meneghini, E. Morvan, A. Nakajima, E.M.S. Narayanan, T. Stephen Oliver, D. Palacios, M. Piedra, R. Plissonnier, M. Reddy, I. Sun, A.T. Thayne, V. Nicola Trivellin, M.J. Unni, M.V. Uren, D.J. Hove, J. Wallis, J. Wang, S. Xie, S.Y. Yagi, C. Youtsey, R. Yu, E. Zanoni, S. Zeltner, Y. Zhang, The 2018 GaN power electronics roadmap. J. Phys D. Appl. Phys **51**(16), 163001 (2018). https://doi.org/10.1088/1361-6463/aaaf9d

3. N. Camara, K. Zekentes, L.P. Romanov, A.V. Kirillov, M.S. Boltovets, K.V. Vassilevski, G. Haddad, Microwave *p-i-n* diodes and switches based on 4H-SiC. IEEE EDL **27**, 108–110 (2006). https://doi.org/10.1109/led.2005.862686
4. K.V. Vassilevski, A.V. Zorenko, K. Zekentes, X-band Silicon Carbide IMPATT oscillator. MRS Online Proc. Lib **680**, 1011 (2001). https://doi.org/10.1557/PROC-680-E10.11
5. K.V. Vassilevski, A.V. Zorenko, K. Zekentes, Experimental observation of microwave oscillations produced by pulsed silicon carbide IMPATT diode. Electron. Lett **37**(7), 466–467 (2001). https://doi.org/10.1049/el:20010285
6. M. Arai, S. Ono, C. Kimura, IMPATT oscillation in SiC p^+-n-n^+ diodes with a guard ring formed by vanadium ion implantation. Electron. Lett **40**, 1026 (2004). https://doi.org/10.1049/el:20045312
7. K.V. Vassilevski, K. Zekentes, A.V. Zorenko, L.P. Romanov, Experimental determination of electron drift velocity in 4H-SiC p^+-n-n^+ avalanche diodes. IEEE EDL **21**, 485–487 (2000). https://doi.org/10.1109/55.870609
8. P.A. Ivanov, A.S. Potapov, T.P. Samsonova, I.V. Grekhov, Current–voltage characteristics of high-voltage 4H-SiC p^+–n_0–n^+ diodes in the avalanche breakdown mode. Semiconductors **51**, 374–378 (2017). https://doi.org/10.1134/s1063782617030095
9. K. Vassilevski, Silicon carbide diodes for microwave applications. Int J. High Speed Electron. Syst. **15**(04), 899–930 (2005). https://doi.org/10.1142/s0129156405003454
10. A. Salemi, Silicon carbide technology for high- and ultra-high-voltage bipolar junction transistors and PiN diodes, PhD dissertation, KTH, Sweden, 2017
11. T. Tsuji, A. Kinoshita, N. Iwamuro, K. Fukuda, K. Tezuka, T. Tsuyuki, H. Kimura, Experimental demonstration of 1200V SiC-SBDs with lower forward voltage drop at high temperature. Mater. Sci. Forum **717**, 917–920 (2012)
12. C.E. Weitzel, J.W. Palmour, C.H. Carter, K. Moore, K.K. Nordquist, S. Allen, C. Thero, M. Bhatnagar, Silicon carbide high-power devices. IEEE Trans. Electron Dev **43**(10), 1732–1741 (1996)
13. G. Chen, S. Bai, A. Liu, L. Wang, R.H. Huang, Y.H. Tao, Y. Li, Fabrication and application of 1.7 KV SiC-Schottky diodes. Mater. Sci. Forum **821**, 579–582 (2015)
14. Q. Wahab, T. Kimoto, A. Ellison, C. Hallin, M. Tuominen, R. Yakimova, A. Henry, J.P. Bergman, E. Janzen, A 3 kV Schottky barrier diode in 4H-SiC. Appl. Phys. Lett **72**(4), 445–447 (1998)
15. K. Vassilevski, I. Nikitina, A. Horsfall, N. Wright, C.M. Johnson, 4.6 kV, 10.5 mOhm.cm^2 Nickel Silicide Schottky diodes on commercial 4H-SiC Epitaxial Wafers. Mater. Sci. Forum **645-648**, 897 (2010)
16. R. Singh, J.A. Cooper, M.R. Melloch, T.P. Chow, J.W. Palmour, SiC power Schottky and PiN diodes. IEEE Trans. Electron Dev **49**(4), 665–672 (2002) vol. 49(12), pp. 2308–2316, 2002
17. K. Vassilevski, I.P. Nikitina, A.B. Horsfall, N.G. Wright, K.P. Hilton, A.G. Munday, A.J. Hydes, M.J. Uren, C.M. Johnson, et al., High voltage silicon Carbide Schottky diodes with single zone junction termination extension. Mater. Sci. Forum **556**, 873–876 (2007)
18. J.H. Zhao, P. Alexandrov, X. Li, Demonstration of the first 10-kV 4H-SiC Schottky barrier diodes. IEEE Electron Dev. Lett **24**(6), 402–404 (2003)
19. F. Roccaforte, G. Brezeanu, P. Gammon, F. Giannazzo, S. Rascunà, M. Saggio, Schottky contacts to silicon carbide: Physics, technology and applications, in *Advancing Silicon Carbide Electronics Technology, I: Metal Contacts to Silicon Carbide: Physics, Technology, Applications*, (Materials Research Forum LLC, 2018) ISSN 2471-8890 (Print) ISSN 2471-8904 (Online)
20. D. Stephani, Status, prospects and commercialization of SiC power devices. Device research conference. Conference digest (Cat. No.01TH8561), p. 14, 2001. https://doi.org/10.1109/DRC.2001.937852
21. B.M. Wilamowski, Schottky diodes with high breakdown voltages. Solid-State Electron **26**(5), 491–493 (1983). https://doi.org/10.1016/0038-1101(83)90106-5
22. K. Tone, J.H. Zhao, M. Weiner, M. Pan, Fabrication and testing of 1,000 V - 60 A 4H-SiC MPS diodes in an inductive half-bridge circuit. Mater. Sci. Forum **338-342**, 1187–1190 (2000). https://doi.org/10.4028/www.scientific.net/msf.338-342.1187

23. Toshiba Corp, Improved JBS structure to reduce the leakage current and increase the surge current capability. https://toshiba.semicon-storage.com/eu/semiconductor/product/diodes/sic-schottky-barrier-diodes/articles/improved-jbs-structure-to-reduce-the-leakage-current-and-increase-the-surge-current-capability.html
24. T. Kimoto, J.A. Cooper, *Fundamentals of Silicon Carbide Technology: Growth, Characterization, Devices and Applications* (Wiley, 2014)
25. Y. Jiang, W. Sung, X. Song, H. Ke, S. Liu, B.J. Baliga, A.Q. Huang, E. Van Brunt, 10 kV SiC MPS diodes for high temperature applications. 28th Int. symposium on power semiconductor devices and ICs (ISPSD), pp. 43–46, 2016. https://doi.org/10.1109/ispsd.2016.7520773
26. F. Dahlquist, C.M. Zetterling, Mikael Östling, and K Rottner., Junction barrier Schottky diodes in 4H-SiC and 6H-SiC. Mater. Sci. Forum **264**, 1061–1064 (1998)
27. T. Yamamoto, J. Kojima, T. Endo, E. Okuno, T. Sakakibara, S. Onda, 1200-V JBS diodes with low threshold voltage and low leakage current. Mater. Sci. Forum **600**, 939–942 (2009)
28. N. Ren, J. Wang, K. Sheng, Design and experimental study of 4H-SiC trenched junction barrier Schottky diodes. IEEE Trans. Electron Dev **61**(7), 2459–2465 (2014)
29. F. Dahlquist, J.O. Svedberg, C.M. Zetterling, M. Östling, B. Breitholtz, H. Lendenmann, A 2.8 kV, forward drop JBS diode with low leakage. Mater. Sci. Forum **338**, 1179–1182 (2000)
30. J. Hu, L.X. Li, P. Alexandrov, X.H. Wang, J.H. Zhao, 5 kV, 9.5 A SiC JBS diodes with non-uniform Guard Ring Edge termination for high power switching application. Mat. Sci. Forum **600**, 947–950 (2009)
31. R.H. Huang, G. Chen, S. Bai, R. Li, Y. Li, Y. Hong Tao, Simulation, fabrication and characterization of 4500V 4H-SiC JBS diode. Mater. Sci. Forum **778**, 800–803 (2014)
32. B.A. Hull, J.J. Sumakeris, Q.J. Zhang, J. Richmond, A.R. Powell, M.J. Paisley, V.F. Tsvetkov, A. Hefner, A. Rivera, et al., Development of large area (up to 1.5 cm^2) 4H-SiC 10 kV junction barrier Schottky rectifiers. Mater. Sci. Forum **600**, 931–934 (2009)
33. Information on https://www.genesicsemi.com
34. O. Kordina, C. Hallin, R.C. Glass, E. Janzen, A novel hot-wall CVD reactor for SiC epitaxy. Inst. Phys. Conf. Ser. **137**, 41–44 (1994)
35. P. Grivickas, A. Galeckas, J. Linnros, M. Syväjärvi, R. Yakimova, V. Grivickas, J.A. Tellefsen, Carrier lifetime investigation in 4H–SiC grown by CVD and sublimation epitaxy. Mater. Sci. Semiconduct. Proc **4**, 191 (2001). https://doi.org/10.1016/s1369-8001(00)00133-5
36. L. Storasta, H. Tsuchida, Reduction of traps and improvement of carrier lifetime in 4H-SiC epilayers by ion implantation. Appl. Phys. Lett **90**(6), 062116 (2007). https://doi.org/10.1063/1.2472530
37. T. Hiyoshi, T. Kimoto, Reduction of deep levels and improvement of carrier lifetime in n-Type 4H-SiC by thermal oxidation. Appl. Phys. Exp **2**, 041101 (2009). https://doi.org/10.1143/apex.2.041101
38. S.H. Ryu, D.J. Lichtenwalner, M. O'Loughlin, C. Capell, J. Richmond, E. van Brunt, C. Jonas, Y. Lemma, A. Burk, B. Hull, M. McCain, S. Sabri, H. O'Brien, A. Ogunniyi, A. Lelis, J. Casady, D. Grider, S. Allen, J.W. Palmour, 15 kV n-GTOs in 4H-SiC. Mater. Sci. Forum **963**, 651–654 (2019). https://doi.org/10.4028/www.scientific.net/MSF.963.651
39. H. Lendenmann, F. Dahlquist, N. Johansson, R. Söderholm, P.Å. Nilsson, P. Bergman, P. Skytt, Long term operation of 4.5 kV PiN and 2.5 kV JBS diodes. Mater. Sci. Forum **353-356**, 727–730 (2001). https://doi.org/10.4028/www.scientific.net/msf.353-356.727
40. N. Camara, A. Thuaire, E. Bano, K. Zekentes, Forward-bias degradation in 4H-SiC p+nn+ diodes: Influence of the mesa etching. Phys. State Sol. (a) **202**(4), 660–664 (2005)
41. T. Ohno, H. Yamaguchi, S. Kuroda, K. Kojima, T. Suzuki, K. Arai, Influence of growth conditions on basal plane dislocation in 4H-SiC epitaxial layer. J. Cryst. Growth **271**, 1–7 (2004). https://doi.org/10.1016/j.jcrysgro.2004.04.044
42. J.J. Sumakeris, J.R. Jenny, A.R. Powell, Bulk crystal growth, epitaxy, and defect reduction in silicon carbide materials for microwave and power devices. MRS Bullet **30**(4), 280–286 (2005). https://doi.org/10.1557/mrs2005.74
43. J.J. Sumakeris, P. Bergman, M.K. Das, C. Hallin, B.A. Hull, E. Janzén, H. Lendenmann, M.J. O'Loughlin, M.J. Paisley, S.Y. Ha, M. Skowronski, J.W. Palmour, C.H. Carter Jr., Techniques for minimizing the basal plane dislocation density in SiC epilayers to reduce V_f drift in SiC

bipolar power devices. Mater. Sci. Forum **527-529**, 141–146 (2006). https://doi.org/10.4028/www.scientific.net/msf.527-529.141
44. W. Chen, M.A. Capano, Growth and characterization of 4H-SiC epilayers on substrates with different off-cut angles. J. Appl. Phys **98**(11), 114907 (2005). https://doi.org/10.1063/1.2137442
45. Y. Bu, H. Yoshimoto, N. Watanabe, A. Shima, Fabrication of 4H-SiC PiN diodes without bipolar degradation by improved device processes. J. Appl. Phys **122**(24), 244504 (2017). https://doi.org/10.1063/1.5001370
46. H. Niwa, J. Suda, T. Kimoto, 21.7 kV 4H-SiC PiN diode with a space-modulated junction termination extension. Appl. Phys. Exp **5**, 064001 (2012). https://doi.org/10.1143/apex.5.064001
47. B.J. Baliga, *Power Semiconductor Devices* (PWS Publishing Company, 1996)
48. D.C. Sheridan, G. Niu, J.N. Merrett, J.D. Cressler, C. Ellis, C.C. Tin, Design and fabrication of planar guard ring termination for high- voltage SiC diodes. Solid State Electron. **44**(8), 1367–1372 (2000)
49. M.C. Tarplee, V.P. Madangarli, Q. Zhang, S. Sudardhan, De- sign rules for field plate edge termination in SiC Schottky diodes. IEEE Trans. Electron. Dev **48**(12), 2659–2664 (2001)
50. T. Hiyoshi, T. Hori, J. Suda, T. Kimoto, Simulation and experimental study on the junction termination structures for high voltage 4H-SiC pin diodes. IEEE Trans. Electron Dev **55**(8), 1841–1846 (2008)
51. W. Sung, J. Baliga, A.Q. Huang, Area-efficient bevel-edge termination techniques for SiC high-voltage devices. IEEE Trans. Electron. Dev **63**(4), 1630–1636 (2016)
52. J. Bardeen, W.H. Brattain, The transistor, a semi-conductor triode. Phys. Rev **74**(2), 230 (1948)
53. B. Jayant, Baliga., *Power Semiconductor Devices* (PWS Publishing Company, 1996)
54. W.V. Muench et al., Silicon carbide filed-effect and bipolar transistors. IEEE electron devices meeting, 1977 international, volume 23, pages 337–339. IEEE, 1977
55. S.H. Ryu, A.K. Agarwal, J.W. Palmour, M.E. Levinshtein, 1.8 kV, 3.8 A bipolar junction transistors in 4H-SiC. IEEE 13th international symposium on power semiconductor devices & IC's (ISPSD). IEEE, p. 37–40, 2001
56. Y. Gao, A.Q. Huang, A.K. Agarwal, Q. Zhang, Theoretical and experimental analyses of safe operating area (soa) of 1200-V 4H-SiC BJT. IEEE Trans. Electron Dev **55**(8), 1887–1893 (2008)
57. J.W. Palmour, Silicon carbide power device development for industrial markets. IEEE international electron devices meeting, pages 1–1. IEEE, 2014
58. A. Salemi, H. Elahipanah, G. Malm, C.M. Zetterling, M. Östling, Area-and efficiency-optimized junction termination for a 5.6 kV SiC BJT process with low ON-resistance, in *Proc. 27th ISPSD*, (2015), pp. 249–252
59. H. Elahipanah, A. Salemi, C.M. Zetterling, M. Östling, 5.8-kV implantation-Free 4H-SiC BJT with multiple-Shallow-Trench junction termination extension. IEEE Electron Dev. Lett. **36**(2), 168–170 (2015)
60. A. Salemi, H. Elahipanah, K. Jacobs, C.M. Zetterling, M. Östling, 15 kV-class implantation-free 4H-SiC BJTs with record high current gain. IEEE Electron Dev. Lett. **39**(1), 63–66 (2018)
61. E. Danielsson, M. Domeij, H.S. Lee, C.M. Zetterling, M. Östling, A. Schöner, C. Hallin, A 4H-SiC BJT with an epitaxially regrown extrinsic base layer. Mater. Sci. Forum **483-485**, 905–908 (2005)
62. Q. Zhang, A. Burk, F. Husna, R. Callanan, A. Agarwal, J. Palmour, R. Stahlbush, C. Scozzie, 4H-SiC bipolar junction transistors: From research to development-a case study: 1200 V, 20 A, stable SiC BJTs with high blocking yield. 21st international symposium on power semiconductor devices & IC's IEEE, pp. 339–342, 2009
63. S.G. Sundaresan, S. Jeliazkov, B. Grummel, R. Singh, Rapidly maturing SiC junction transistors featuring current gain (β)> 130, blocking voltages up to 2700 V and stable long-term operation. Mater. Sci. Forum **778**, 780 (2014)

64. C.F. Huang, J.A. Cooper, 4H-SiC npn bipolar junction transistors VWith BV CEO > 3,200 V. Proc. 14th international symposium on power semiconductor devices and ICs, IEEE, pp. 57–60, 2002
65. S. Balachandran, C. Li, P.A. Losee, I.B. Bhat, T.P. Chow, 6kV 4H-SiC BJTs with specific on-resistance below the unipolar limit using a selectively grown base contact process. Proc. 19th international symposium on power semiconductor devices and IC's IEEE, pp. 293–296, 2007
66. S. Sundaresan, C. Li, P.A. Losee, I.B. Bhat, T.P. Chow, 10 kV SiC BJTs static, switching and reliability characteristics. 25th international symposium on power semiconductor devices & IC's IEEE (ISPSD), pp. 303–306, 2013
67. H. Miyake, T. Okuda, H. Niwa, T. Kimoto, J. Suda, 21-kV SiC BJTs with space-modulated junction termination extensioN. IEEE Electron Dev. Lett **33**(11), 1598–1600 (2012)
68. H. Elahipanah, S. Kargarrazi, A. Salemi, C.M. Zetterling, M. Östling, 500 °C high current 4H-SiC lateral BJTs for high-temperature integrated circuits. IEEE Electron Dev. Lett **38**(10), 1429–1432 (2017)
69. M.W. Hussain, H. Elahipanah, J.E. Zumbro, S. Rodriguez, B.G. Malm, H.A. Mantooth, A. Rusu, A SiC BJT-based negative resistance oscillator for high-temperature applications. IEEE J. Electron Dev. Soc **7** (2019)
70. Silicon Carbide Bipolar Junction Transistor. 22nd European conference on power electronics and applications (EPE'20 ECCE Europe), 2020
71. H. Mitlehner, W. Bartsch, K.O. Dohnke, P. Friedrichs, R. Kaltschmidt, U. Weinert, B. Weis, D. Stephani, Dynamic characteristics of high voltage 4H-SiC vertical JFETs. Presented at the 11th international symposium on power semiconductor devices and ICs, ISPSD'99 proceedings, 1999
72. D. Stephani, P. Friedrichs, Silicon carbide junction field effect transistors. Int. J. High Speed Electron. Syst **16**(3), 825–854 (2006)
73. S.-H. Ryu, S. Krishnaswami, B.A. Hull, B. Heath, F. Husna, J. Richmond, A. Agarwal, J. Palmour, J. Scofield, A comparison of high temperature performance of SiC DMOSFETs and JFETs. Mater. Sci. Forum **556-557**, 775–778 (2007)
74. J.H. Zhao, K. Tone, X. Li, P. Alexandrov, L. Fursin, M. Weiner, 3.6 mΩ cm2, 1726V 4H-SiC normally-off trenched and-implanted vertical JFETs and circuit applications. IEE Proc.-Circuits Dev. Syst. **151**(3), 231–237 (2004)
75. V. Veliadis, Silicon carbide junction field effect transistors (SiC – JFETs), in *Wiley Encyclopedia of Electrical and Electronics Engineering*, (2014), p. 1–37
76. K. Vamvoukakis, D. Stefanakis, A. Stavrinidis, K. Vassilevski, G. Konstantinidis, M. Kayambaki, K. Zekentes, Channel width effect on the operation of 4H-SiC vertical JFETs. Phys. Status Solidi A **214**(4), 1600452 (2017). https://doi.org/10.1002/pssa.201600452
77. K. Zekentes, A. Stavrinidis, G. Konstantinidis, M. Kayambaki, K. Vamvoukakis, E. Vassakis, K. Vassilevski, A.B. Horsfall, N.G. Wright, P. Brosselard, S. Niu, M. Lazar, D. Planson, D. Tournier, N. Camara, 4H-SiC VJFETs with self-aligned contacts. Mater. Sci. Forum **821-823**, 793–796 (2015)
78. A. Stavrinidis, G. Konstantinidis, K. Vamvoukakis, K. Zekentes, Salicide-like process for the formation of gate and source contacts in 4H-SiC TSI-VJFET. Mater. Sci. Forum **897**, 407–410 (2017)
79. V. Veliadis, T. McNutt, M. McCoy, H. Hearne, P. Potyraj, C. Scozzie, Large area silicon carbide VJFETs for 1200 V cascode switch operation. Int. J. Power Manag. Electron **2008**, ID. 523721 (2008)
80. V. Veliadis, H. Hearne, T. McNutt, M. Snook, P. Potyraj, C. Scozzie, VJFET based all-SiC normally-off cascode switch for high temperature power handling applications. Mater. Sci. Forum **615-617**, 711–714 (2009)
81. V. Veliadis, B. Steiner, K. Lawson, S.B. Bayne, D. Urciuoli, H.C. Ha, N. El-Hinnawy, S. Gupta, P. Borodulin, R.S. Howell, C. Scozzie, Reliable operation of SiC JFET subjected to over 2.4 million 1200-V/115-A hard switch stressing events at 150 °C. IEEE Electron Dev. Lett. **34**(3), 384–386 (2013)

82. P.G. Neudeck, D.J. Spry, L. Chen, N.F. Prokop, M.J. Krasowski, Demonstration of 4H-SiC digital integrated circuits above 800 °C. IEEE Electron Dev. Lett **38**, 1082–1085 (2017). https://doi.org/10.1109/led.2017.2719280
83. P.G. Neudeck, D.J. Spry, C. Liang-Yu, G.M. Beheim, R.S. Okojie, C.W. Chang, R.D. Meredith, T.L. Ferrier, L.J. Evans, M.J. Krasowski, N.F. Prokop, Stable electrical operation of 6H-SiC JFETs and ICs for thousands of hours at 500C. IEEE Electron Dev. Lett **29**(5), 456–459 (2008)
84. P.G. Neudeck, S.L. Garverick, D.J. Spry, L.-Y. Chen, G.M. Beheim, M.J. Krasowski, M. Mehregany, Extreme temperature 6H-SiC JFET integrated circuit technology. Phys. Status Solidi (a) **206**(10), 2329–2345 (2009)
85. https://unitedsic.com/
86. J.P. Russell, A.M. Goodman, L.A. Goodman, J.M. Nielson, The COMFET: a new high conductance MOS gated device. IEEE Electron Dev. Lett. **EDL-44**(3), 63–65 (1983)
87. B.J. Baliga, M.S. Adler, R.P. Love, P.V. Gray, N. Zommer, The insulated gate transistor : A new three-terminal MOS-controlled bipolar power device. IEEE Trans. Electron Dev **ED-31**(6), 821–828 (1984)
88. T. Fujihira, Theory of semiconductor superjunction devices. Jpn. J.Appl. Phys **36**(part 1, 10), 6254–6262 (1997)
89. Infineon Technologies, https://www.infineon.com/cms/en/product/power/mosfet/n-channel/500v-950v. Accessed 23 Dec 2021.
90. J. Palmour, SiC devices: Powering the next generation of electric vehicles. Presented at WIPDA 2019, Raleigh, NC, Oct 29–Oct 31, 2019
91. A. Agarwal, S. Ryu, J. Palmour, Power MOSFETs in 4H-SiC: Device design and technology, in *Silicon Carbide, Recent Major Advances*, ed. by W. J. Choyke, H. Matsunami, G. Pensl, (Springer-Verlag, 2004), pp. 785–811. ISBN 3-540-40459-9
92. D.J. Lichtenwalner, B. Hull, V. Pala, E. van Brunt, S. Ryu, J.J. Sumakeris, M.J. O'Loughlin, A.A. Burk, S.T. Allen, J.W. Palmour, Performance and Reliability of SiC Power MOSFETs. MRS Adv. **1**(02), 81–89 (2016)
93. S. Ryu, S. Krishnaswami, M. Das, B. Hull, J. Richmond, B. Heath, A. Agarwal, J. Palmour, J. Scofield, 10.3 mΩ-cm^2, 2kV power DMOSFETs in 4H-SiC. Proceedings of the 17th international symposium on power semiconductor devices & IC's, Santa Barbra, CA, pp. 275–278, 23–26 May, 2005
94. D. Okamoto, H. Yano, K. Hirata, T. Hatayama, T. Fuyuki, Improved inversion channel mobility in 4H-SiC MOSFETs on Si face utilizing phosphorus-doped gate oxide. IEEE Electron Dev. Lett. **31**(7), 710–712 (2010)
95. D. Okamoto, M. Sometani, S. Harada, R. Kosugi, Y. Yonezawa, H. Yano, Improved channel mobility in 4H-SiC MOSFETs by Boron passivation. IEEE Electron Dev. Lett. **35**(12), 1176–1178 (2014)
96. V. Soler, M. Cabello, J. Montserrat, J. Rebollo, J. Millan, P. Godignon, M. Berthou, E. Bianda, A. Mihaila, 4.5 kV SiC MOSFET with Boron doped gate dielectric. Proceedings of the 28th international symposium on power semiconductor devices and ICs, Prague, Czech Republic, pp. 283–286, 12–16 June, 2016
97. D.J. Lichtenwalner, L. Cheng, S. Dhar, A. Agarwal, J.W. Palmour, High mobility 4H-SiC (0001) transistors using alkali and alkaline earth interface layers. Appl. Phys. Lett. **105**, 182107 (2014). https://doi.org/10.1063/1.4901259
98. T. Kobayashi, Y. Matsushita, Structure and energetics of carbon defects in SiC(0001)/SiO_2 systems at realistic temperatures: Defects in SiC, SiO2, and at their interface. J. Appl. Phys. **126**, 145302 (2019)
99. T. Kimoto, M. Kaneko, T. Tachiki, K. Ito, R. Ishikawa, X. Chi, D. Stefanakis, H. Tanaka, Physics and innovative technologies in SiC power devices. 2021 IEDM technical digest, pp. 36.1.1–36.1.4, San Francisco, CA, Dec 11–Dec 15, 2021
100. T. Kobayashi, T. Okuda, K. Tachiki, K. Ito, Y. Matsushita, T. Kimoto, Design and formation of SiC (0001)/SiO_2 interfaces via Si deposition followed by low-temperature oxidation and high temperature nitridation. Appl. Phys. Exp **13**, 091003 (2020)

101. K. Tachiki, M. Kaneko, T. Kimoto, Mobility improvement of 4H-SiC (0001) MOSFETs by a three-step process of H_2 etching, SiO_2 deposition, and interface nitridation. Appl. Phys. Exp **14**, 031001 (2021)
102. M. Wang, M. Yang, W. Liu, S. Yang, C. Han, L. Geng, Y. Hao, Toward High Performance 4H-SiC MOSFETs Using Low Temperature Annealing Process with Supercritical Fluid. 2021 IEDM technical digest, pp. 36.2.1 – 36.2.4, San Francisco, CA, Dec 11–Dec 15, 2021
103. D. Heer, D. Domes, D. Peters, Switching performance of a 1200 V SiC-Trench-MOSFET in a low-power module. Proceedings of PCIM, Nuremberg, pp. 1–7, 10–12 May, 2016
104. J.W. Palmour, J.A. Edmond, H.S. Kong, C.H. Carter Jr., 6H-silicon carbide devices and applications. Physica B **185**, 461–465 (1993)
105. J.W. Palmour, S.T. Allen, R. Singh, L.A. Lipkin, D.G. Waltz, 4H-silicon carbide power switching devices, in *Silicon Carbide and Related Materials 1995, Institute of Phys. Conf. Series No. 142*, ed. by S. Nakashima, H. Matsunami, S. Yoshida, H. Harima, (Inst. Of Phys. Publ, Bristol, 1996), pp. 813–816
106. H. Yano, H. Nakao, T. Hatayama, Y. Uraoka, T. Fuyuki, Increased channel mobility in 4H-SiC UMOSFETs using on-axis substrates. Mater. Sci. Forum **556-557**, 807–811 (2007)
107. J. Tan, J.A. Cooper Jr., M.R. Melloch, High-voltage accumulation-layer UMOSFET's in 4H-SiC. IEEE Electron Dev. Lett. **19**(12), 487–489 (1998)
108. H. Takaya, T. Misumi, H. Fujiwara, T. Ito, 4H-SiC Trench MOSFET with low on-resistance at high temperature. Proceedings of ISPSD 2020, Vienna, Austria, pp. 118–121, Sept 13–18, 2020
109. S. Kyogoku, K. Tanaka, K. Ariyoshi, R. Iijima, Y. Kobayashi, S. Harada, Role of trench bottom shielding region on switching characteristics of 4H-SiC double-trench MOSFETs. Mater. Sci. Forum **924**, 748–751 (2018)
110. T. Nakamura, Y. Nakano, M. Aketa, R. Nakamura, S. Mitani, H. Sakairi, Y. Yokotshuji, High performance SiC trench devices with ultra-low Ron. IEEE IEDM Tech. Dig., Washington DC, USA, pp. 26.5.1–26.5.3, 5–7 Dec 2011
111. D. Peters, R. Siemieniec, T. Aichinger, T. Basler, R. Esteve, W. Bergner, D. Kueck, Performance and ruggedness of 1200V SiC – Trench – MOSFET. Proceedings of the 29th international symposium on power semiconductor devices & ICs, Sapporo, Japan, pp. 239–242, 28 May–1 June 2017
112. G. Rescher, G. Pobegen, T. Aichinger, T. Grasser, Preconditioned BTI on 4H-SiC: Proposal for a nearly delay time-independent measurement technique. IEEE Trans. Electron Dev **65**(4), 1419–1426 (2018)
113. R. Kosugi, Y. Sakuma, K. Kojima, S. Itoh, A. Nagaka, T. Yatsuo, T. Tanake, H. Okumura, First experimental demonstration of SiC superjunction (SJ) structure by multi-epitaxial growth method. International symposium on power semiconductor devices and ICs, Waikoloa, USA, pp. 346–349, 15–19 June, 2014
114. S. Harada, Y. Kobayashi, S. Kyogoku, T. Morimoto, T. Tanaka, M. Takei, H. Okumura, First demonstration of dynamic characteristics for SiC superjunction MOSFET realized using multi-epitaxial growth method. Proceedings of IEDM 2018, San Francisco, CA, USA, pp. 8.2.1–8.2.4, 3–5 Dec 2018
115. Y. Kobayashi, S. Kyogoku, T. Morimoto, T. Kumazawa, Y. Yamashiro, M. Takei, S. Harada, High-temperature performance of 1.2 kV-class SiC super junction MOSFET. Proceedings of the 31st international symposium on power semiconductor devices & ICs, Shanghai, China, pp. 31–34, 19–23 May 2019
116. https://www.aist.go.jp/aist_e/list/latest_research/2020/20200108/en20200108.html
117. https://www.fujielectric.com/company/tech/pdf/66-04/FER66-04-237-2020.pdf
118. R. Kosugi, S. Ji, K. Mochizuki, K. Adachi, S. Segawa, Y. Kawada, Y. Yonezawa, H. Okumura, Breaking the theoretical limit of 6.5 kV-class 4H-SiC Super-Junction (SJ) MOSFETs by trench-filling epitaxial growth. Proceedings of the 31st international symposium on power semiconductor devices & ICs, Shanghai, China, pp. 39–42, 19–23 May 2019
119. S.H. Ryu, L. Cheng, S. Dhar, C. Capell, C. Jonas, J. Clayton, M. Donofrio, M.J. O'Loughlin, A.A. Burk, A.K. Agarwal, J.W. Palmour, Development of 15 kV 4H-SiC IGBTs. Mater. Sci. Forum **717–720**, 1135–1138 (2012)

120. V. Pala, E.V. Brunt, L. Cheng, M. O'Loughlin, J. Richmond, A. Burk, S.T. Allen, D. Grider, J.W. Palmour, 10 kV and 15 kV silicon carbide power MOSFETs for next-generation energy conversion and transmission systems. Proc. ECCE 2014 (Pittsburgh, PA, Sept 14–18), pp. 449–454, 2014
121. S. Ryu, C. Capell, E. Van Brunt, C. Jonas, M. O'Loughlin, J. Clayton, K. Lam, V. Pala, B. Hull, Y. Lemma, Ultra high voltage MOS controlled 4H-SiC power switching devices. Semicond. Sci. Tech. **30**, 084001 (2015)
122. S. Ryu, C. Capell, C. Jonas, M.J. O'Loughlin, J. Clayton, E. van Brunt, K. Lam, J. Richmond, A. Kadavelugu, S. Bhattacharya, A.A. Burk, A. Agarwal, D. Grider, S.T. Allen, J.W. Palmour, 20 kV 4H-SiC n-IGBTs. Mater. Sci. Forum **778-780**, 1030–1033 (2014)
123. E. van Brunt, L. Cheng, M.J. O'Loughlin, J. Richmond, V. Pala, J.W. Palmour, C.W. Tipton, C. Scozzie, 27 kV, 20A 4H-SiC n-IGBTs. Mater. Sci. Forum **821-823**, 847–850 (2015)
124. A. Koyama, Y. Kiuchi, T. Mizushima, K. Takenaka, S. Matsunaga, M. Sometani, K. Nakayama, H. Ishimori, A. Kimoto, M. Takei, T. Kato, Y. Yonezawa, H. Okumura, 20 kV-class ultra-high voltage 4H-SiC n-IE-IGBTs. Mater. Sci. Forum **1004**, 899–904 (2020)
125. N. Iwamuro, T. Laska, IGBT history, state-of -the-art, and future prospects. IEEE Trans. Electron Dev **64**(3), 741–752 (2017)
126. M. Kitagawa, I. Omura, S. Hasegawa, T. Inoue, A. Nakagawa, A 4500V Injection enhanced insulated gate bipolar transistor (IEGT) operating in a mode similar to a thyristor, in *IEEE IEDM Technical Digest*, (1993), pp. 679–682
127. K. Eikyu, A. Sakai, H. Matsuura, Y. Nakazawa, Y. Akiyama, Y. Yamaguchi, M. Inuishi, On the scaling limit of the Si-IGBTs with very narrow mesa structures. Proceedings of ISPSD 2015, pp. 211 – 214, 2015
128. H. Takahashi, H. Haruguchi, H. Hagino, T. Yamada, Carrier stored trench gate bipolar transistor (CSTBT) – A novel power device for high voltage application. Proceedings of ISPSD 1996, pp. 349 – 352, 1996
129. S. Ryu, C. Capell, C. Jonas, M. O'Loughlin, J. Clayton, K. Lam, E. Van Brunt, Y. Lemma, J. Richmond, D. Grider, S. Allen, J.W. Palmour, An analysis of forward conduction characteristics of ultra high voltage 4H-SiC n-IGBTs, in *Materials Science Forum, 858*, (Trans Tech Publications, 2016), pp. 945–948
130. A. M. Research, "GaN power device market is expected to reach $1.24 Billion by 2027, at 35.4% CAGR," GlobeNewswire News Room, May 11, 2020. http://www.globenewswire.com/news-release/2020/05/11/2031230/0/en/GaN-Power-Device-Market-Is-Expected-to-Reach-1-24-Billion-by-2027-at-35-4-CAGR.html. Accessed 15 Feb 2021.
131. Y. Zhang, A. Dadgar, T. Palacios, Gallium nitride vertical power devices on foreign substrates: a review and outlook. J. Phys D Appl. Phys **51**(27), 273001 (2018). https://doi.org/10.1088/1361-6463/aac8aa
132. O. Ambacher, J. Smart, J.R. Shealy, N.G. Weimann, K. Chu, M. Murphy, W.J. Schaff, L.F. Eastman, R. Dimitrov, L. Wittmer, Two-dimensional electron gases induced by spontaneous and piezoelectric polarization charges in N-and Ga-face AlGaN/GaN heterostructures. J. Appl. Phys **85**(6), 3222–3233 (1999)
133. J.P. Ibbetson, P.T. Fini, K.D. Ness, S.P. DenBaars, J.S. Speck, U.K. Mishra, Polarization effects, surface states, and the source of electrons in AlGaN/GaN heterostructure field effect transistors. Appl. Phys. Lett **77**(2), 250–252 (2000). https://doi.org/10.1063/1.126940
134. K. Shinohara, D.C. Regan, T. Yan, A.L. Corrion, D.F. Brown, J.C. Wong, J.F. Robinson, H.H. Fung, A. Schmitz, T.C. Oh, S.J. Kim, P.S. Chen, R.G. Nagele, A.D. Margomenos, M. Micovic, Scaling of GaN HEMTs and Schottky diodes for submillimeter-wave MMIC applications. IEEE Trans. Electron Dev **60**(10), 2982–2996 (2013). https://doi.org/10.1109/TED.2013.2268160
135. G.C. Barisich, S. Pavlidis, C.A.D. Morcillo, O.L. Chlieh, J. Papapolymerou, E. Gebara, An X-band GaN HEMT hybrid power amplifier with low-loss Wilkinson division on AlN substrate. Presented at the IEEE international conference on microwaves, communications, antennas and electronics systems (COMCAS), Tel Aviv, Israel, Oct 2013. https://doi.org/10.1109/COMCAS.2013.6685285

136. K. Hoo Teo, Y. Zhang, N. Chowdhury, S. Rakheja, R. Ma, Q. Xie, E. Yagyu, K. Yamanaka, K. Li, T. Palacios, Emerging GaN technologies for power, RF, digital, and quantum computing applications: Recent advances and prospects. J. Appl. Phys **130**(16), 160902 (2021). https://doi.org/10.1063/5.0061555
137. Y. Uemoto, M. Hikita, H. Ueno, H. Matsuo, H. Ishida, M. Yanagihara, T. Ueda, T. Tanaka, D. Ueda, Gate injection transistor (GIT)—A normally-off AlGaN/GaN power transistor using conductivity modulation. IEEE Trans. Electron Dev **54**(12), 3393–3399 (2007). https://doi.org/10.1109/TED.2007.908601
138. Q. Song, R. Zhang, J.P. Kozak, J. Liu, Q. Li, Y. Zhang, Robustness of cascode GaN HEMTs in unclamped inductive switching. IEEE Trans. Power Electron **37**(4), 4148–4160 (2022). https://doi.org/10.1109/TPEL.2021.3122740
139. Q. Song, J.P. Kozak, M. Xiao, Y. Ma, B. Wang, R. Zhang, R. Volkov, K. Smith, T. Baksht, Y. Zhang, Evaluation of 650V, 100A direct-drive GaN power switch for electric vehicle powertrain applications, in *2021 IEEE 8th Workshop on Wide Bandgap Power Devices and Applications (WiPDA)*, (Nov 2021), pp. 28–33. https://doi.org/10.1109/WiPDA49284.2021.9645143
140. Y. Zhang, A. Zubair, Z. Liu, M. Xiao, J.A. Perozek, Y. Ma, T. Palacios, GaN FinFETs and trigate devices for power and RF applications: review and perspective. Semicond. Sci. Technol **36**(5), 054001 (2021). https://doi.org/10.1088/1361-6641/abde17
141. B. Lu, E. Matioli, T. Palacios, Tri-gate normally-off GaN power MISFET. IEEE Electron Dev. Lett **33**(3), 360–362 (2012). https://doi.org/10.1109/LED.2011.2179971
142. Y. Ma, M. Xiao, Z. Du, X. Yan, K. Cheng, M. Clavel, M.K. Hudait, I. Kravchenko, H. Wang, Y. Zhang, Tri-gate GaN junction HEMT. Appl. Phys. Lett **117**(14), 143506 (2020). https://doi.org/10.1063/5.0025351
143. M. Xiao, Y. Ma, K. Cheng, K. Liu, A. Xie, E. Beam, Y. Cao, Y. Zhang, 3.3 kV multi-channel AlGaN/GaN Schottky barrier diodes with P-GaN termination. IEEE Electron Dev. Lett **41**(8), 1177–1180 (2020). https://doi.org/10.1109/LED.2020.3005934
144. J. Ma, G. Kampitsis, P. Xiang, K. Cheng, E. Matioli, Multi-channel tri-gate GaN power Schottky diodes with low ON-resistance. IEEE Electron Dev. Lett **40**(2), 275–278 (2019). https://doi.org/10.1109/LED.2018.2887199
145. M. Xiao, Y. Ma, Z. Du, X. Yan, R. Zhang, K. Cheng, K. Liu, A. Xie, E. Beam, Y. Cao, H. Wang, Y. Zhang, 5 kV multi-channel AlGaN/GaN power Schottky barrier diodes with junction-Fin-anode, in *2020 IEEE International Electron Devices Meeting (IEDM)*, (Dec 2020), pp. 5.4.1–5.4.4. https://doi.org/10.1109/IEDM13553.2020.9372025
146. M. Xiao, Y. Ma, K. Liu, K. Cheng, Y. Zhang, 10 kV, 39 mΩ·cm2 multi-channel AlGaN/GaN Schottky barrier diodes. IEEE Electron Dev. Lett **42**(6), 808–811 (2021). https://doi.org/10.1109/LED.2021.3076802
147. M. Xiao, Y. Ma, V. Pathirana, K. Cheng, A. Xie, E. Beam, Y. Cao, F. Udrea, H. Wang, Y. Zhang, Multi-channel monolithic-Cascode HEMT (MC2-HEMT): A new GaN power switch up to 10 kV, in *2021 IEEE International Electron Devices Meeting (IEDM)*, pp. 5.5.1–5.5.4
148. C. Sukwon, E.R. Heller, D. Dorsey, R. Vetury, S. Graham, The impact of bias conditions on self-heating in AlGaN/GaN HEMTs. IEEE Trans. Electron Dev **60**(1), 159–162 (2013). https://doi.org/10.1109/TED.2012.2224115
149. S. Pavlidis, A.C. Ulusoy, W.T. Khan, O.L. Chlieh, E. Gebara, J. Papapolymerou, A feasibility study of flip-chip packaged gallium nitride HEMTs on organic substrates for wideband RF amplifier applications, in *IEEE Electronic Components and Technology Conference (ECTC)*, (May 2014), pp. 2293–2298. https://doi.org/10.1109/ECTC.2014.6897625
150. Wide Bandgap Power Semiconductors | JEDEC. https://www.jedec.org/category/technology-focus-area/wide-bandgap-power-semiconductors-gan-sic. Accessed 6 Jan 2022.
151. "TP65H015G5WS," Transphorm. https://www.transphormusa.com/en/product/tp65h015g5ws/. Accessed 6 Jan 2022.
152. S. Pavlidis, A.C. Ulusoy, J. Papapolymerou, A 5.4W X-band Gallium Nitride (GaN) power amplifier in an encapsulated organic package, in *European Microwave Conference (EuMC)*, (2015), pp. 789–792. https://doi.org/10.1109/EuMC.2015.7345882

153. R. Reiner, B. Weiss, D. Meder, P. Waltereit, T. Gerrer, R. Quay, C. Vockenberger, O. Ambacher, PCB-embedding for GaN-on-Si power devices and ICs, in *CIPS 2018; 10th International Conference on Integrated Power Electronics Systems*, (Mar 2018), pp. 1–6
154. GaN Systems, GN002: thermal design for packaged GaNPX® Devices, Application Note, Aug 2020.
155. S. Pavlidis, G. Alexopoulos, A.Ç. Ulusoy, M.K. Cho, J. Papapolymerou, Encapsulated organic package technology for wideband integration of heterogeneous MMICs. IEEE Trans. Microw. Theory Techniq **65**(2), 438–448 (2017). https://doi.org/10.1109/TMTT.2016.2630067
156. J.A. del Alamo, E.S. Lee, Stability and reliability of lateral GaN power field-effect transistors. IEEE Trans. Electron Dev **66**(11), 4578–4590 (2019). https://doi.org/10.1109/TED.2019.2931718
157. G. Zulauf, M. Guacci, J.W. Kolar, Dynamic on-resistance in GaN-on-Si HEMTs: Origins, dependencies, and future characterization frameworks. IEEE Trans. Power Electron **35**(6), 5581–5588 (2020). https://doi.org/10.1109/TPEL.2019.2955656
158. R. Zhang, J.P. Kozak, M. Xiao, J. Liu, Y. Zhang, Surge-energy and overvoltage ruggedness of P-gate GaN HEMTs. IEEE Trans. Power Electron **35**(12), 13409–13419 (2020). https://doi.org/10.1109/TPEL.2020.2993982
159. R. Zhang, J.P. Kozak, Q. Song, M. Xiao, J. Liu, Y. Zhang, Dynamic breakdown voltage of GaN Power HEMTs, in *2020 IEEE International Electron Devices Meeting (IEDM)*, (Dec 2020), pp. 23.3.1–23.3.4. https://doi.org/10.1109/IEDM13553.2020.9371904
160. J.P. Kozak, R. Zhang, Q. Song, J. Liu, W. Saito, Y. Zhang, True breakdown voltage and overvoltage margin of GaN power HEMTs in hard switching. IEEE Electron Dev. Lett **42**(4), 505–508 (2021). https://doi.org/10.1109/LED.2021.3063360
161. R. Zhang, J. Liu, Q. Li, S. Pidaparthi, A. Edwards, C. Drowley, Y. Zhang, Breakthrough short circuit robustness demonstrated in vertical GaN Fin JFET. IEEE Trans. Power Electron, 1–1 (2021). https://doi.org/10.1109/TPEL.2021.3138451
162. H. Li, X. Li, X. Wang, X. Lyu, H. Cai, Y.M. Alsmadi, L. Liu, S. Bala, J. Wang, Robustness of 650-V enhancement-mode GaN HEMTs under various short-circuit conditions. IEEE Trans. Indus. Appl **55**(2), 1807–1816 (2019). https://doi.org/10.1109/TIA.2018.2879289
163. J. Liu, M. Xiao, R. Zhang, S. Pidaparthi, C. Drowley, L. Baubutr, A. Edwards, H. Cui, C. Coles, Y. Zhang, Trap-mediated Avalanche in large-area 1.2 kV vertical GaN p-n diodes. IEEE Electron Dev. Lett **41**(9), 1328–1331 (2020). https://doi.org/10.1109/LED.2020.3010784
164. J. Liu, R. Zhang, M. Xiao, S. Pidaparthi, H. Cui, A. Edwards, L. Baubutr, C. Drowley, Y. Zhang, Surge current and Avalanche ruggedness of 1.2-kV vertical GaN p-n diodes. IEEE Trans. Power Electron **36**(10), 10959–10964 (2021). https://doi.org/10.1109/TPEL.2021.3067019
165. J. Liu, M. Xiao, Y. Zhang, S. Pidaparthi, H. Cui, A. Edwards, L. Baubutr, W. Meier, C. Coles, C. Drowley, 1.2 kV vertical GaN Fin JFETs with robust Avalanche and fast switching capabilities, in *2020 IEEE International Electron Devices Meeting (IEDM)*, (Dec 2020), pp. 23.2.1–23.2.4. https://doi.org/10.1109/IEDM13553.2020.9372048
166. J. Liu, M. Xiao, R. Zhang, S. Pidaparthi, H. Cui, A. Edwards, M. Craven, L. Baubutr, C. Drowley, Y. Zhang, 1.2-kV vertical GaN Fin-JFETs: High-temperature characteristics and Avalanche capability. IEEE Trans. Electron Dev **68**(4), 2025–2032 (2021). https://doi.org/10.1109/TED.2021.3059192
167. J. Liu, R. Zhang, M. Xiao, S. Pidaparthi, H. Cui, A. Edwards, C. Drowley, Y. Zhang, Tuning Avalanche path in vertical GaN JFETs by gate driver design. IEEE Trans. Power Electron, 1–1 (2021). https://doi.org/10.1109/TPEL.2021.3132906
168. M. Giandalia, J. Zhang, T. Ribarich, 650 V AllGaN™ power IC for power supply applications, in *2016 IEEE 4th Workshop on Wide Bandgap Power Devices and Applications (WiPDA)*, (Nov 2016), pp. 220–222. https://doi.org/10.1109/WiPDA.2016.7799941
169. R. Chu, Y. Cao, M. Chen, R. Li, D. Zehnder, An experimental demonstration of GaN CMOS technology. IEEE Electron Dev. Lett **37**(3), 269–271 (2016). https://doi.org/10.1109/LED.2016.2515103

170. N. Chowdhury, J. Lemettinen, Q. Xie, Y. Zhang, N.S. Rajput, P. Xiang, K. Cheng, S. Suihkonen, H.W. Then, T. Palacios, p-channel GaN transistor based on p-GaN/AlGaN/GaN on Si. IEEE Electron Dev Lett **40**(7), 1036–1039 (2019). https://doi.org/10.1109/LED.2019.2916253
171. N. Chowdhury, Q. Xie, M. Yuan, K. Cheng, H.W. Then, T. Palacios, Regrowth-free GaN-based complementary logic on a Si substrate. IEEE Electron Dev. Lett **41**(6), 820–823 (2020). https://doi.org/10.1109/LED.2020.2987003
172. H.W. Then, S. Dasgupta, M. Radosavljevic, P. Agababov, I. Ban, R. Bristol, M. Chandhok, S. Chouksey, B. Holybee, C.Y. Huang, B. Krist, K. Jun, K. Lin, N. Nidhi, T. Michaelos, B. Mueller, R. Paul, J. Peck, W. Rachmady, D. Staines, T. Talukdar, N. Thomas, T. Tronic, P. Fischer, W. Hafez, 3D heterogeneous integration of high performance high-K metal gate GaN NMOS and Si PMOS transistors on 300mm high-resistivity Si substrate for energy-efficient and compact power delivery, RF (5G and beyond) and SoC applications, in *2019 IEEE International Electron Devices Meeting (IEDM)*, (Dec 2019), pp. 17.3.1–17.3.4. https://doi.org/10.1109/IEDM19573.2019.8993583
173. H.W. Then, M. Radosavljevic, P. Agababov, I. Ban, R. Bristol, M. Chandhok, S. Chouksey, B. Holybee, C.Y. Huang, B. Krist, K. Jun, P. Koirala, K. Lin, T. Michaelos, R. Paul, J. Peck, W. Rachmady, D. Staines, T. Talukdar, N. Thomas, T. Tronic, P. Fischer, W. Hafez, GaN and Si transistors on 300mm Si(111) enabled by 3D monolithic heterogeneous integration, in *2020 IEEE Symposium on VLSI Technology*, (Jun 2020), pp. 1–2. https://doi.org/10.1109/VLSITechnology18217.2020.9265093
174. Y. Zhang, M. Sun, Z. Liu, D. Piedra, H.S. Lee, F. Gao, T. Fujishima, T. Palacios, Electrothermal simulation and thermal performance study of GaN vertical and lateral power transistors. IEEE Trans. Electron Dev **60**(7), 2224–2230 (2013). https://doi.org/10.1109/TED.2013.2261072
175. Y. Zhang, T. Palacios, (Ultra)Wide-bandgap vertical power FinFETs. IEEE Trans. Electron Dev **67**(10), 3960–3971 (2020). https://doi.org/10.1109/TED.2020.3002880
176. M. Xiao, T. Palacios, Y. Zhang, ON-resistance in vertical power FinFETs. IEEE Trans. Electron Dev **66**(9), 3903–3909 (2019). https://doi.org/10.1109/TED.2019.2928825
177. Y. Zhang, M. Sun, D. Piedra, J. Hu, Z. Liu, Y. Lin, X. Gao, K. Shepard, T. Palacios, 1200 V GaN vertical fin power field-effect transistors, in *2017 IEEE International Electron Devices Meeting (IEDM)*, (Dec 2017), pp. 9.2.1–9.2.4. https://doi.org/10.1109/IEDM.2017.8268357
178. Y. Zhang, M. Sun, J. Perozek, Z. Liu, A. Zubair, D. Piedra, N. Chowdhury, X. Gao, K. Shepard, T. Palacios, Large-area 1.2-kV GaN vertical power FinFETs with a record switching of merit. IEEE Electron Dev. Lett **40**(1), 75–78 (2019). https://doi.org/10.1109/LED.2018.2880306
179. I. Ben-Yaacov, Y.-K. Seck, U.K. Mishra, S.P. DenBaars, AlGaN/GaN current aperture vertical electron transistors with regrown channels. J. Appl. Phys **95**(4), 2073–2078 (2004). https://doi.org/10.1063/1.1641520
180. D. Ji, A. Agarwal, W. Li, S. Keller, S. Chowdhury, Demonstration of GaN current aperture vertical electron transistors with aperture region formed by ion implantation. IEEE Trans. Electron Dev **65**(2), 483–487 (2018). https://doi.org/10.1109/TED.2017.2786141
181. D. Ji, A. Agarwal, H. Li, W. Li, S. Keller, S. Chowdhury, 880 V/2.7 mΩ-cm2 MIS gate trench CAVET on bulk GaN substrates. IEEE Electron Dev. Lett **39**(6), 863–865 (2018). https://doi.org/10.1109/LED.2018.2828844
182. D. Shibata, R. Kajitani, M. Ogawa, K. Tanaka, S. Tamura, T. Hatsuda, M. Ishida, T. Ueda, 1.7 kV/1.0 mΩcm2 normally-off vertical GaN transistor on GaN substrate with regrown p-GaN/AlGaN/GaN semipolar gate structure, in *IEDM Tech. Dig*, (Dec 2016), pp. 10.1.1–10.1.4. https://doi.org/10.1109/IEDM.2016.7838385
183. T. Oka, T. Ina, Y. Ueno, J. Nishii, Over 10A operation with switching characteristics of 1.2 kV-class vertical GaN trench MOSFETs on a bulk GaN substrate, 459–462 (2016). https://doi.org/10.1109/ISPSD.2016.7520877
184. R. Li, Y. Cao, M. Chen, R. Chu, 600 V/1.7 Ω normally-off GaN vertical trench metal–oxide–semiconductor field-effect transistor. IEEE Electron Dev. Lett **37**(11), 1466–1469 (2016). https://doi.org/10.1109/LED.2016.2614515

185. C. Gupta, C. Lund, S.H. Chan, A. Agarwal, J. Liu, Y. Enatsu, S. Keller, U.K. Mishra, In situ oxide, GaN interlayer-based vertical trench MOSFET (OG-FET) on bulk GaN substrates. IEEE Electron Dev. Lett **38**(3), 353–355 (2017). https://doi.org/10.1109/LED.2017.2649599
186. R. Tanaka, S. Takashima, K. Ueno, H. Matsuyama, M. Edo, K. Nakagawa, Mg implantation dose dependence of MOS channel characteristics in GaN double-implanted MOSFETs. Appl. Phys. Exp **12**(5), 054001 (2019). https://doi.org/10.7567/1882-0786/ab0c2c
187. H. Ohta, K. Hayashi, F. Horikiri, M. Yoshino, T. Nakamura, T. Mishima, 5.0 kV breakdown-voltage vertical GaN p–n junction diodes. Japanese J. Appl. Phys **57**(4S), 04FG09 (2018). https://doi.org/10.7567/JJAP.57.04FG09
188. K. Nomoto, Z. Hu, B. Song, M. Zhu, M. Qi, R. Yan, V. Protasenko, E. Imhoff, J. Kuo, N. Kaneda, T. Mishima, T. Nakamura, D. Jena, H.G. Xing, GaN-on-GaN p-n power diodes with 3.48 kV and 0.95 m???-cm2: A record high figure-of-merit of 12.8 GW/cm2, in *2015 IEEE International Electron Devices Meeting (IEDM)*, (Dec 2015), pp. 9.7.1–9.7.4. https://doi.org/10.1109/IEDM.2015.7409665
189. Y. Zhang, M. Sun, Z. Liu, D. Piedra, M. Pan, X. Gao, Y. Lin, A. Zubair, L. Yu, T. Palacios, Novel GaN trench MIS barrier Schottky rectifiers with implanted field rings, in *2016 IEEE International Electron Devices Meeting (IEDM)*, (Dec 2016), pp. 10.2.1–10.2.4. https://doi.org/10.1109/IEDM.2016.7838386
190. Y. Zhang, Z. Liu, M.J. Tadjer, M. Sun, D. Piedra, C. Hatem, T.J. Anderson, L.E. Luna, A. Nath, A.D. Koehler, H. Okumura, J. Hu, X. Zhang, X. Gao, B.N. Feigelson, K.D. Hobart, T. Palacios, Vertical GaN junction barrier Schottky rectifiers by selective ion implantation. IEEE Electron Dev. Lett **38**(8), 1097–1100 (2017). https://doi.org/10.1109/LED.2017.2720689
191. T. Hayashida, T. Nanjo, A. Furukawa, M. Yamamuka, Vertical GaN merged PiN Schottky diode with a breakdown voltage of 2 kV. Appl. Phys. Exp **10**(6), 061003 (2017). https://doi.org/10.7567/APEX.10.061003
192. K. Hasegawa, G. Nishio, K. Yasunishi, N. Tanaka, N. Murakami, T. Oka, Vertical GaN trench MOS barrier Schottky rectifier maintaining low leakage current at 200 °C with blocking voltage of 750 V. Appl. Phys. Exp **10**(12), 121002 (2017). https://doi.org/10.7567/APEX.10.121002
193. Y. Zhang, M. Sun, D. Piedra, M. Azize, X. Zhang, T. Fujishima, T. Palacios, GaN-on-Si vertical Schottky and p-n diodes. IEEE Electron Dev. Lett **35**(6), 618–620 (2014). https://doi.org/10.1109/LED.2014.2314637
194. Y. Zhang, D. Piedra, M. Sun, J. Hennig, A. Dadgar, L. Yu, T. Palacios, High-performance 500 V quasi- and fully-vertical GaN-on-Si pn diodes. IEEE Electron Dev. Lett **38**(2), 248–251 (2017). https://doi.org/10.1109/LED.2016.2646669
195. S. Mase, T. Hamada, J.J. Freedsman, T. Egawa, Effect of drift layer on the breakdown voltage of fully-vertical GaN-on-Si p-n diodes. IEEE Electron Dev. Lett **38**(12), 1720–1723 (2017). https://doi.org/10.1109/LED.2017.2765340
196. Y. Zhang, M. Yuan, N. Chowdhury, K. Cheng, T. Palacios, 720-V/0.35-mΩ•cm^2 fully vertical GaN-on-Si power diodes by selective removal of Si substrates and buffer layers. IEEE Electron Dev. Lett **39**(5), 715–718 (2018). https://doi.org/10.1109/LED.2018.2819642
197. R.A. Khadar, C. Liu, R. Soleimanzadeh, E. Matioli, Fully vertical GaN-on-Si power MOSFETs. IEEE Electron Dev. Lett **40**(3), 443–446 (2019). https://doi.org/10.1109/LED.2019.2894177
198. R. Xu, P. Chen, M. Liu, J. Zhou, Y. Yang, Y. Li, C. Ge, H. Peng, B. Liu, D. Chen, Z. Xie, R. Zhang, Y. Zheng, 1.4-kV quasi-vertical GaN Schottky barrier diode with reverse p-n junction termination. IEEE J. Electron Dev. Soc **8**, 316–320 (2020). https://doi.org/10.1109/JEDS.2020.2980759
199. Y. Zhang, H.Y. Wong, M. Sun, S. Joglekar, L. Yu, N.A. Braga, R.V. Mickevicius, T. Palacios, Design space and origin of off-state leakage in GaN vertical power diodes, in *2015 IEEE International Electron Devices Meeting (IEDM)*, (Dec 2015), p. 35.1.1–35.1.4. https://doi.org/10.1109/IEDM.2015.7409830
200. F. Udrea, G. Deboy, T. Fujihira, Superjunction power devices, history, development, and future prospects. IEEE Trans. Electron Dev **64**(3), 720–734 (2017). https://doi.org/10.1109/TED.2017.2658344

201. H. Ishida, D. Shibata, M. Yanagihara, Y. Uemoto, H. Matsuo, T. Ueda, T. Tanaka, D. Ueda, Unlimited high breakdown voltage by natural super junction of polarized semiconductor. IEEE Electron Dev. Lett **29**(10), 1087–1089 (2008). https://doi.org/10.1109/LED.2008.2002753
202. M. Xiao, Z. Du, J. Xie, E. Beam, X. Yan, K. Cheng, H. Wang, Y. Cao, Y. Zhang, Lateral p-GaN/2DEG junction diodes by selective-area p-GaN trench-filling-regrowth in AlGaN/GaN. Appl. Phys. Lett **116**(5), 053503 (2020). https://doi.org/10.1063/1.5139906
203. M. Xiao, R. Zhang, D. Dong, H. Wang, Y. Zhang, Design and simulation of GaN superjunction transistors with 2-DEG channels and fin channels. IEEE J. Emerg. Select. Topics Power Electron **7**(3), 1475–1484 (2019). https://doi.org/10.1109/JESTPE.2019.2912978
204. D. Khachariya, D. Szymanski, P. Reddy, E. Kohn, Z. Sitar, R. Collazo, S. Pavlidis, (Invited) A path toward vertical GaN superjunction devices. ECS Trans **98**(6), 69 (2020). https://doi.org/10.1149/09806.0069ecst
205. D. Szymanski, D. Khachariya, T.B. Eldred, P. Bagheri, S. Washiyama, A. Chang, S. Pavlidis, R. Kirste, P. Reddy, E. Kohn, L. Lauhon, R. Collazo, Z. Sitar, GaN lateral polar junction arrays with 3D control of doping by supersaturation modulated growth: A path toward III-nitride superjunctions. J. Appl. Phys **131**(1), 015703 (2022). https://doi.org/10.1063/5.0076044
206. B.J. Baliga, *Fundamentals of Power Semiconductor Devices*, 2nd edn. (Springer International Publishing AG, Boston, 2019)
207. K. Fu, H. Fu, X. Deng, P.-Y. Su, H. Liu, K. Hatch, C.-Y. Cheng, D. Messina, R.V. Meidanshahi, P. Peri, C. Yang, T.-H. Yang, J. Montes, J. Zhou, X. Qi, S.M. Goodnick, F.A. Ponce, D.J. Smith, R. Nemanich, Y. Zhao, The impact of interfacial Si contamination on GaN-on-GaN regrowth for high power vertical devices. Appl. Phys. Lett **118**(22), 222104 (2021). https://doi.org/10.1063/5.0049473
208. S. Kotzea, A. Debald, M. Heuken, H. Kalisch, A. Vescan, Demonstration of a GaN-based vertical-channel JFET fabricated by selective-area regrowth. IEEE Trans. Electron Dev **65**(12), 5329–5336 (2018). https://doi.org/10.1109/TED.2018.2875534
209. C. Yang, H. Fu, V.N. Kumar, K. Fu, H. Liu, X. Huang, T.-H. Yang, H. Chen, J. Zhou, X. Deng, J. Montes, F.A. Ponce, D. Vasileska, Y. Zhao, GaN vertical-channel junction field-effect transistors with regrown p-GaN by MOCVD. IEEE Trans. Electron Dev, 1–6 (2020). https://doi.org/10.1109/TED.2020.3010183
210. M. Xiao, X. Yan, J. Xie, E. Beam, Y. Cao, H. Wang, Y. Zhang, Origin of leakage current in vertical GaN devices with nonplanar regrown p-GaN. Appl. Phys. Lett **117**(18), 183502 (2020). https://doi.org/10.1063/5.0021374
211. H.W. Choi, M.A. Rana, S.J. Chua, T. Osipowicz, J.S. Pan, Surface analysis of GaN decomposition. Semicond. Sci. Technol **17**(12), 1223–1225 (2002). https://doi.org/10.1088/0268-1242/17/12/304
212. M.J. Tadjer, B.N. Feigelson, J.D. Greenlee, J.A. Freitas, T.J. Anderson, J.K. Hite, L. Ruppalt, C.R. Eddy, K.D. Hobart, F.J. Kub, Selective p-type doping of GaN:Si by Mg ion implantation and multicycle rapid thermal annealing. ECS J. Solid State Sci. Technol **5**(2), P124–P127 (2016). https://doi.org/10.1149/2.0371602jss
213. V. Meyers, E. Rocco, T.J. Anderson, J.C. Gallagher, M.A. Ebrish, K. Jones, M. Derenge, M. Shevelev, V. Sklyar, K. Hogan, B. McEwen, F. Shahedipour-Sandvik, p-type conductivity and damage recovery in implanted GaN annealed by rapid gyrotron microwave annealing. J. Appl. Phys **128**(8), 085701 (2020). https://doi.org/10.1063/5.0016358
214. S.R. Aid, T. Uneme, N. Wakabayashi, K. Yamazaki, A. Uedono, S. Matsumoto, Carrier activation in Mg implanted GaN by short wavelength Nd:YAG laser thermal annealing. Phys. Status Solidi (A) **214**(10), 1700225 (2017). https://doi.org/10.1002/pssa.201700225
215. M.H. Breckenridge, J. Tweedie, P. Reddy, Y. Guan, P. Bagheri, D. Szymanski, S. Mita, K. Sierakowski, M. Boćkowski, R. Collazo, Z. Sitar, High Mg activation in implanted GaN by high temperature and ultrahigh pressure annealing. Appl. Phys. Lett **118**(2), 022101 (2021). https://doi.org/10.1063/5.0038628

216. H. Sakurai, M. Omori, S. Yamada, Y. Furukawa, H. Suzuki, T. Narita, K. Kataoka, M. Horita, M. Bockowski, J. Suda, T. Kachi, Highly effective activation of Mg-implanted p-type GaN by ultra-high-pressure annealing. Appl. Phys. Lett **115**(14), 142104 (2019). https://doi.org/10.1063/1.5116866
217. D. Khachariya, M.H. Breckenridge, W. Kim, A. Klump, K. Wang, S. Mita, J. Tweedie, S. Stein, P. Reddy, M. Bockowski, Z. Sitar, R. Collazo, S. Pavlidis, 1 kV GaN-on-GaN PN diode using Mg implantation. Presented at the IEEE device research conference (DRC), Virtual, 2020
218. D. Khachariya, D. Szymanski, R. Sengupta, P. Reddy, E. Kohn, Z. Sitar, R. Collazo, S. Pavlidis, Chemical treatment effects on Schottky contacts to metalorganic chemical vapor deposited n-type N-polar GaN. J. Appl. Phys **128**(6), 064501 (2020). https://doi.org/10.1063/5.0015140
219. B. Romanczyk, X. Zheng, M. Guidry, H. Li, N. Hatui, C. Wurm, A. Krishna, E. Ahmadi, S. Keller, U.K. Mishra, W-band power performance of SiN-passivated N-Polar GaN deep recess HEMTs. IEEE Electron Dev. Lett **41**(3), 349–352 (2020). https://doi.org/10.1109/LED.2020.2967034
220. O.S. Koksaldi, J. Haller, H. Li, B. Romanczyk, M. Guidry, S. Wienecke, S. Keller, U.K. Mishra, N-Polar GaN HEMTs exhibiting record breakdown voltage over 2000 V and low dynamic on-resistance. IEEE Electron Dev. Lett **39**(7), 1014–1017 (2018). https://doi.org/10.1109/LED.2018.2834939
221. J.Y. Tsao, S. Chowdhury, M.A. Hollis, D. Jena, N.M. Johnson, K.A. Jones, R.J. Kaplar, S. Rajan, C.G. Van de Walle, E. Bellotti, C.L. Chua, R. Collazo, M.E. Coltrin, J.A. Cooper, K.R. Evans, S. Graham, T.A. Grotjohn, E.R. Heller, M. Higashiwaki, M.S. Islam, P.W. Juodawlkis, M.A. Khan, A.D. Koehler, J.H. Leach, U.K. Mishra, R.J. Nemanich, R.C.N. Pilawa-Podgurski, J.B. Shealy, Z. Sitar, M.J. Tadjer, A.F. Witulski, M. Wraback, J.A. Simmons, Ultrawide-bandgap semiconductors: Research opportunities and challenges. Adv. Electron. Mater **4**(1), 1600501 (2018). https://doi.org/10.1002/aelm.201600501
222. R. Dalmau, B. Moody, R. Schlesser, S. Mita, J. Xie, M. Feneberg, B. Neuschl, K. Thonke, R. Collazo, A. Rice, J. Tweedie, Z. Sitar, Growth and characterization of AlN and AlGaN epitaxial films on AlN single crystal substrates. J. Electrochem. Soc **158**(5), H530 (2011). https://doi.org/10.1149/1.3560527
223. D. Khachariya, S. Mita, P. Reddy, S. Dangi, P. Bagheri, M.H. Breckenridge, R. Sengupta, E. Kohn, Z. Sitar, R. Collazo, S. Pavlidis, Al0.85Ga0.15N/Al0.6Ga0.4N high electron mobility transistors on native AlN substrates with >9 MV/cm Mesa breakdown fields, in *2021 Device Research Conference (DRC)*, (Jun. 2021), pp. 1–2. https://doi.org/10.1109/DRC52342.2021.9467186
224. I. Abid, J. Mehta, Y. Cordier, J. Derluyn, S. Degroote, H. Miyake, F. Medjdoub, AlGaN channel high electron mobility transistors with Regrown Ohmic contacts. Electronics **10**(6), Art. no. 6 (2021). https://doi.org/10.3390/electronics10060635
225. H. Okumura, S. Suihkonen, J. Lemettinen, A. Uedono, Y. Zhang, D. Piedra, T. Palacios, AlN metal–semiconductor field-effect transistors using Si-ion implantation. Japanese J. Appl. Phys **57**(4S), 04FR11 (2018). https://doi.org/10.7567/JJAP.57.04FR11

Chapter 3
Flexible and Printed Electronics

Benjamin Iñiguez

1 Introduction

The field of flexible, printed and organic electronics has progressed enormously in the last years.

Printed electronics is one of the most promising fields in electronics [1–4]. It is based on creating electronic devices by printing on a variety of substrates, some of them being flexible. Inks for printed electronics are usually made of carbon-based compounds. In particular, inkjet printed electronics has progressed very fast during the last years, and nowadays inkjet printers are capable of printing electrical circuits very quickly and inexpensively. At an industrial level, high-quality printed electronics are already being used to produce flexible keyboards, conformable antennas, flexible screens, interactive books and posters, electronic skin patches and more with industrial processing such as flexography or screen printing. Indeed, there are already a high number of printed electronics products in the market. Some of the latest progress in printed electronics were the smart packaging, ranging from RFID labels to RFID sensing labels. The market for printed electronics is growing [1, 2] because the Internet of Things is expanding and require slow-cost, lightweight technology that can sense and store information securely and transmit data.

A number of materials may be more adequate for flexible and wearable electronics than silicon, whether used in combination with silicon or not. The more promising of these alternative materials are the organic and oxide materials. In particular, the combination of organic and inorganic materials with printing technologies allows thin, lightweight, eco-friendly and very cost-efficient electronic systems.

B. Iñiguez (✉)
Department of Electronic, Electrical and Auromatic Control Engineering, Universitat Rovira i Virgili, Tarragona, Spain
e-mail: benjamin.iniguez@urv.cat

F. Iacopi, F. Balestra (eds.), *More-than-Moore Devices and Integration for Semiconductors*, https://doi.org/10.1007/978-3-031-21610-7_3

On the other hand, despite the significant performances already achieved by devices based on oxide and organic materials, more research is still needed to further improve their electrical characteristics, increase mobility and threshold voltage stability and reduce bias and light stress instability and reduce the voltage operating range [1, 2].

We review the recent developments, current status and challenges in the field of flexible electronics. We address the main device technologies, materials synthesis, fabrication techniques and simulation and modelling and design techniques. In addition, we also highlight the current and expected opportunities and applications of flexible electronics.

2 Materials and Processes for Flexible and Printed Electronics

Although in rigid integrated circuits and displays the mainstream material is silicon (either in crystalline or in amorphous form), there are more suitable materials for flexible and printed electronics, such as organic and amorphous oxide semiconductors. Indeed, flexible silicon electronics [5, 6] still faces several obstacles: make the silicon substrate containing the devices thinner, transfer to the suitable encapsulation materials, placement, interconnection, etc.

Alternative and naturally flexible materials for hybrid electronics are 2D materials (graphene, two-dimensional dichalcogenide materials) and 1D materials (carbon nanotubes, nanowires). Anyway, the management of data obtained by IoE or IoE sensors is still carried out by rigid materials electronics [6]. Therefore, it is still very challenging to develop fully flexible electronic systems. Obviously, the materials proposed for flexible electronics must not lose the advantages of their rigid counterparts for data management.

2.1 Printing Processes

Several printing methods have been used for printed electronics, including gravure, inkjet, reverse offset, aerosol jet printing and others [1, 4].

The two major printing approaches are contact and noncontact printing. In contact printing, the patterned structures with inked surfaces are put in physical contact with the substrate. In noncontact printing, the solution is dispensed through by means of openings, and structures are defined by moving the substrate holder in a pre-programmed pattern. The contact-based printing technologies include gravure printing, gravure-offset printing, flexographic printing and roll-to-roll (R2R) printing. The main noncontact printing techniques include screen printing, slot-die coating and inkjet printing.

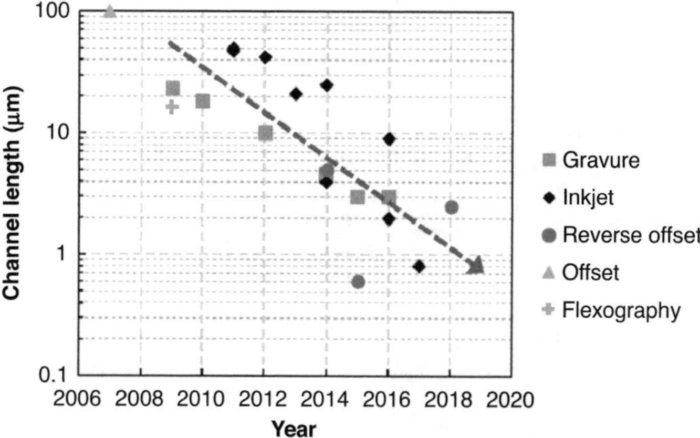

Fig. 3.1 Evolution of the reduction of printed TFT channel length for several printing methods [7]

The noncontact printing techniques have received greater attractions due to their simplicity, affordability, speed and adaptability to the fabrication process. Anyway, there have been important progress in some noncontact printing techniques, such as R2R. Recently, polymeric stamp-based printing methods such as nanoimprint, micro-contact printing and transfer printing have also attracted considerable interest, in particular for semiconductor-based flexible electronics [4].

For the advance of printed electronics, new inks are needed in order to achieve higher mobility thin-film transistors (TFTs) with improved uniformity and reliability. Silver or other metallic oxide-based inks seem promising.

Size downscaling is beneficial for the performance of electron devices in many applications. It needs high-resolution printing. Printed thin-film transistors (TFTs) require downscaled electrodes to achieve a sufficiently high switching frequency for wireless communication [7]. Besides, channel length reduction leads to increased on-current in active-matrix displays or image sensors would need space. It is shown in Fig. 3.1 that channel length in printed TFTs has exponentially been reduced.

On the other hand, printed current collectors with narrower linewidth blocking less light increase solar cell efficiency [8]. The sensitivity of printed sensors can also be improved by downscaling, for example, using interdigitated electrodes [9]. Generally, the layers that require the most aggressive downscaling are conductive electrodes.

One problem for miniaturization is that comparable printed features between commercial ink jet printers (40–50 micron) and lithography (1–3 micron) are still difficult to achieve [10]. A thin uniform defect-free gate dielectric is particularly challenging.

Further progress in printing methods requires an increase of the understanding of the underlying physics, usually the related microscale fluid mechanics, in order to optimize printing parameters and ink formulations (Fig. 3.2).

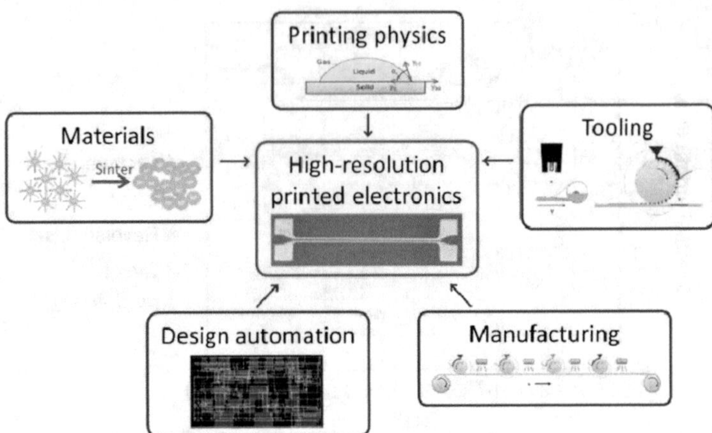

Fig. 3.2 Overview of areas requiring innovation in high-resolution printed electronics [1]

In particular, the scaling of fluid mechanical forces, and maybe the possibility to manipulate electrostatic forces too, will be instrumental to improve printing resolution. Dry transfer, photopatterning or self-assembly can also be incorporated. Besides, the progress needs to be translated into real manufacturing. In order to take advantage of the high-resolution printing techniques, new design rules and electronic design automation (EDA) tools need to be created.

More well-controlled equipment will also improve layer-to-layer registration in addition to other approaches including minimizing flexible substrate deformation, self-alignment or misalignment tolerant device structures such as fully overlapped transistors [1]. A more innovative approach to optimize printing parameters and ink properties could be machine learning [11].

The roll-to-roll (R2R) gravure (Fig. 3.3) has a high potential to fabricate inexpensive, wearable and large-area electronic devices [13]. However, there are some important challenges and constraints to print logic gates and active matrix. The implementation of the roll-to-roll R2R gravure foundry faces several challenges, such as the overlay printing registration accuracy (OPRA), the nanoscale consistency in printed layers and design rule [14]. The OPRA must be less than ± 30 μm. Overlay printings require at least four or more layers to print a thin-film transistor (TFT) with a sufficient performance [14].

Fast and low-cost technology is very much restricted by the substrates. In the case of stretchable devices, during the roll-to-roll processing, materials may be deformed, which would cause huge problems in roll-to-roll processing. Multilayer making with roll-to-roll is prone to high mismatch.

The design rules must take into account that printed TFTs are vulnerable to trapped charges [Noh]. The gate width and channel length of TFT should contain a channel aligned on the top of gate in order to obtain a yield higher than 90% can be achieved. In order to print the electronic devices, the printing speed should be more than 6 m/min.

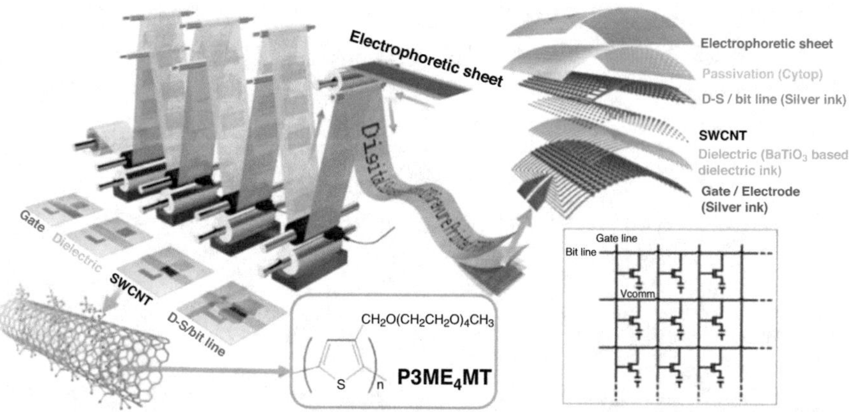

Fig. 3.3 Schematic description of R2R gravure printing process to print TFT-active matrix using carbon nanotube as semiconducting material [12]

2.2 3D Printed Electronics

In the last years, new opportunities have been opened to 3D printing electronics [1, 15], and the sector is rapidly growing, but most of these 3D printed electronic components are based on passive electronic materials such as conductors and dielectrics.

The growth of the IoT is strongly dependent on the emergence of flexible devices which can conform to their three-dimensional dynamic features. 3D printing is an emerging option for consumer wearable electronics, bioelectronics and personalized healthcare.

Several challenges must be overcome. They are related to issues such as types of printable materials, device performances and printing processes. These problems can lead to negative impacts on device performances such as high leakage currents, low device-to-device reliability and dielectric breakdown. It is particularly critical to achieve uniformity of the printed layers. Besides, 3D printing technologies often have low process yields. We need to optimize the materials and ink formulations to achieve the desirable performances.

Once these problems are solved, 3D printing platforms can become more mobile and ubiquitous. It is foreseen that in a near future, electronic devices will be able to be printed from one's own cell phone.

Besides, there is a need for strategies to improve the interfaces and interactions between the different material layers in the device, and between the device and the substrate. These actions can include leveraging the development of high-performance conducting polymers [16], polymers with continuously tunable stiffnesses or reconfigurable soft electronics with programmed ferromagnetic domains [17] in the development of next-generation 3D printed devices.

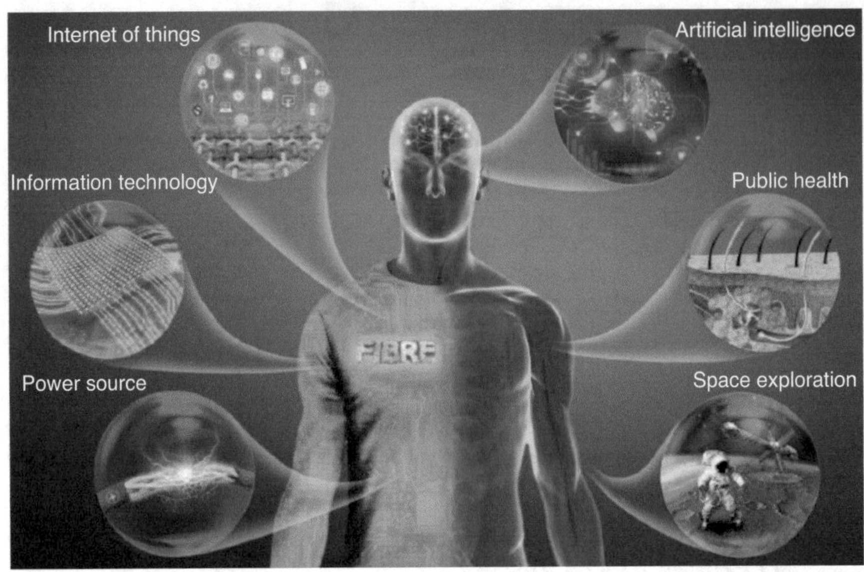

Fig. 3.4 Potential applications of electronic textiles: power sources, information technology, the Internet of Things, artificial intelligence, public health and space exploration [18]

2.3 Electronic Textiles

Research efforts in electronic (e-textiles), which started by simply attaching rigid electronic devices onto the surface of textiles, have advanced to the level of developing methods to integrate electronics into textiles in ways compatible to the maintenance of the softness and stretchability demanded by users [1]. Progress in e-textiles has been driven by the evolution of electronic devices from rigid 3D structures to flexible 2D films and recently to 1D fibres.

Challenges for e-textiles are often related to the curved surfaces of textile fibres and the 3D, porous structures of textiles (Fig. 3.4).

E-fibres are building blocks for e-textile devices [18]. Integration of e-fibres to form e-textiles is a challenging development issue. Very efficient technologies will be required to interconnect large numbers of e-fibres. A promising approach is to knit together e-fibre device components, such as electrodes and electroactive fibres, to produce e-textile devices during the process of textile manufacturing. Appropriate design of the knitted or woven structure and geometry can improve the device performance [19, 20].

There is also research focusing on the development of scalable deposition methods that are compatible with 3D, porous textile structures [21].

3 Devices for Flexible and Printed Electronics

Organic light-emitting diodes (OLEDs) and organic photovoltaics (OPV) have already reached a certain level of maturity. Anyway, for its use in printing technologies on flexible substrates, stability is still an important challenge. On the other hand, printed TFT processes and performances have considerably improved greatly, and they have a high potential as essential devices in printed displays and integrated circuits, but still they are far from being a robust technology [1–3].

In addition, the progress of printed electronics is still limited by the lack of appropriate electron design automation (EDA) tools as well as suitable compact models incorporated to those tools.

3.1 OLED Technologies

The first practical OLED display was made by C. Tang et al. [22]. In the last years, OLED displays and lighting technologies have been improving their efficacy, lifetime and colour quality [1, 2]. Flexible OLEDs have been demonstrated as a promising technique for display and lighting applications with smart cell phones as the main application. According to Sigmaintell Consulting, about 470 million OLED panels for cell phones (290 million rigid and 180 million flexible) were shipped in 2019 and expected to reach 39% of all cell phones in 2020. Although the cost of OLED display and lighting products is higher than LCDs and LEDs, some OLED display and lighting products are available in affordable mobile phones, TV and automotive lighting applications.

The production of an OLED screen requires the matching selection of light-emitting/electronic and hole injection and transport materials, patterning technologies, backplane technologies and encapsulation technologies Novel materials have been instrumental for the development of OLEDs, as shown in Fig. 3.5. Fluorescent materials show a low internal quantum efficiency (IQE) of 25%, whereas the phosphorescent materials can achieve 100% IQE [23]. New fluorescent organic materials, such as thermally assisted delayed fluorescence (TADF) [24], can achieve an EQE of 20%.

Flexible OLED displays are mainly manufactured by means of sublimation in a high vacuum system at a high cost, limiting the area. Solution processing has long been anticipated as the manufacturing technology for future OLED displays.

Anyway, flexible OLED technology faces several challenges. Low-efficiency blue light-emitting fluorescent materials are still widely used in the production of OLEDs, because high efficiency ones have operational lifetimes of only a few thousand hours. Therefore, higher-performance blue EL materials need to be developed [24]. TADF materials can be a solution. Double-doped polymers can also lead to an increase of efficiency [2]. On the other hand, simplified OLED structures, with a lower number of layers, can be helpful to increase the yield and the EL efficacy and to reduce costs.

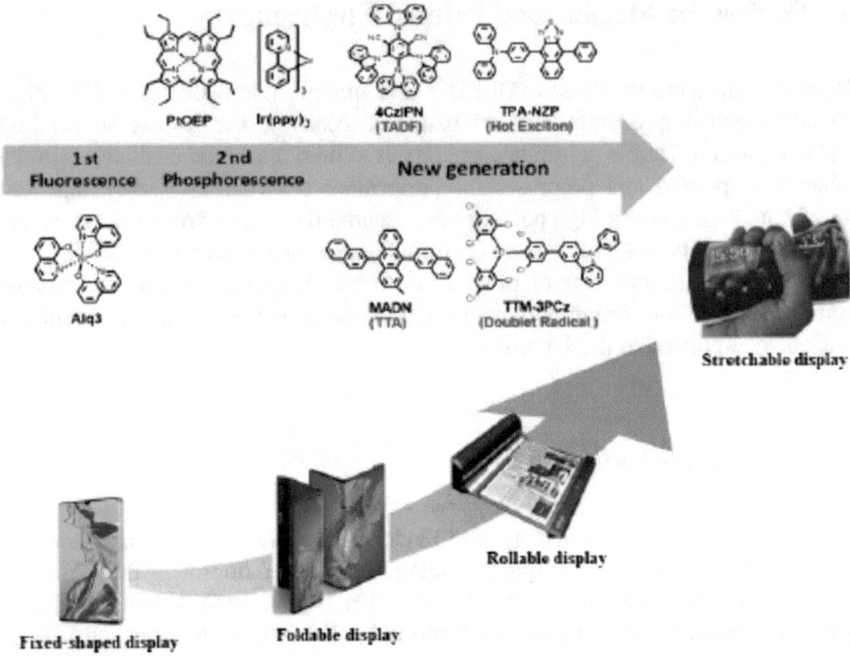

Fig. 3.5 Roadmap of OLED emission materials and flexible OLED displays

Vacuum thermal evaporation allows for high-quality film fabrication, because of its good thickness control and flexible multilayer design. An alternative and promising method is inkjet printing, but it needs a better understanding of the ink formulation, droplet jetting and spreading, solvent evaporation and fusion control. Roll-to-roll manufacturing of print layers is challenging and will require printing methods for fully printed cathodes.

Metal oxide TFT (such as indium gallium zinc oxide) [9] is a promising technology to drive OLED displays, but improvements are needed in the electron mobility and photoelectric stability [25]. Doped oxide TFTs may be a solution.

Further improvements in flexible OLED technology will also require a deeper understanding of several physical mechanisms in these devices, in particular with new materials, such as charge recombination, electron excited state processes, carrier transfer/transport process, new light-emitting and matched electron/hole injection and transport materials.

3.2 *Organic Photovoltaics*

OPV is a very promising technology for indoor and outdoor integration on flat and curved surfaces, like glasses, windows or facades, due to its properties of semitransparency, flexibility and compatibility with digital printing (Fig. 3.6).

Fig. 3.6 Top: Visionary concept of the OPV product portfolio. Flexible and semitransparent modules are integrated indoors as well as outdoors into windows, facades, installations, greenhouses, urban mobility concepts or mobile applications. Bottom: "Real-world" integration of OPV modules in glass construction elements [1]

The first organic photovoltaics (OPV) products were put in the market by Konarka in 2008–2009. They are series of P3HT:PCBM-based solar modules with a nominal peak power between 1 and 40 Wp. Efficiencies of around 20%, a guaranteed lifetime of more than 25 years and costs between 0.3 and 0.5 €/Wp have been obtained. Forecasts are predicting OPV costs at GW level as low as 5 €ct/Wp [27].

OPV efficiencies have already surpassed the performance of older technologies like amorphous silicon (a-Si:H) or dye-sensitized solar cells (DSSC) [28]. Lifetime is progressing quickly. OPV has been proven to be a light stable technology which can operate for tens of thousands of equivalent sun-hours if protected from oxygen and humidity [29, 30].

Anyway, high-performance materials with a low BoM (bill of materials) are still a challenging subject of research. The BoM of the current flexible OPV technology is dominated by the costs for the active material, followed by packaging costs and electrode costs. Few organic semiconductors like P3HT, PCBM, etc. already fulfil these requirements, but despite good stability data, their efficiency is a factor 3–5 too low for most products. Non-fullerene acceptors (NFAs) have an excellent performance, but it is still challenging to reduce their associated costs.

OPV allows low-temperature and low-cost solution coating and printing processes, which offers a high reliability. However, commercial OPV products have much lower efficiency than modules processed in the lab (about 5% vs 13%).

To obtain higher performances, more research is needed to develop new semiconductors and semiconductor inks which can be fully compatible to environmental and green processing, as well as interface and charge extraction layers which can form long-time stable contacts. On the other hand, high-resolution patterning processes with feature sizes of 100 micron or lower are required. Lamination and packaging processes which operate below 140 °C are also crucial.

Commercial OPV modules are currently processed by slot-die coating with shims, with lateral resolutions in the mm regime. Laser patterning on roll-to-roll pilot machines has been shown-web resolution down to 100 microns. Digital printing like roll-to-roll ink jet printing is envisaged to produce free patterns of solar cells [31].

3.3 Printed TFTs

Mobility of thin-film transistors (TFTs) has increased from 10^{-5} cm^2/V to 1–10 cm^2/Vs (see Fig. 3.7) [1, 2]. Small molecules use to have higher mobility than polymers, although many cannot be deposited from solution. Many TFT properties (mobility, threshold voltage, subthreshold slope, off current) depend on the type of gate dielectric material and the fabrication method. Flexible organic TFTs have already been achieved by deposition on a wide range of plastic substrates. Instability to a gate bias voltage and ambient humidity is still a challenge in organic TFTs, although it has considerably improved [33].

Silicon hydrogenated (a-Si:H), laser-recrystallized polysilicon (LTPS) and amorphous oxide semiconductors, especially InGaZnO (IGZO), are used as active layers in flexible TFTs for products such as LCDs, OLED displays and x-ray detectors. Backplanes are fabricated on a thin polyimide film released from a glass carrier after processing. A-Si:H and IGZO can be deposited below 200C and are compatible with other plastic substrates. Besides, IGZO can be printed from a sol-gel solution with annealing at about 400 °C. IGZO flexible microprocessors have been demonstrated [34, 35] (Fig. 3.8).

Other materials include perovskites, carbon nanotubes, graphene and other two-dimensional materials [36]. Electrolyte-gated and electrochemical (EC) TFTs use a liquid or solid electrolyte gate dielectric and operate by transferring charge from the gate dielectric directly to the semiconductor, often PEDOT. ECTFTs typically have high current but slow response and have applications for chemical sensing [37].

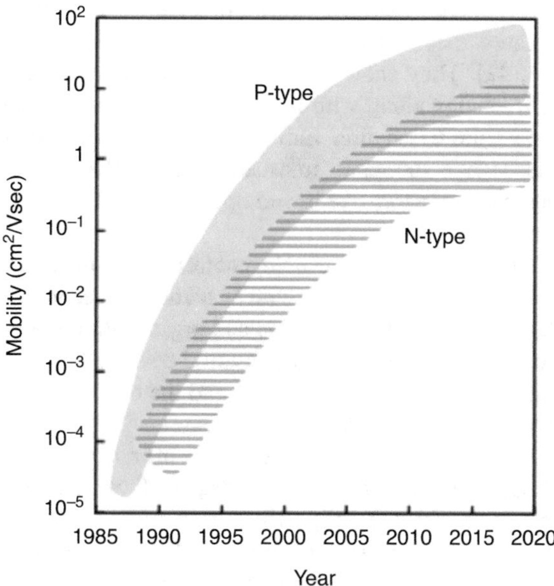

Fig. 3.7 Organic TFT mobility trend over three decades [32]

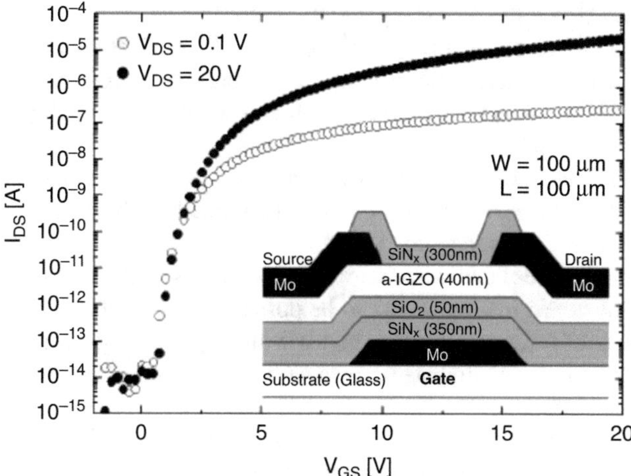

Fig. 3.8 Schematics of an IGZO TFT with a SiO_2/SiN_x dielectric and transfer characteristics [26]

Injection-controlled devices such as the source-gated thin-film transistor (SGT) [38] and the multimodal transistor (MMT) [39] have been shown to achieve a high voltage gain with organic and oxide materials but present complementary challenges. Here, the nature of the charge control mechanism brings important functional trade-offs in terms of saturation voltage, intrinsic gain, transconductance, tolerance to fabrication variability and bias stress behaviour [40].

Perovskite materials have attracted huge interest of the research community because they combine the advantages of both organic and inorganic materials [41, 42]. They show excellent optical properties, high mobilities and long diffusion lengths along with the low-cost and low-temperature fabrication techniques. Perovskite-based thin film transistors can potentially operate at high switching speeds like inorganic transistors and at the same time combine advantages of organic transistors like low-temperature and low-cost processibility and mechanical flexibility.

To address the threshold control issue, dual-gate transistors, consisting of a single thin-film transistor with an additional second gate and second dielectric, have been developed both in oxide and organic structure [43, 44]. Since the electrostatic potential and the carrier density in the whole film become a function of the second gate bias, the threshold voltage and the off-state current can be easily tuned. Both organic and oxide dual-gate transistors with a steeper subthreshold slope improved carrier mobility, and an increased on/off ratio has been demonstrated.

On the other hand, scalability is favoured by the organic permeable base transistor (OPBT) [45], which is a truly vertical device with a semiconductor thickness in the order of 100 nm. Due to the extraordinarily short channel and a reduced influence of the contact resistance, the OPBT can drive very large current densities above 1 kA cm^{-2} and reaches record-high transition frequencies of 40 MHz, making it the fastest organic transistor to date.

Printed TFTs are especially promising for IOT devices, such as disposable flexible tags with an Internet link. Its growth is foreseen as IOT grows. But before reaching the TFT backplane market, problems related to stability and process integration must be solved.

Organic TFT technology shows lower mobility and less uniformity inorganic TFTs, and mobility is even lower when fabricated with a solution-deposited dielectric on a flexible substrate. Mobilities can be increased by means of new poly(benzothiadiazole-naphthalenediimide) derivatives and fine-tuning the material's backbone conformation. This can be possible by the introduction of vinylene bridges capable of forming hydrogen bonds with neighbouring fluorine and oxygen atoms. Introducing these vinylene bridges required a technical feat so as to optimize the reaction conditions.

Oxide TFTs are limited in their use in backplane drivers by the lack of good p-type materials, which is an important gap to fill.

Vacuum-deposited and lithographically patterned flexible TFTs are already being produced. Challenges in printed TFTs have been discussed before. Comparable printed features between commercial ink jet printers (40–50 micron) and lithography (1–3 micron) are still difficult to achieve [10]. The roll-to-roll processing must be improved too.

Better inks must be developed in order to achieve higher mobility. Silver or other metallic-based inks seem promising. A reduction of the costs of inks would also help in the extension of printed electronics applications.

Technological improvements are needed in order to increase mobility values and threshold voltage stability in thin-film transistors (TFTs) and reduce the voltage operating range, bias and light stress instability and the voltage operating range. High-k dielectrics are helpful to reduce the bias operating range. There is research to develop oxides including p-type materials that can be processed at low temperature with stability and high mobility. A suitable complementary TFT technology is still a challenge, but can be achieved by using an n-type oxide TFT and a p-type organic TFT. Another possible solution is the control of doping to balance charge transport in complementary circuits. This approach can be combined with dedicated circuit design and the development of advanced patterning techniques such as high-resolution printing. This can allow TFTs with reduced parasitic capacitances and high transconductance.

Large TFT device variabilities can arise from nonuniform thicknesses of the materials, variability in the size of the devices during fabrication or inhomogeneities within the thin-film materials leading to variable mobilities. Variabilities can be reduced by circuit design for device performance compensation and dynamic performance (threshold voltage, etc.) control using top-gate or floating-gate structure.

Hybrid circuits can overcome some of the performance limitations [46, 47]. The integration of flexible electronics into traditional products will require high mobility TFTs, complementary circuits, new and improved inks and high-resolution printing process.

It is still a challenge for ICs to become thin and flexible for plastic substrates. On the other hand, there is an increasing research activity in microwave flexible electronics. Substrates with high thermal conductivity are necessary. Single crystalline nanomembranes can implement high-frequency flexible transistors due to their transferability, scalability and relatively low cost. The frequency figure of merit of nanomembrane-based flexible transistors has already reached 100 GHz.

Besides, it is still needed a development of 3D printing electronics with the same precision as a 2D printing technology and the improvement as well. 3D integration can help achieve high-density circuits. Initial progress in 3D integration needs to be developed into a robust technology [48]. 3D printing flexible electronics is especially useful in the healthcare sector, where 3D printers can be used for direct printing of biomedical devices onto human skin and can facilitate the manufacturing of flexible electronic sensors of body pressure. 3D printed flexible electronics has also applications in the field of prosthetic organs for the disable.

To carry out the manufacturing of CMOS systems with acceptable costs and reliability, a possible solution is the use of non-planar coin-like 3D architecture with some components placed in the outer sides of both planes, and other elements remain in the middle which are also physically flexible.

3.4 Device Modelling and Design Automation

Thin-film transistors (TFTs) in various semiconductor technologies, including amorphous silicon (a-Si), polycrystalline silicon (poly-Si), amorphous oxide semiconductor (AOS) and organic semiconductor (OSC), provide abundant choices to accommodate diverse flexible integrated electronics applications. Implementation of

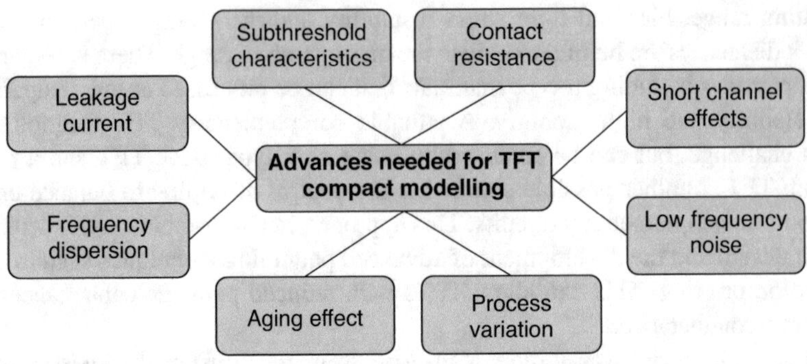

Fig. 3.9 Required advances for TFT compact modelling

these applications, including active-matrix reflective or emissive displays, imagers, radio-frequency identification and wearable sensor systems, needs different considerations in terms of performance, cost, area and mechanical flexibility.

To enable efficient design of circuits and systems, compact models are needed for those TFTs to accurately describe their electrical characteristics and be incorporated into circuit simulators to perform circuit-level simulations (Fig. 3.9). The earliest developed TFT compact models were for a-Si and LTPS TFTs by Shur et al. in 1997 [49]. The models are threshold voltage-based and define the field effect mobility as the usual crystalline silicon carrier mobility scaled by the ratio of the free carrier density to the induced total carrier density. Both current-voltage and capacitance models were developed, and for the poly-Si TFT model, both the kink effect and the short channel effect were taken into account. The models are named as RPI a-Si TFT and poly-Si TFT models and have been widely adopted in commercial circuit simulators for practical circuit simulations.

The design and optimization of large-scale printed and flexible electronic circuits and systems require efficient and accurate device compact models. Due to the limited availability of organic and oxide semiconductor devices' compact models today, circuit and system designers often rely on empirical behavioural macromodels and/or use existing silicon device compact models based on the conventional understanding of transport processes. However, neither approach provides a fully adequate device description under all operation conditions, nor the quantitative predictive quality required for the accurate production quality design. The problem is of course much more serious in emerging TFT structures, such as the ones we will target in this project.

In OTFT the main transport mechanism was assumed to be the variable range hopping [50]. The density of states (DOS) has a Gaussian form, which makes an analytical electrostatic model challenging to develop. Using two exponential density of states (DOS) function, the compact model for organic TFT could be developed, no free carriers considered [51]. Anyway, it was shown that an accurate compact model for OTFTs can be developed by assuming only one exponential DOS (which is a reasonable approximation in the practical range of applied gate voltages). The resulting model, which has a similar formulation as the RPI one, allows to apply direct methods for parameter extraction [52], such as an integral function to extract key parameters. More recently, compact model for OTFTs with Gaussian DOS was presented in [53]. This model considers a power-law approach for mobility and contact resistance.

On the other hand, a few works [54] present quasi-static charge-based compact OTFT model for organic and oxide TFTs. In [55] it was shown that in OTFTs frequency dispersion up to 10 KHz can be incorporated to a quasi-static capacitance model by means of a frequency-dependent dielectric permittivity.

In [56] a charge-based compact model for OTFT has been presented which is based on a power-law expression for the mobility and introduces one-piece current equations for all regions of operation. The model has been extended to include short-channel effects as threshold voltage shift and drain-induced barrier lowering (DIBL) in submicron devices and nonlinear injection at the source contact [57] for the case of staggered and coplanar device architectures. From the expression for the accumulated channel charge, a closed-form model for the drain-current variability due to carrier-number and correlated mobility fluctuations was derived, relating these statistical variations to the trap density in the channel [58].

Furthermore, by applying a partitioning scheme, charges are explicitly attributed to the source.

A compact AOS TFT model assuming a two exponential DOS (corresponding to deep and tail states, respectively), with analytical expressions for current, charges and capacitances was presented in [59]. Parameters of this model can be extracted applying direct methods to both the deep state and the tail state dominated regimes. In [60] authors develop a compact model for TFT accounting for the charge transport by percolation path, where both trap-controlled transport and free carrier movements are included. On the other hand, it has been reported that in mature IGZO TFT technologies, the deep DOS is negligible, and an accurate model can be developed assuming only the tail DOS [61].

In [62] the RPI model was extended to become a surface potential based, which allows us to avoid the threshold voltage problems in TFT and showed good symmetry for circuit design. On the basis of surface potential, a variety of different improvements have been reported, with more physical effect and parameter extraction method [63].

These models addressed particular types of OTFT and AOS TFT structures which were considered among the most matures ones at that time. However, these technologies have evolved a lot in the last 3 years (together with the growth of flexible electronics), and novel and promising OTFT and AOS TFT structures have appeared, where those models are not sufficiently accurate (since they were developed for other structures), and new physical effects need to be considered.

Furthermore, even for "classical" OTFT and AOS TFT structures, the model still needs to become more physical, more accurate and more easily implemented and to have less computation cost [64]. This would be important for industrial applications (Fig. 3.10). For example, most of the compact models need to know the trap density of states' distribution or mobility parameters in advance; however, the

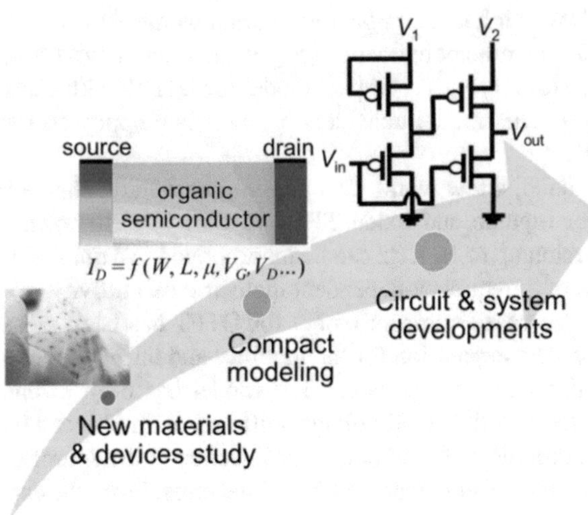

Fig. 3.10 Conceptual representation of the general research and development steps toward organic circuits and systems [53]

correct extraction method is still under discussion. Besides, a particular challenge in flexible electronics is that an adequate model must take into account the TFT channel length variation (and possible changes in other parameters) as a substrate is stretched and bent.

On the other hand, more work is needed to incorporate non-quasi-static effects in the structure of a compact model. A frequency-dependent dielectric permittivity can be accurate up to 10 kHz, but more effects need to be accounted for at higher frequencies. Most of the approaches at high frequencies are based on an equivalent circuit, but a fast compact model needs to take into account these effects in an analytical way in the core model structure.

Novel flexible printed electronics (FPE) applications [65, 66] will need a design automation framework and EDA tools to carry out system simulation and design verification. A big challenge to device modellers and flexible printed electronics designers is the performance variations and the degradation due to mechanical bending, stretching and twisting during the use.

Some of the most popular flexible substrates are plastic films such as polyimide (PI), polyethylene terephthalate (PET) or thermoplastic polyurethane (TPU). The process temperature is limited by the melting temperature of the plastic films that is usually lower than 200°.

Additive manufacturing, such as screen printing, ink jet printing or roll-to-roll imprinting, lowers the manufacturing cost, but limits the minimum feature sizes and the FPE circuit performance.

Design optimization including multi-physics modelling and simulation is essential for meeting the performance target under the usage scenarios. A PDK for FPE and flexible hybrid electronics (FHE) has been developed recently [67] in order

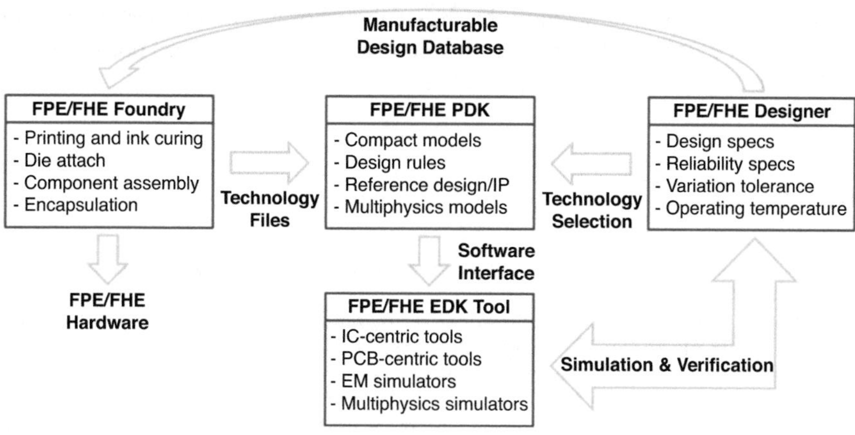

Fig. 3.11 FPE/FHE ecosystem enabled by PDK [1]

to allow the design of complex circuits manufacturable at large quantities. FHE enhances FPE through introducing heterogeneous integration of thinned silicon chips (e.g. <50 μm thick) with FPE elements on a flexible substrate. The FPE/FHE ecosystem that can be allowed by the PDK is illustrated in Fig. 3.11.

The FPE/FHE designers can perform several simulations under target operating temperatures and bending radii and in turn produce manufacturable design database using EDA tools and PDK [68].

Besides, customized place-and-route (P&R) algorithms for physical design flow are also required in order to accommodate the bending use cases for TFT circuits.

The study in [69] suggested inclusion of both mechanical strain and temperature drift's impacts on TFT circuit's performance in layout optimization. For bending [70] or other use cases that require mechanical deformation or thermal cycles, FPE/FHE multi-physics models for electrical, mechanical and thermal interactions are necessary to derive useful information from multi-physics simulation.

4 Conclusions

Flexible electronics is foreseen to be used in an increasing electronics number of products as the related manufacturing technologies continue to progress, due to the expansion of the Internet of Things, which requires low-cost, lightweight, flexible and wearable technologies. The market of flexible and printed electronics is expected to reach over $73 billion by 2025. As substrates become thinner, devices become thin, light and flexible. The performances of the components used in flexible and printed electronics, such as TFTs, OLEDs and OPVs, have improved in the last years. However, for a further expansion of flexible electronics, it is required to overcome several technological challenges, such as improvements in mobility, environmental stability, biodegradable substrates and incorporation in crystalline and rigid electronics. Anyway, several solutions have been proposed, and performances are expected to improve in the coming years.

References

1. Y. Bonnassieux et al., The flexible and printed electronics roadmap. Flex. Print. Electron. **6**(2), 023001 (2021)
2. Wearable, printed and flexible electronics, in *International Roadmap for Devices and Systems*. 2021 Update. Moore than Moore White Paper, pp. 35–39.
3. A. Nathan et al., Flexible electronics: The next ubiquitous platform. Proc. IEEE **100**, 1486–1517 (May 2012)
4. S. Khan et al., Technologies for printing sensors and electronics over large flexible substrates: A review. IEEE Sensors J. **15**(6), 3126–3185 (2015)
5. A.M. Hussain, M.M. Hussain, CMOS-technology-enabled flexible and stretchable electronics for internet of everything applications. Adv. Mater. **28**(22), 4219–4249 (2016)
6. J.A. Rogers, X. Chen, X. Feng, Flexible hybrid electronics. Adv. Mater. **32**(15), 1905590 (2020)
7. G. Grau, V. Subramanian, Fully high-speed gravure printed, low-variability, high-performance organic polymer transistors with sub-5 V operation. Adv. Electron. Mater. **2**(4), 1500328 (2016). https://doi.org/10.1002/aelm.201500328
8. International Technology Roadmap for Photovoltaic (ITRPV), *2019 Results*, vol 2020, 11th edn. (VDMA)
9. G. Mattana, D. Briand, Recent advances in printed sensors on foil. Mater. Today **19**(2), 88–99 (2016). https://doi.org/10.1016/j.mattod.2015.08.001
10. K. Fukuda, T. Someya, Recent progress in the development of printed thin-film transistors and circuits with high-resolution printing technology. Adv. Mater. **29**, 1602736 (2017)
11. Y. Liu, T. Zhao, W. Ju, S. Shi, Materials discovery and design using machine learning. J. Mater. **3**(3), 159–177 (2017). https://doi.org/10.1016/j.jmat.2017.08.002
12. J. Sun et al., Fully R2R-printed carbon-nanotube-based limitless length of flexible active-matrix for electrophoretic display application. Adv. Electron. Mater. **1901431**, 1–9 (2020). https://doi.org/10.1002/aelm.201901431
13. H. Koo, W. Lee, Y. Choi, J. Sun, J. Noh, V. Subramanian, Y. Azuma, Y. Majima, G. Cho, Scalability of carbon-nanotube-based thin film transistors for flexible electronic devices manufactured using an all roll-To-roll gravure printing system. Sci. Rep. **5**(August), 1–11 (2015). https://doi.org/10.1038/srep14459
14. J. Noh, D. Yeom, C. Lim, H. Cha, J. Han, J. Kim, Y. Park, V. Subramanian, G. Cho, Scalability of roll-to-roll gravure-printed electrodes on plastic foils. IEEE Trans. Electron. Packag. Manuf. **33**(4), 275–283 (2010). https://doi.org/10.1109/TEPM.2010.2057512
15. M. Mahto, B. Sniderman, *3D Opportunity for Electronics. Additive Manufacturing Powers Up*. Available: https://www2.deloitte.com/insights/us/en/focus/3d-opportunity/additive-manufacturing-3d-printed-electronics.html (2017)
16. H. Yuk, B. Lu, S. Lin, K. Qu, J. Xu, J. Luo, et al., 3D printing of conducting polymers. Nat. Commun. **11**, 1–8 (2020)
17. Y. Kim, H. Yuk, R. Zhao, S.A. Chester, X. Zhao, Printing ferromagnetic domains for untethered fast-transforming soft materials. Nature **558**, 274–279 (2018)
18. X. Xu, S. Xie, Y. Zhang, H. Peng, The rise of fiber electronics. Angew. Chem. Int. Ed. **131**(39), 13778–13788 (2019)
19. A. Levitt et al., 3D knitted energy storage textiles using MXene-coated yarns. Mater. Today **34**, 17–29 (2020)

20. P. Liu et al., Polymer solar cell textiles with interlaced cathode and anode fibers. J. Mater. Chem. A **6**(41), 19947–19953 (2018)
21. H. Jin, N. Matsuhisa, S. Lee, M. Abbas, T. Yokota, T. Someya, Enhancing the performance of stretchable conductors for e-textiles by controlled ink permeation. Adv. Mater. **29**(21), 1605848 (2017)
22. C. Tang, S.A. VanSlyke, Organic electroluminescent diodes. Appl. Phys. Lett. **51**, 913–915 (1987)
23. Y. Ma, H. Zhang, J. Shen, C. Che, Electroluminescence from triplet metal—Ligand charge-transfer excited state of transition metal complexes. Synth. Met. **94**, 245–248 (1998)
24. H. Uoyama, K. Goushi, K. Shizu, H. Nomura, C. Adachi, Highly efficient organic light-emitting diodes from delayed fluorescence. Nature **492**, 234–238 (2012)
25. K. Nomura, H. Ohta, A. Takagi, T. Kamiya, M. Hirano, H. Hosono, Room-temperature fabrication of transparent flexible thin-film transistors using amorphous oxide semiconductors. Nature **432**, 488–492 (2004)
26. S. Lee, D. Striakhilev, S. Jeon, A. Nathan, Unified analytic model for current–voltage behavior in amorphous oxide semiconductor TFTs. IEEE Electron. Device Lett. **35**(1), 84–86 (2014)
27. F. Machui et al., Cost analysis of roll-to-roll fabricated ITO free single and tandem organic solar modules based on data from manufacture. Energy Environ. Sci. **7**(9), 2792 (2014)
28. M.A. Green, E.D. Dunlop, J. Hohl-Ebinger, M. Yoshita, N. Kopidakis, A.W.Y. Ho-Baillie, Solar cell efficiency tables (Version 55). Prog. Photovoltaics Res. Appl. **28**, 3 (2020)
29. X. Du et al., Efficient polymer solar cells based on non-fullerene acceptors with potential device lifetime approaching 10 years. Joule **3**(1), 215–226 (2019)
30. Q. Burlingame, X. Huang, X. Liu, C. Jeong, C. Coburn, S.R. Forrest, Intrinsically stable organic solar cells under high-intensity illumination. Nature **573**(7774), 394–397 (2019)
31. P. Maisch, K.C. Tam, P. Schilinsky, H.J. Egelhaaf, C.J. Brabec, Shy organic photovoltaics: Digitally printed organic solar modules with hidden interconnects. Sol. RRL **2**, 1800005 (2018)
32. A.F. Paterson, S. Singh, K.J. Fallon, T. Hodsden, Y. Han, B.C. Schroeder, H. Bronstein, M. Heeney, I. McCulloch, T.D. Anthopoulos, Recent progress in high-mobility organic transistors: A reality check. Adv. Mater. **30**, 1801079 (2018)
33. M. Kettner, M. Zhou, J. Brill, P.W.M. Blom, R.T. Weitz, Complete suppression of bias-induced threshold voltage shiftbelow 273 K in solution-processed high-performance organic transistors. Appl. Mater. Interfaces **10**, 35449–35454 (2018)
34. K. Myny, The development of flexible integrated circuits based on thin-film transistors. Nat. Electron. **1**, 30–39 (2018)
35. E. Carlos, J. Leppäniemi, A. Sneck, A. Alastalo, J. Deuermeier, R. Branquinho, R. Martins, E. Fortunato, Printed, highly stable metal oxide thin-film transistors with ultra-thin high-κ oxide dielectric. Adv. Elect. Mater. **6**, 1901071 (2020)
36. L. Xiang, H. Zhang, Y. Hu, L.-M. Peng, Carbon nanotube-based flexible electronics. J. Mater. Chem. C **6**, 7714–7727 (2018)
37. J. Rivnay, S. Inal, A. Salleo, M. Róisín, Owens, M. Berggren, G. Malliaras, Organic electrochemical transistors. Nat. Rev. Mater **3**, 17086 (2018)
38. R.A. Sporea et al., Intrinsic gain in self-aligned polysilicon source-gated transistors. IEEE Trans. Electron. Devices **57**(10) (2010)
39. E. Bestelink et al., Versatile thin-film transistor with independent control of charge injection and transport for mixed signal and analog computation. Adv. Intell. Sys., 2000199 (2021)
40. R.A. Sporea, K.M. Niang, A.J. Flewitt, S.R.P. Silva, Novel tunnel-contact-controlled IGZO thin-film transistors with high tolerance to geometrical variability. Adv. Mat. **1902551** (2019)
41. S.D. Stranks et al., Electron-hole diffusion lengths exceeding 1 micrometer in an organometal trihalide perovskite absorber. Science **342**, 341–344 (2013)
42. C. Wehrenfennig et al., High charge carrier mobilities and lifetimes in organolead trihalide perovskites. Adv. Mater. **26**, 1584–1589 (2014)
43. L.L. Chua et al., Organic double-gate field-effect transistors: Logic-AND operation. Appl. Phys. Lett. **87**, 253512 (2005)

44. R. Pfattner et al., Dual-gate organic field-effect transistor for pH sensors with tunable sensitivity. Adv. Electron. Mater. **5**, 1800381 (2019)
45. F. Dollinger et al., Electrically stable organic permeable base transistors for display applications. Adv. Electron. Mater. **5**, 1900576 (2019)
46. G. Tong et al., Flexible hybrid electronics: Reviews and challenges. Proc. of ISCAS 2018, Florence (Italy) (2018)
47. D.E. Schwartz, J. Rivnay, G.L. Whiting, P. Mei, Y. Zhang, B. Krusor, S. Kor, G. Daniel, S.E. Ready, J. Veres, R.A. Street, Flexible hybrid electronic circuits and systems. IEEE JETCAS7 **27** (2016)
48. J. Kwon, Y. Takeda, R. Shiwaku, S. Tokito, K. Cho, S. Jung, Three-dimensional monolithic integration in flexible printed organic transistors. Nat. Commun. **10**, 54 (2019)
49. B. Iñíguez, T. Fjeldly, T. Ytterdal, M.S. Shur, Thin film transistor modeling, in *Silicon and beyond: Advanced Circuit Simulators and Device Models*, ed. by M. S. Shur, T. A. Fjeldly, (World Scientific Publishers, Singapore, 2000), pp. 703–723
50. M.C.J.M. Vissenberg, M. Matters, Theory of the field-effect mobility in amorphous organic transistors. Physical Review B **57**(20), 13 (1998)
51. L. Li, H. Marien, J. Genoe, M. Steyaert, P. Heremans, Compact model for organic thin-film transistor. IEEE Electron. Dev. Lett. **31**, 210–212 (2010)
52. C.H. Kim et al., A compact model for organic field-effect transistors with improved output asymptotic behaviors. IIEEE Trans. Electron Devices **60**(3), 1136–1141 (2013)
53. S. Jung et al., Advances in compact modeling of organic field-effect transistors. IEEE Trans. Electron Devices **66**(11), 4894–4900 (2019)
54. O. Moldovan et al., A compact model and direct parameters extraction techniques for amorphous gallium-indium-zinc-oxide thin film transistors. IEEE Electron Device Letters **40**(5), 730–733 (2019)
55. H. Cortés-Ordóñez, S. Jacob, F. Mohamed, G. Ghibaudo, B. Iñiguez, Analysis and compact modeling of gate capacitance in organic thin-film transistors. IEEE Trans. Electron Devices **65**(5), 2370–2374 (2019)
56. F. Hain, M. Graef, B. Iniguez, A. Kloes, F. Hain, M. Graef, B. Iñiguez, A. Kloes, Charge based, continuous compact model for the channel current in organic thin film transistors in all regions of operation. Solid State Electron. **133**, 17–24 (2017)
57. J. Prüfer et al., Compact modeling of nonlinear contact effects in short-channel coplanar and staggered organic thin-film transistors. IEEE Trans. Electron Devices **68**(8), 3843–3850 (2021)
58. A. Nikolaou et al., Charge-based model for the drain-current variability in organic thin-film transistors due to carrier-number and correlated-mobility fluctuation. IEEE Trans. Electron Devices **67**(11), 4667–4671 (2020)
59. O. Moldovan et al., A complete charge based capacitance model for IGZO TFTs. Solid State Electron. **269**, 81–86 (2016)
60. X. Cheng et al., TFT compact modelling. IEEE J. Display Technol. **12**(9), 898–906 (2016)
61. M. Estrada et al., Crystalline-like temperature dependence of the electrical characteristics in amorphous Indium- Gallium-Zinc-Oxide thin film transistors. Solid State Electron. **135**, 43–48 (2017)
62. Z. Zong et al., A new surface potential-based compact model for a-IGZO TFTs in RFID applications. IEEE IEDM Dig. Tech., 35.5.1–35.5.4 (2014)
63. Y.H. Barrios et al., An insight to mobility parameters for AOSTFTs, when the effect of both, localized and free carriers, must be considered to describe the device behavior. Solid State Electron. **149**, 32–37 (2018)
64. B. Iñiguez et al., New compact modeling solutions for organic and amorphous oxide TFTs. IEEE J. Electron Devices Soc. **9**, 911–932 (September 2021)
65. T. Someya, Z. Bao, G.G. Malliaras, The rise of plastic bioelectronics. Nature **540**(7633), 379–385 (2016)
66. T. Lei et al., Low-voltage high-performance flexible digital and analog circuits based on ultrahigh-purity semiconducting carbon nanotubes. Nat. Commun. **10**(1), 1–10 (2019)

67. F. Rasheed et al., Predictive modeling and design automation of inorganic printed electronics, in *2019 Design, Automation & Test in Europe Conf. & Exhibition (DATE)*, (IEEE, 2019)
68. L. Shao et al., Ultra-thin skin electronics for high quality and continuous skin-sensor-silicon interfacing, in *Proceedings of the 56th Annual Design Automation*, (2019)
69. J.-L. Lin, P.-H. Wu, T.-Y. Ho, A novel cell placement algorithm for flexible TFT circuit with mechanical strain and temperature consideration. ACM J. Emerg. Technol. Comput. Syst. **11**(1), 1 (2014)
70. H. Gleskova, S. Wagner, W. Soboyejo, J. Suo, Electrical response of amorphous silicon thin-film transistors under mechanical strain. J. Appl. Phys. **92**, 6224–6229 (2002)

Chapter 4
Terahertz Metasurfaces, Metawaveguides, and Applications

Wendy S. L. Lee, Shaghik Atakaramians, and Withawat Withayachumnankul

1 Introduction

Terahertz waves refer to the electromagnetic waves propagating in the frequency range from 0.1 to 10 THz. An overview of the electromagnetic spectrum positions the terahertz frequency regime between the microwave and infrared ranges is depicted in Fig. 4.1. On the lower end of the frequency spectrum, there exists an overlap with the millimeter-wave (MMW) range between 30 GHz and 300 GHz. The terahertz frequency region encapsulates the sub-millimeter-wave (sub-MMW) region, from 300 GHz to 3 THz [1]. Toward the higher end of the frequency spectrum, the terahertz wave overlaps with the far infrared region, ranging from 3 to 20 THz [2].

W. S. L. Lee
Department of Electrical and Electronic Engineering, University of Melbourne, Melbourne, VIC, Australia
e-mail: wendy.lee@unimelb.edu.au

S. Atakaramians (✉)
School of Electrical Engineering and Telecommunications, University of New South Wales, Sydney, NSW, Australia
e-mail: s.atakaramians@unsw.edu.au

W. Withayachumnankul
School of Electrical and Electronic Engineering, The University of Adelaide, Adelaide, SA, Australia
e-mail: withawat@adelaide.edu.au

F. Iacopi, F. Balestra (eds.), *More-than-Moore Devices and Integration for Semiconductors*, https://doi.org/10.1007/978-3-031-21610-7_4

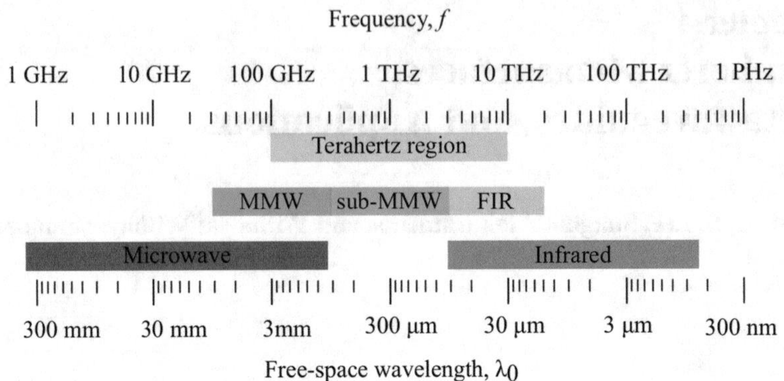

Fig. 4.1 The terahertz frequency band within the electromagnetic spectrum and its overlap with the microwave and infrared bands

1.1 Terahertz Gap

As one of the underutilized sections of the electromagnetic spectrum, the terahertz region is often referred to as the "terahertz gap." This is due to the underdevelopment of technology to generate and detect terahertz waves. Currently, the common approaches of generating and detecting broadband terahertz radiation are photoconductive antennas [3, 4], organic crystals [5], and non-organic crystals [6] combined with ultrafast lasers. A more recent technology to generate and detect terahertz radiation is by an electronic method, where a microwave signal is amplified by signal extension modules to generate terahertz waves [7]. However, these terahertz systems are expensive and bulky, and thus extensive research into low-cost, miniaturized, and efficient systems is needed. Nevertheless, the terahertz gap is still worth exploring as this frequency range shows promise for a myriad of applications in sensing, imaging, and communications.

1.2 Principles of Metamaterials

Conventional components such as lenses, prisms, beam splitters, and wave plates manipulate electromagnetic waves based on the principles of reflection and refraction. As a wave travels through a medium, the electric field components undergo a phase delay that corresponds to the refractive index of the medium. Thus, by engineering the refractive indices of a material, the phase and amplitude of electromagnetic waves can be tailored to achieve intricate manipulations to the outgoing electromagnetic waves. Unfortunately, the existing conventional techniques such as bulk size, limited naturally available materials, and complicated fabrication are not suitable for on-chip integration and other miniature devices. Metamaterials, on the other hand, have been shown to be a promising tool for exotic wave manipulation [8]. This is because metamaterials have been shown to have full control over the permittivity and permeability of materials, applications such as negative refraction [9–11], cloaking [12–14], and field enhancements [15–17] just to name a few. Two-dimensional metamaterials are commonly known as metasurfaces [18], which consist of miniaturized metallic or dielectric resonators arranged in a subwavelength periodicity. These resonators, otherwise known as meta-atoms, are the key toward shaping the outgoing waveform. Additionally, metasurfaces are ultrathin, which lowers loss that arises from wave propagation in bulk substrates. Planar metasurfaces can also be readily manufactured using the existing fabrication techniques.

1.3 Meta-atoms

Metasurfaces consist of building blocks known as meta-atoms. These individual meta-atoms are responsible for the phase, polarization, and amplitude of the outgoing wavefront from the metasurface. By tuning the size, shape, and arrangements of these meta-atoms, specific electromagnetic responses can be generated. In order to generate exotic wavefronts, a few criteria should be achieved depending on the intended outcome. Typically, these meta-atoms can induce either an electric or magnetic resonance or even both. These resonances will provide abrupt phase shifts across the metasurface, which will mold the outgoing wavefront. The following section will discuss two of the most common forms of these meta-atoms, namely metallic and dielectric resonators.

1.3.1 Metallic Resonators

An example of metallic meta-atoms that can provide the electric response is thin metal wire arrays [19, 20] as illustrated in Fig. 4.2a. Upon excitation from electric fields polarized parallel to the wires, these meta-atoms behave as a Drude metal as shown by the following equation:

$$\epsilon(\omega) = \epsilon_0\left(1 - \frac{\omega_p^2}{\omega^2 - \gamma_e \omega j}\right), \tag{4.1}$$

where its plasma frequency ω_p^2 is given by

$$\omega_p^2 = \frac{Ne^2}{m_e \epsilon_0}, \tag{4.2}$$

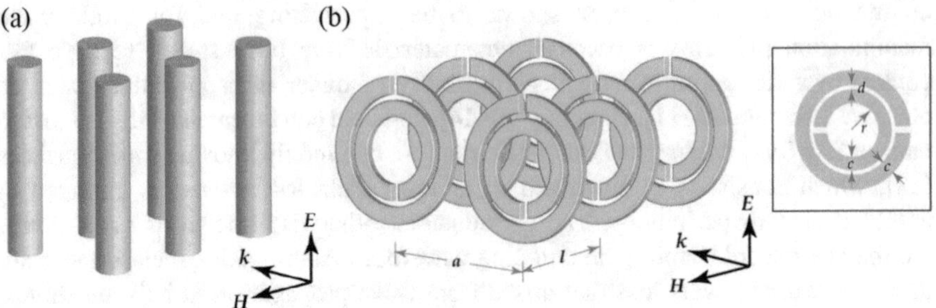

Fig. 4.2 Schematic of metamaterial geometry for a (**a**) wire array and (**b**) split-ring resonator. Adopted from [22]

where N is the density of electrons, e is the electron charge, and m_e is the effective mass of the electron. In the case of thin wire arrays, the effective plasma frequency depends on the lattice geometry, which can be approximated by the following equation:

$$\omega_p^2 = \frac{2\pi c^2}{a^2 \ln \frac{a}{r}}, \tag{4.3}$$

where c is the speed of light in vacuum, r is the radius of the thin wire, and a is the spacing in between the unit cells. Thus, based on Eq. 4.1, the plasma frequency of a metamaterial can be tuned according to the parameters of the thin wire array, namely the radius of the wire r and the lattice spacing a.

On the other hand, a magnetic response can be provided by meta-atoms that can support current loops, which generates a magnetic dipole moment such as split-ring resonators (SRRs) [21] as illustrated in Fig. 4.2b. This is analogous to an LC circuit, where the resonance frequency can be given by $\omega_{m0} = \sqrt{1/LC}$, where L and C are the inductance and capacitance of the meta-atom, respectively. When magnetic fields polarized parallel to the SRRs axis propagate through the SRR, the meta-atoms display a resonance that amplifies the magnetic effect. The effective magnetic permeability equation can be described by the Lorentzian model [21],

$$\mu(\omega) = \epsilon_0 \left(1 - \frac{F\omega^2}{\omega^2 - \omega_{m0}^2 + \Gamma\omega j}\right), \tag{4.4}$$

where F is the fill factor of the SRR given by

$$F = \frac{\pi r^2}{a^2}, \tag{4.5}$$

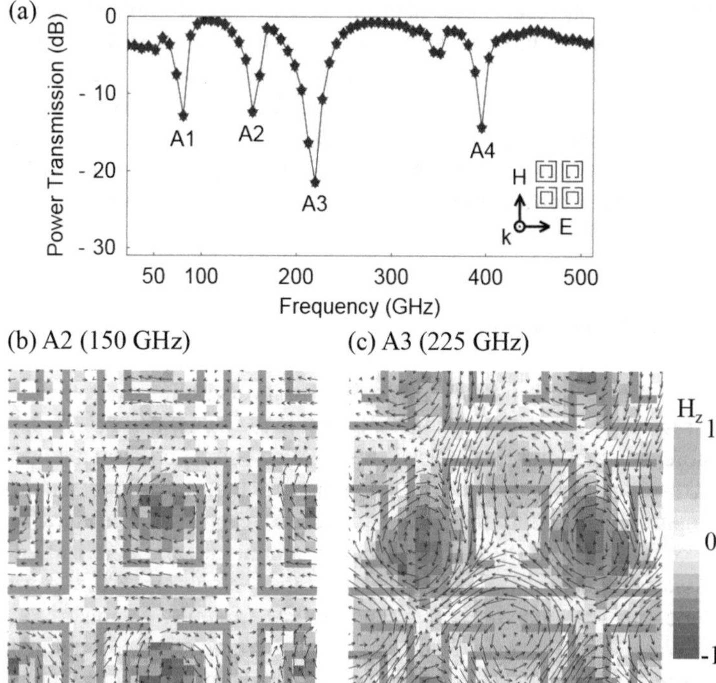

Fig. 4.3 (**a**) Power transmission spectrum of the SRR array sample. Experimentally determined electromagnetic near fields for the resonances (**b**) A2 and (**c**) A3. The arrows indicate the electric in-plane field vectors in the xy plane. Adapted with permission from [23] ©The Optical Society

and ω_{m0} is the resonance frequency given by

$$\omega_{m0} = \sqrt{\frac{3lc^2}{\pi^2 r^3}}, \tag{4.6}$$

and Γ is the damping factor of the SRR, where l is the distance between the rings of the SRR and r is the radius of the outer ring. Thus, by tailoring the dimensions of the SRRs, the meta-atoms can provide either a positive or negative response to the magnetic field.

The geometry and material properties of the SRR can be specifically designed to support both electric and magnetic resonances at particular frequencies [23]. To demonstrate the nature of the resonances, both far-field and near-field measurements were done. The following example shows a copper SRR on a polytetrafluoroethene (PTFE) substrate. Figure 4.3a shows the far-field transmission of the SRR array with four dominant resonances (A1–A4) identified from their minimum transmission peaks. The terahaertz near-field terahertz microscope can then be used to characterize these particular resonances, namely, to determine if they are an electric or magnetic resonance. Figure 4.3b shows the measured near-field distribution of the resonance A2, where the oscillating fields are rotating toward the middle of the SRR. These modes correspond to the LC resonances of the outer and inner split ring where the circulating currents in the metal structures form a magnetic moment normal to the SRR plane. For Fig. 4.3c, the magnetic field vectors are pointing toward and field vectors are travelling toward and away from the inner resonator on all four corners, which indicate an electric quadrupole resonance.

1.3.2 Dielectric Resonators

In contrast to metallic resonators that require two layers to support both electric and magnetic resonances, dielectric resonators are able to do this with a single layer [24]. Furthermore, the operation of dielectric resonators is based on resonant oscillations of displacement current; this alleviates Ohmic loss that metallic resonators suffer from. Similarly, with proper magnetic and electric resonances, dielectric resonator metasurfaces can exhibit extraordinary phase, amplitude, and polarization control of output waves. It is noteworthy that dielectric meta-atoms can take various shapes that include spheres, cuboids, and cylinders. For example, a terahertz dielectric resonator array is constructed from periodically arranged silicon cylinders on a quartz substrate in order to characterize the near-field properties [25]. The resonator height and diameter are designed to show isolated electric and magnetic dipole resonances. As dielectric resonators are based on oscillation of displacement currents within the resonator, the size of the resonator dictates the field confinement within the boundary. Consequently, the resonant frequency of the dielectric resonator is dependent on the size of the resonator. The electric and magnetic dipole resonance modes are classified by examining their electric and magnetic fields, as revealed by cross-sectional views in Fig. 4.4a,b obtained through numerical computation. The electric dipole resonance in Fig. 4.4a shows an electric field that is oscillating in the direction of the incident wave's polarization ($\mathbf{E}_y$). Magnetic dipole resonances in Fig. 4.4b can be identified with a circulating electric field around the center of the resonator. For further confirmation of the positions of the electric dipole resonance and the magnetic dipole resonance, the far-field reflection phase and transmission amplitude spectrum can be studied as shown in Fig. 4.4c and d, respectively. The simulated reflection phase curve in Fig. 4.4c exhibits a π radian crossing at 0.58 THz, which is the location of the electric dipole resonance, and a zero crossing at 0.61 THz, which is the location of the magnetic dipole resonance. Additionally, the dips at 0.58 THz and 0.61 THz are observed in Fig. 4.4d, which correspond to the electric dipole resonance and magnetic dipole resonance, respectively. This is in line with the expected π phase difference between electric and magnetic dipole resonances [26].

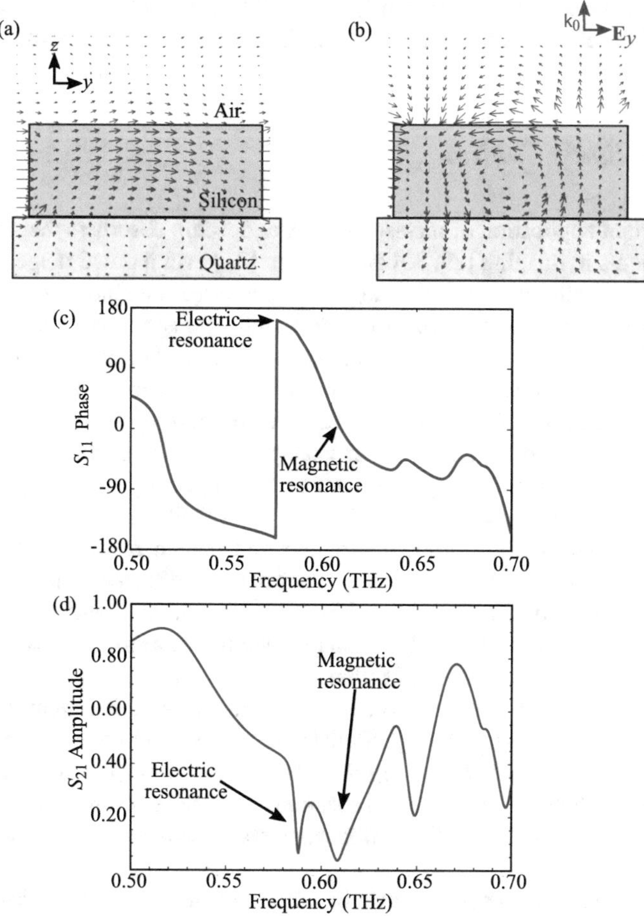

Fig. 4.4 Cross-sectional view of a single dielectric resonator. The instantaneous electric fields are represented by blue arrows. (**a**) Electric dipole resonance and (**b**) magnetic dipole resonance. (**c**) Reflection phase response of the dielectric resonator array. (**d**) Normalized transmission spectra of the dielectric resonator array. The red and blue lines refer to simulated and measured results, respectively. The labels indicate the positions of the electric dipole resonance and the magnetic dipole resonance. Adapted with permission from [25] ©The Optical Society

2 Metasurfaces

Metasurfaces are two-dimensional planar structures consisting of meta-atoms arranged periodically. The mechanism behind wavefront control of metasurfaces is in principle the same as that known in reflectarrays [27] and transmitarrays [28], which are well-established concepts at microwave frequencies. Metasurfaces that operate in transmission are known as transmitarrays, while those that operate in reflection are known as reflectarrays. The following subsections detail typical functionalities of metasurfaces such as amplitude, phase, and polarization control.

2.1 Amplitude Control

One important application of metasurfaces is to develop absorbers that can be used as components in imaging [29, 30] and sensing [31] over a broad range of frequencies [32]. Typically, metamaterial absorbers comprise metallic resonators resting on a ground plane that function to remove reflections and enhance absorption. Key characteristics in absorber designs include bandwidth, overall thickness, and stability of the absorbance over oblique angles of incidence [33]. One potential solution to address stability of absorbers under oblique incident angles is to miniaturize the unit cell elements. For example, a narrowband terahertz absorber consisting of three metallic layers separated by a cyclic olefin copolymer (COC) dielectric spacer is shown in Fig. 4.5a. An equivalent circuit model is used to analytically determine the properties of the unit cell in relation to the resonant frequency. In order to obtain a high-Q absorber, the effective capacitance of the unit cell is increased. This is achieved by providing an additional capacitive layer via a patch array sandwiched in between the metallic patterns. The layer allows for destructive interference in the reflection path as the circuit works as an absorber if the input impedance matches the free-space impedance. Measurements utilizing a typical THz-TDS set-up were performed with various oblique incidence angles ranging from 30° to 60°. From Fig. 4.5b and c, it is observed that the absorber works well for both TE and TM polarizations. It is noteworthy that this absorber has a high-Q factor and has a stable frequency response for various oblique incident angles.

Dielectric semiconductors can also be used to create near-perfect absorbers. For example, cavity arrays that can absorb nearly 100% of incident terahertz energy at resonance have been demonstrated by [35]. Scanning electron micrograph images of the array and a single cavity structure can be seen in Fig. 4.6a and b, respectively. The resonant cavities can efficiently trap energy through dimensions that satisfy the critical coupling condition. The linearly polarized terahertz waves normally incident on the surface will be diffracted by the annular gap. The diffraction leads to phase matching to surface plasmon polaritons (SPPs) that is sustained along the sidewalls of the cavity. The excited SPPs can propagate back and forth along the cavity axis. This motion will result in a resonance from a standing wave.

2.2 Phase Control

Generally, the designed meta-atoms that make up a metasurface induce a phase discontinuity with the incoming electromagnetic wave to mold the outgoing wavefront.

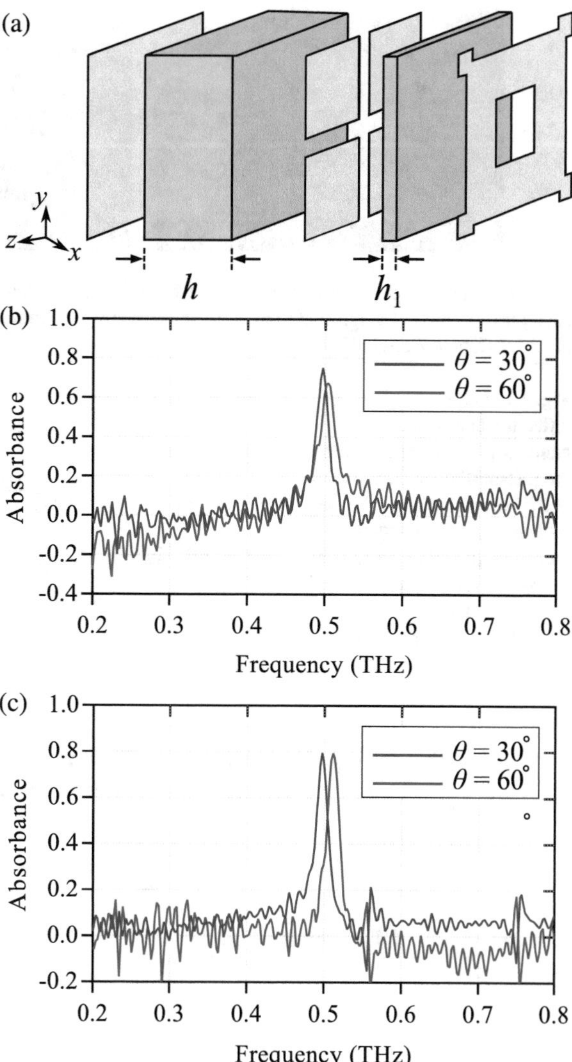

Fig. 4.5 Unit cell of the proposed narrowband absorber. (**a**) 3-D view of the unit cell showing the three metallic layers and dielectric spacers. (**b**) Power absorption for the TE polarization and (**c**) TM polarization for varying incidence angles [34]

In order to achieve such a function, the designed meta-atoms need to cover a full-phase cycle of 2π. The meta-atoms introduce these phase shifts through resonances, whether electric, magnetic, or a combination of both. These abrupt phase shifts constitute a spatial phase variation that enables the metasurface to collectively reshape the direction of propagation of incoming electromagnetic waves. Figure 4.7 shows a conceptual illustration of a linear gradient metasurface at the boundary between two mediums. The meta-atoms are arranged subwavelengths apart, with a unit cell size of a. There exists a fixed phase difference between adjacent meta-atoms of $\Delta\phi = \phi_1 - \phi_0$. When the distance between the unit cells is constant, the

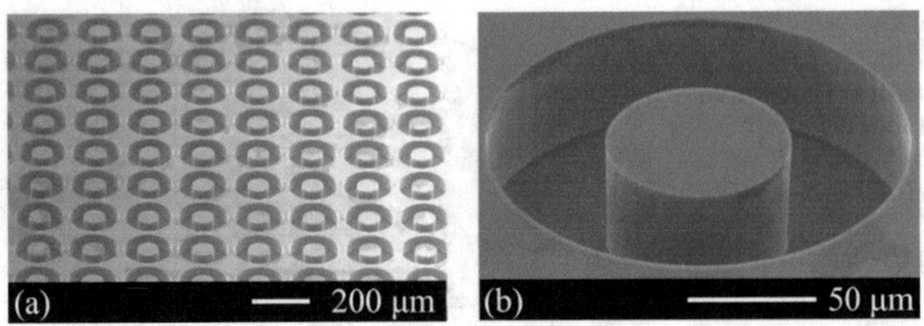

Fig. 4.6 Scanning electron micrographs of the fabricated micro-cavities. (**a**) Partial view of the cavity arrays and (**b**) magnified view of a single cavity. The micrographs are taken at the tilt angle of 45°. Adopted from [35]

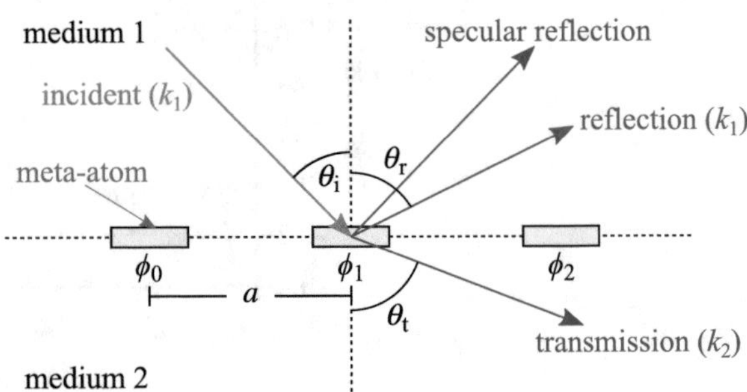

Fig. 4.7 Generalized law of reflection and refraction. The incident wave impinges onto the metasurface at angle θ_i. θ_r and θ_t depict the angle of the reflected and transmitted waves, respectively. k_1 and k_2 are the wavenumbers for mediums 1 and 2, respectively. Adopted from [52]

wavevector that is imparted to the incident wave is $\frac{\Delta\phi}{a}$. Herein, the angle of the reflected wave θ_r can be determined by the following equation:

$$k_1 \sin\theta_i + \frac{\Delta\phi}{a} = k_1 \sin\theta_r, \tag{4.7}$$

and the angle of the transmitted wave, θ_t, can be determined by the following equation:

$$k_1 \sin\theta_i + \frac{\Delta\phi}{a} = k_2 \sin\theta_t. \tag{4.8}$$

From Eqs. 4.7 and 4.8, θ_i refers to the angle of incidence, while k_1 and k_2 are wavenumbers in mediums 1 and 2, respectively. In Eqs. 4.7 and 4.8, the reflected and transmitted electromagnetic waves can be steered into a predetermined direction given a phase gradient that is imposed by the metasurface through resonating meta-atoms. As the nature of the phase gradient is influenced by the arrangement of meta-atoms in the lattice, a wide variety of phase gradients can be designed to tailor the outgoing electromagnetic wave. Among such applications include beam splitting [36–40], beam focusing [41–44], beam steering [45–48], and generation of vortex beams [49–51].

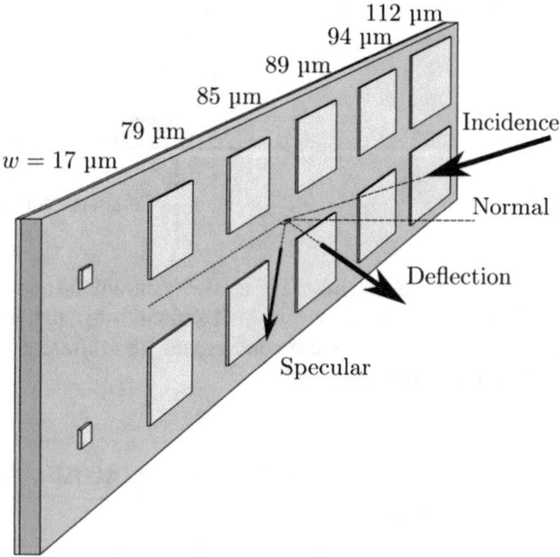

Fig. 4.8 Two subarrays with arrows indicating the incident, deflected, and specular beams. Adapted from [57]

Through variation of the dimensions of the metallic resonating element, the local reflection phase response can be varied. A requirement for effective outgoing wavefront control is a phase coverage of near 2π radians. Inspired by their microwave [53] and millimeter-wave [54–56] implementations, terahertz reflectarrays essentially share some features of phased arrays and reflector antennas [27].

The earliest demonstration of terahertz reflectarrays was by [46] and showed a device that worked as an isotropic deflector. Figure 4.8 shows the intended beam deflecting operation of the metasurface. The terahertz wave is incident at 45° on the metasurface and is then deflected away from the specular reflection. A square patch was employed as the resonating element (meta-atom) with a gradual increase in size across the subarray as shown in Fig. 4.8a. This progressive increment in size is related to the reflection phase response. Thus, the width increment will encompass the near 2π phase coverage required to effectively deflect the incident beam.

For normal incidence, the incident angle θ_i is 0°. As such, Eq. 4.7 is simplified to

$$\frac{\Delta\phi}{a} = k_1 \sin\theta_r, \tag{4.9}$$

where the wavenumber, k_1, can be expressed as $\frac{2\pi}{\lambda_0}$, where λ_0 is the operational wavelength. Hence, the deflection angle of the outgoing wave can be calculated by using the following equation [46, 57]:

$$\theta_r = \arcsin\frac{\Delta\phi\lambda_0}{2\pi a}, \tag{4.10}$$

where a is the unit cell size and $\Delta\phi$ is the phase difference between two adjacent unit cells. The progressive phase shift, $\Delta\phi$, is fixed to 60°, which sets the number of meta-atoms in one linear subarray to six, as this covers one full 2π phase cycle

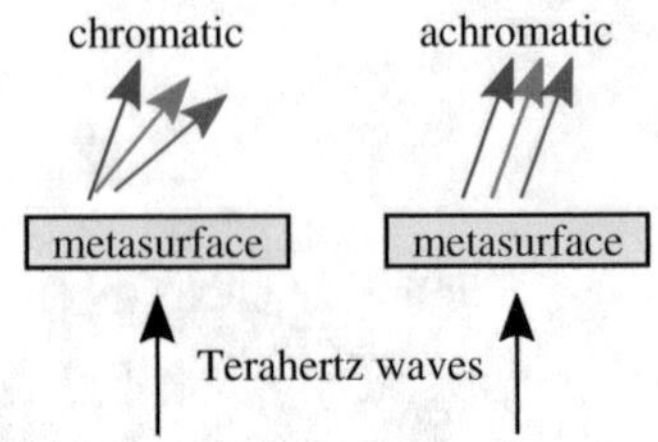

Fig. 4.9 The schematic of the terahertz chromatic and achromatic deflectors. The steering angles are constant in the whole target bandwidth for the achromatic beam deflector, while they are different for the chromatic counterpart. The different colored arrows represent the output beam at different frequencies

in a periodic manner. Hence, for $a = 140$ μm, $f_0 = 1$ THz, and $\Delta\phi = 60\,°$, the angle of reflection is calculated to be 20.5°.

One of the major challenges in metasurfaces with regard to phase control is their applicability in a achromatic scenario. The phase gradient utilized to deflect waves into predetermined directions is locked to a specific wavelength. In order to achieve achromatic deflection, the engineered metasurface has to impart both the phase gradient and dispersion simultaneously [58]. Based on Eq. 4.10, the progressive phase shift, which can be referred to as the phase gradient, should hold a linear relation between ϕ and the position of the unit cell (a and ω) across the metasurface, a. Hence, the required transmission phase can be defined as the following:

$$\frac{\Delta\phi}{\Delta\omega} = \frac{a \sin\theta_r}{c}, \tag{4.11}$$

where ω is the angular frequency and c is the speed of the light in vacuum. The phase gradient imparted by the metasurface should equal to $k \sin\theta_r$ based on Eq. 4.10. Additionally, for θ_r to remain the same across the frequency range, the unit cell needs a linear dispersion whose slope is given as $\frac{a \sin\theta_r}{c}$. Arrays of silicon pillars and gratings with carefully designed geometry and placement have been shown to be able to achieve simultaneous control of both the phase and dispersion at terahertz frequencies [59]. Two beam deflectors are combined as shown in Fig. 4.9 to obtain an achromatic beam deflector.

2.3 Polarization Control

A fundamental functionality of various optical components is polarization control that includes but is not limited to polarizers, polarizing beam splitters, and polarization converters. Conventional optical components that utilize polarization control operate on the strong anisotropy of naturally available materials at a given frequency. At the higher end of the terahertz frequency spectrum, polarization

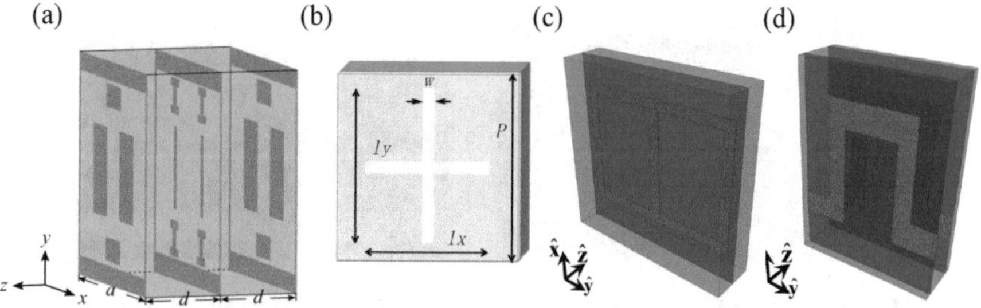

Fig. 4.10 Unit cell configurations of various quarter-wave plates. (**a**) Multilayer quarter-wave plate. Adopted from [67]. (**b**) Babinet-inverted resonator array consisting of cross-shaped slots. Adapted with permission from [75] ©The Optical Society. (**c**) Electric split-ring resonator variant and (**d**) meanderline quarter-wave plate. Adapted with permission from [76] ©The Optical Society

control devices are typically manufactured from crystalline materials [60–62], while at lower terahertz frequencies, materials such as wood [63] and paper [64] were used to provide polarization control. These aforementioned materials possess birefringence, where the material would portray a different refractive index that aligns with the propagation and polarization of an incoming electromagnetic wave. Alternatively, birefringence can be created with periodic structures in the form of reflectarrays [36, 46, 65, 66], multilayered materials [67–70], or gratings [71–74].

Among the key components in terahertz systems that would benefit from broadband, designable birefringences are the quarter- and half-wave plates, which operate in transmission mode. Quarter-wave plates introduce a $\frac{\pi}{2}$ phase difference between the two orthogonal electric field components of an incident wave, while the amplitude responses are equal. This allows conversion from linearly polarized waves to circularly polarized waves and vice versa. Several examples of planar quarter-wave plates made up of metallic resonators were demonstrated at terahertz frequencies and are shown in Fig. 4.10.

A variant of wave plates that operate in reflection instead of transmission can be referred to as birefringent mirrors. A potential solution to overcome low radiation efficiencies of metallic resonators at higher frequencies is to incorporate dielectric resonators in metasurfaces. It is also noteworthy that dielectric resonators exhibit a smoother phase variation as compared to their metallic counterparts. The phase gradient as a function of the frequency is dependent on the radiation quality factor of the resonators, Q_{rad}, where a higher value would result in a larger phase gradient. The lower Q_{rad} and thus smoother phase gradient of the dielectric resonators are due to their higher radiation loss or better coupling with free-space waves [77]. This feature benefits the polarization conversion purity of resulting wave plates and mirrors. The polarization conversion purity refers to the conversion efficiency between the initial polarization and the conversion. Several examples of quarter-wave plates made up of dielectric materials at terahertz frequencies are shown in Fig. 4.11.

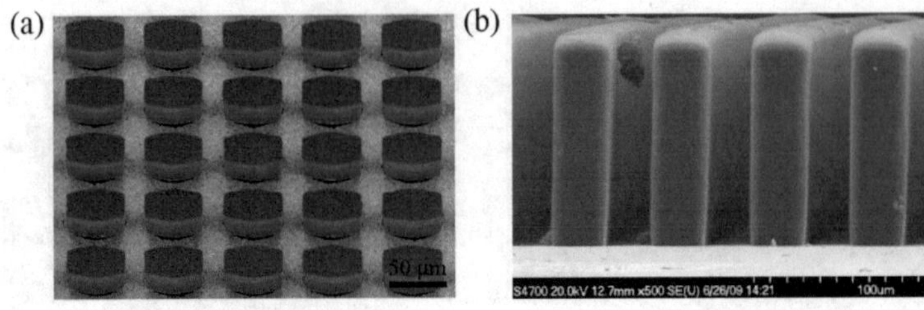

Fig. 4.11 (**a**) False colored scanning electron micrograph of the quarter-wave mirror consisting of silicon resonators. Adapted with permission from [25] ©The Optical Society. (**b**) A vertical grating quarter-wave plate with two high-density polyethylene (HDPE) plates alternating with air. Adapted with permission from [78] ©The Optical Society

Unlike quarter-wave plates, half-wave plates introduce a π phase difference between the two orthogonal field components. As a consequence, they can rotate 45° incident linearly polarized waves by 90°. For example, in order to determine the conversion efficiency of a half-wave mirror, we then proceed to calculate the polarization conversion ratio (PCR) [79] that is defined as

$$\mathrm{PCR} = \frac{|E_{\mathrm{cr}}|^2}{|E_{\mathrm{cr}}|^2 + |E_{\mathrm{co}}|^2}, \tag{4.12}$$

where E_{cr} represents cross-polarized amplitude and E_{co} represents co-polarized amplitude.

At terahertz frequencies, half-wave plates made up of metallic resonators have also been demonstrated [72, 80–83]. Similar to quarter-wave plates, the performance of half-wave plates can be increased by multilayered structures. As demonstrated by Cong et al. [82], a tri-layer metasurface is capable of enhancing bandwidth and polarization conversion efficiency. The operational mechanism of the structure is shown in Fig. 4.12a. Each layer essentially contains a Fabry–Perot cavity that improves the efficiency of the metasurface. The wire grids at the top and bottom of each layer allow for the cross-polarization wave to pass through the structure and suppress the backpropagating co-polarized waves. Hence, only the cross-polarized wave is transmitted through this multilayer metasurface. Aside from using metallic resonators, a dielectric variant operating in reflection mode has been demonstrated utilizing silicon resonators as shown in Fig 4.12b. The large difference in length and width in this design is essential to obtain the required phase response from the orthogonal electric field components. This allows for the required π phase difference between the two orthogonal field components.

A polarization beam splitter is a device that can split an arbitrarily polarized terahertz wave into its two orthogonal polarization components. Polarization beam splitters are particularly useful in terahertz communications as incoming waves can be manipulated independently, which allows for multiplexing of signals in one communication channel [84]. Aside from that, polarizing beam splitters can benefit

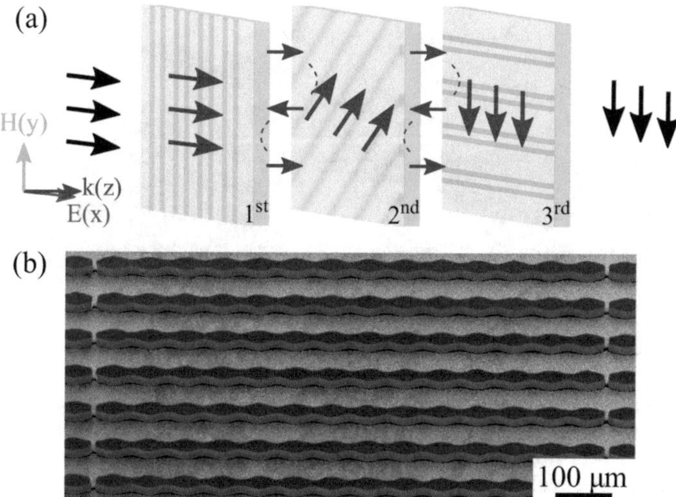

Fig. 4.12 (**a**) Schematic of a multilayer metasurface half-wave plate. The electric field distribution is shown by black arrows. The Fabry–Perot interference between the layers enhances transmission, as indicated by blue arrows [82]. (**b**) False colored scanning electron micrograph of the half-wave mirror consisting of silicon resonators. Adapted with permission from [25] ©The Optical Society

applications in imaging [85] and polarimetric devices [86]. In the terahertz region, wire-grid structures [74, 87–89], metamaterials [90], dielectric bi-layers [91], and stacked metal plates [86] have been shown as functional polarizing beam splitters. Wire-grid polarizers are limited to linear polarization beam splitting and cannot perform circular-polarization beam splitting or beam forming. Thus, it is essential to search for a promising route for structures capable of exotic polarization beam splitting functions.

A wire-grid polarizer consisting of two layers has been demonstrated by [92]. A subarray of this design is shown in Fig. 4.13a. The top layer consists of smaller patterned striplines that function as the resonating elements. Owing to the design and geometry of the resonating elements, this polarization-dependent metasurface has dual functionality. First, it passes TM-polarized waves through the structure. Second, it can operate as a beamforming reflectarray for TE-polarized waves. This configuration is more commonly known as a combination of a reflectarray and a wire-grid polarizer. Splitting circularly polarized waves is desirable for high-data-rate wireless communications and study of molecular chirality at terahertz frequencies. Typically, this functionality is achieved using bulk optical systems with limitations in material availability, bandwidth, and efficiency. As an alternative, we employ metasurfaces with spatially varying broadband birefringence to attain the same functionality. This is in contrast to metasurfaces where the phase discontinuity is introduced by tailoring the meta-atom's geometry a different technique, namely the Pancharatnam–Berry phase [93, 94]. The Pancharatnam–Berry (PB) phase can achieve full-phase coverage of cross-polarized electromagnetic waves by rotating

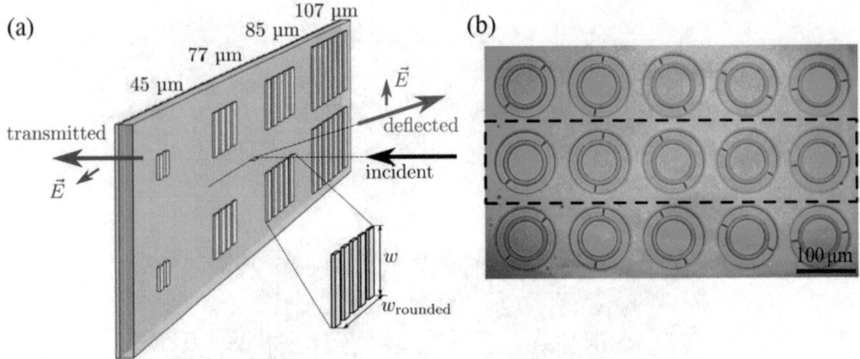

Fig. 4.13 (**a**) Subarray of a metasurface that consists of subwavelength metallic striplines as the ground plane and the resonating elements. This structure passes TM polarization but reflects TE polarization. Adopted from [57, 92]. (**b**) An optical micrograph of a portion of the fabricated broadband terahertz circular-polarization beam splitter is shown. The dashed rectangle encloses a single subarray consisting of 5 resonating elements rotated by 36° cumulatively. Adopted from [38]

meta-atoms of the same geometry. Mathematically, the relation between the cell rotation and the resultant PB phase for circular polarization can be explained in the formalism of the Jones matrices [95].

The meta-atoms can be loosely considered as a local half-wave mirror, which can be represented by a Jones matrix, $\mathbf{M}(\phi)$, where the fast axis of the unit cell is at ϕ angle with respect to the x-axis. The incident left-handed circularly polarized (LHCP) and right-handed circularly polarized (RHCP) Jones vectors are given by $\mathbf{A}_{\pm}$, where the $+$ and $-$ signs indicate their respective handedness. Thus, the resultant Jones vector can be determined by the following equation:

$$\begin{aligned}\mathbf{J}(\phi) &= \mathbf{M}\mathbf{A}_{\pm},\\ &= \begin{bmatrix} -\cos(2\phi) & -\sin(2\phi) \\ -\sin(2\phi) & \cos(2\phi) \end{bmatrix} \frac{1}{\sqrt{2}} \begin{bmatrix} 1 \\ \pm i \end{bmatrix}, \\ &= \frac{1}{\sqrt{2}} \begin{bmatrix} -1 \\ \pm i \end{bmatrix} \exp(\pm i 2\phi).\end{aligned} \tag{4.13}$$

It is observed from the resultant Jones vector $\mathbf{J}$ that the incident wave retains its polarization handedness upon reflection, given a change in the propagation direction. This is accompanied by a $|2\phi|$ phase discontinuity that is imposed to the reflected wave, which is the PB phase. The sign for this phase discontinuity is dependent on the handedness of the incident polarization as shown by $\mathbf{A}_{\pm}$. For example, the metasurface shown in Fig. 4.13b is designed with gradually rotated birefringent metallic resonators and can deflect normally incident left-handed circularly polarized (LHCP) and right-handed circularly polarized (RHCP) waves into different directions [38].

3 Terahertz Metawaveguides

Terahertz frequency range has proven crucial in a wide range of applications such as non-invasive imaging and sensing of biological and chemical samples, and in broadband wireless communication. The lower part of the spectrum (0.1–0.5 THz) has gained particular interest for beyond 5G high-speed communication, for instance, enabling seamless interconnection (terabit per second) between high-speed wired networks and wireless devices. This breath of application is driving research for integrated terahertz technologies and in particular terahertz waveguides. Waveguides are one of the fundamental building blocks of compact electromagnetic devices. Despite the maturity of waveguides in microwave and optics, there is a critical knowledge gap for terahertz spectrum, which hinders development of terahertz integrated systems. Several waveguide solutions based on technologies from both electronics and photonics have been tested for guiding terahertz radiation [96]. In general, metallic waveguides suffer from high Ohmic losses, while dielectric waveguides suffer from absorption losses in terahertz band.

Discovery of metamaterials and topological insulators in condensed matter has opened the opportunity to enhance the performance of terahertz waveguides in terms of loss, dispersion, flexibility, and single-mode operating bandwidth. These waveguides have been dubbed as metawaveguides as their unique properties are achieved by special engineering of waveguide material. Here we will particularly discuss the two categories of terahertz metawaveguides: metamaterial and topologically protected waveguides.

3.1 Metamaterial Waveguides

Although the first reported metamaterial in the literature was anisotropic, the initial theoretical waveguiding studies were focused on waveguides with isotropic metamaterials [97, 98], and subsequently, anisotropy was added into consideration [99–101]. Overall, a waveguide structure has two sections: core, where the electromagnetic energy mainly travels through, and cladding, the supporting structure around the core. Metamaterial waveguides reported for terahertz are mainly waveguides with metamaterial cladding [102, 103]. This can be attributed to a lack of transparent materials in the terahertz spectrum. Despite of that, terahertz waveguides with metamaterial core have been employed for near-field imaging [104, 105]. An elegant example is the wire metamaterial sub-diffraction-limited endoscope [104], where the metamaterial enables propagation of subwavelength features from the object plane to image plane. Here, waveguides with metamaterial cladding are considered, which are divided to subsections based on the core material (air core and silicon core). It is worth noting that in the metamaterials considered here the periodicity between meta-elements/layers is smaller than operating wavelength.

3.1.1 Air-Core Meta-Clad Waveguides

As stated earlier, air-core waveguides are of particular interest in guiding terahertz waves due to the very low material absorption of air. The air-core waveguides in general have dimensions larger than operating wavelength, which makes them large and rigid at terahertz band. These waveguides have very narrow single-mode operating (metallic waveguides) or are multimode (air-core fibers). Utilizing metamaterials in the cladding of air-core terahertz waveguides has enabled addressing these limitations of conventional air-core waveguides in terahertz as elaborated next. Additionally, the unusual properties of anisotropic metamaterial-clad waveguides have enabled characteristics not observed in convectional air-core waveguides such as subwavelength confinement and guidance due to magnetic and electric resonances.

The exquisite example of air-core waveguide is the hollow-core fibers with wire metamaterial cladding, Fig. 4.14a [103]. It is demonstrated that even using a single layer of metal wires in the cladding is sufficient to confine the guided mode in the air core. The large bandwidth of terahertz pulse has resulted in observation of two different types of guided modes in this meta-clad fiber, Fig. 4.14b: the transverse magnetic (TM) modes, Fig. 4.14d, which are confined in the air core as they cannot couple to the TM modes in wire medium meta-cladding, and surface mode like surface plasmon polaritons (SPP), Fig. 4.14c, which propagates at the core cladding interface. This meta-clad waveguide offers wider single-mode bandwidth (2.3 times) compared to dielectric-coated metallic waveguides. Compactness and flexibility are the other advantages of these waveguides, which are achieved due to reduction of core diameter.

The wire array metamaterial has also been exploited to improve the performance of air-core antiresonance terahertz waveguides [102, 106, 107]. In the antiresonant waveguides, the guided modes are confined in the air core due to the resonance effect of the thin layers of cladding and can couple out only when constructive interference occurs [108]. It has been demonstrated numerically that utilizing wire array metamaterials as inner capillary tubes (Fig. 4.15a) or outer cladding (Fig. 4.15b) can lower the losses and broaden the bandwidth compared to the all dielectric counterpart waveguides.

The electric and magnetic resonances of metamaterials can also be harnessed for guiding terahertz waves. An example is an air-core planar waveguides with metamaterial claddings consisting of arrays of split-ring resonators (Fig. 4.16a) [109]. The guiding mechanism is based on total internal reflection due to the electric/magnetic response. The type of response depends on the relative direction of prorogating wave with respect to the cladding resonators. Another example is an air-core waveguide, where the upper and bottom plates are cascaded resonators embedded in

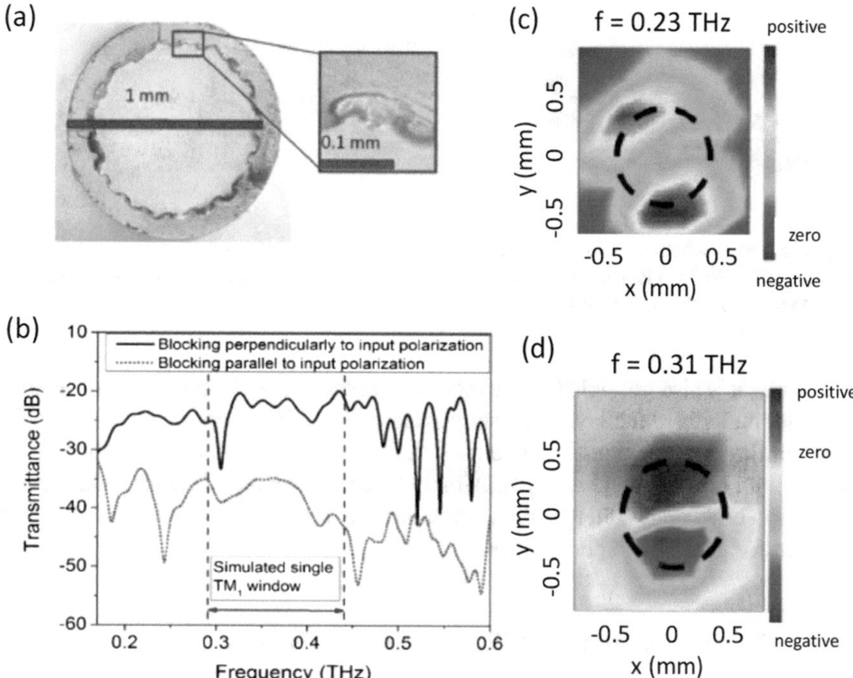

Fig. 4.14 (**a**) Microscope image of the cross-section of air-core metamaterial-clad waveguide. The inset image shows one of the wires. (**b**) Normalized transmissions of TM (black solid curve) and TE (red dotted curve) modes. Measured normalized near-field modal profiles of the waveguide at (**c**) 0.23 THz and (**d**) 0.31 THz. Reprinted with the permission from [103] ©The Optical Society

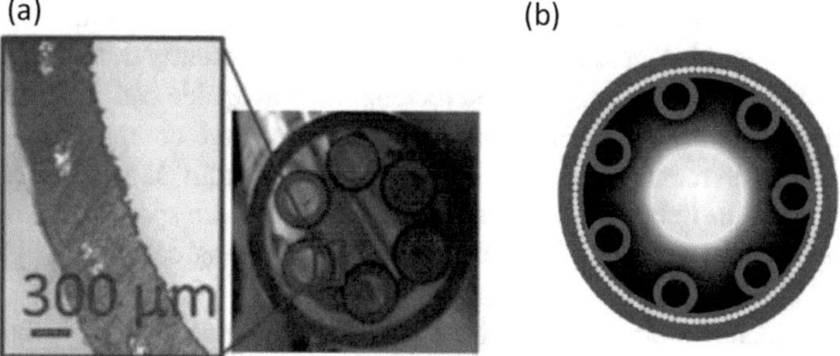

Fig. 4.15 Air-core antiresonance THz waveguides: (**a**) inner capillary tubes contain metal wires (fabricated) and (**b**) outer cladding tube contains metal wires. Reprinted by permission from [106] and [102], respectively

a thin film (Fig. 4.16b) [110]. Apart from guiding terahertz waves, it has also been demonstrated that embedding resonators in the side walls of air-core waveguides, e.g., rectangular metallic waveguides [111], creates functional components.

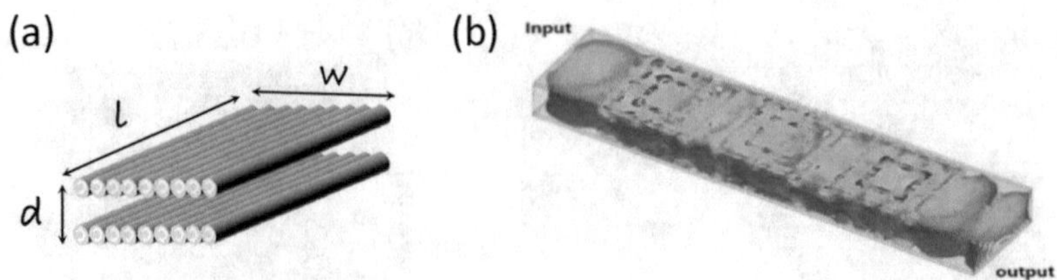

Fig. 4.16 (**a**) A schematic of the air-core waveguide with split-ring resonator arrays as cladding used in [109]. (**b**) The electric field distribution of an air-core waveguide with cascaded resonators as meta-cladding. Reprinted from [110]

Finally, it is also possible to have air-core waveguides with multilayer metamaterial cladding, where the layers are subwavelength relative to operating wavelength. The multilayer metamaterial can be used as the cladding of a slab waveguide [101] or rolled up into a tube [112, 113]. Although such waveguide structures have been demonstrated in optics, there is no waveguide with multilayer metamaterial cladding reported for guiding terahertz waves.

3.1.2 Silicon-Core Meta-Clad Waveguides and Components

One salient feature for the terahertz spectrum is fractional bandwidth (ratio of operation bandwidth to center frequency) that can be enormous compared with that available in other frequency ranges. Indeed, practical applications cannot use the entire band in continuity due to the presence of water vapor absorption bands. However, a number of wide absorption windows exist where extended propagation ranges are supported [114]. Another factor that bounds the bandwidth usage is the banding of rectangular metallic waveguides [115] that currently dominate integrated systems. Nevertheless, a single-moded rectangular waveguide can cover a fractional bandwidth of over 40%, which remains considerably large compared to an optical band of a few percent. Such a vast available bandwidth subsequently informs application designs to benefit the most; communications can span a large bandwidth to increase channel capacity [116], while stand-off sensing can enjoy higher range resolutions [117].

All these demonstrated applications have been demonstrated via bulky free-space optics and blocky metallic waveguides. Toward integrated systems, an unprecedented usage in bandwidth imposes constraints in component designs for broadband operation with low loss and low dispersion. This creates a unique challenge in the most fundamental component—waveguides—that prevails in compact integrated systems. It is possible to either physically upscale a photonic waveguide or downscale microwave transmission lines to a terahertz band. However, these up- or down-scaled interconnects present significant drawbacks. As stated earlier, downscaling metallic transmission lines to a terahertz band exacerbates attenuation through increased Ohmic loss and reduced skin depth. Lending dielectric waveg-

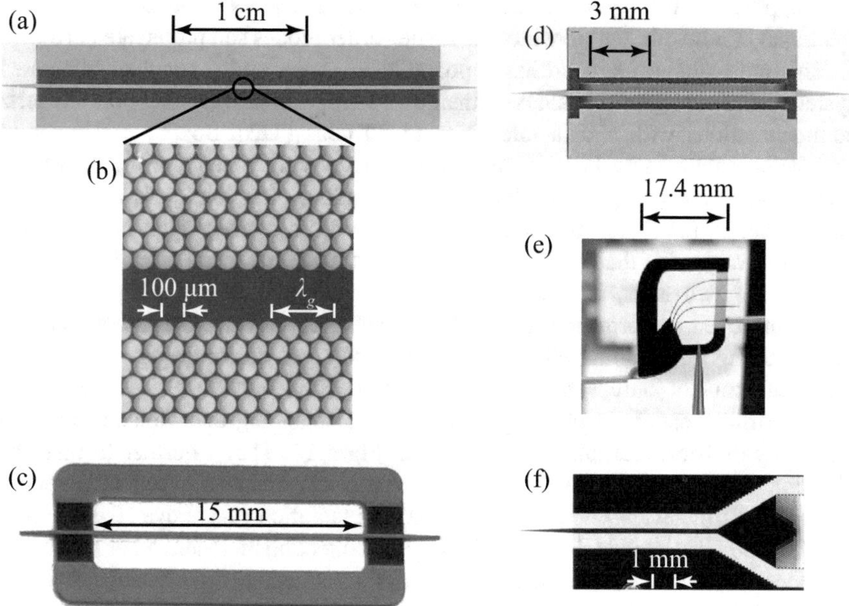

Fig. 4.17 Various components based on the substrateless silicon platform. (**a**) Straight waveguide and (**b**) zoom-in around the core. Adopted from [118]. (**c**) Unclad dielectric waveguide with its ends attached to the frame by effective medium. Adopted from [119]. (**d**) Bragg grating filter [120]. (**e**) Frequency-division multiplexers. Adopted from [121]. (**f**) Lens-integrated horn antenna. Adopted from [122]. All these components operate around 300 GHz and are made of intrinsic float-zone silicon with no other materials. The tapered ends are for coupling to rectangular waveguides for characterization purposes

uides from optics can support this broad terahertz bandwidth, but material choices are of concern. While silicon itself can support guided modes with low loss, an oxide layer or a polymer as a substrate would dissipate significant energy along the path.

To solve this problem, the substrate can be removed completely from dielectric waveguides that are in sub-millimeter scale, while integrability is maintained for system-level construction. This is possible through the concept of effective medium that allows tailoring the material's refractive index with flexibility. As shown in Fig. 4.17a,b, a solid waveguide core is tethered to a solid dielectric plate through an effective medium. The entire structure, including the waveguide, effective medium, and plate, can be made of a single slab of float-zone intrinsic silicon that has an exceptionally low loss with $\tan\delta \approx 0.00002$ and moderate refractive index $n = 3.418$ at 1 THz. A key is air perforation in this silicon slab with a pitch much smaller than a wavelength so that guided waves see this perforated part homogenous with an index average between air and silicon. Effectively, an index contrast between a solid waveguide core and perforated claddings leads to guided modes as a result of total internal reflection. By selecting appropriate dimensions, this waveguide can support two orthogonal modes with low loss and low dispersion across 40% fractional bandwidth, within which no higher-order modes can propagate [118].

This pure-silicon platform has a potential to expand into complete integrated systems. The straight waveguides on their own have been shown to support terahertz communications with a data rate close to 30 Gb/s [123]. So far, a number of peripheral components have been demonstrated, with no supporting substrate. These components range from basic ones including directional couplers, bends, and crossings [118]. A variant of these substrateless waveguides is an unclad dielectric waveguide that is attached to a supporting frame by an effective medium only at the two ends, as shown in Fig. 4.17c [119]. Bragg grating filters can enjoy improved performance in broadband owing to an extra degree of freedom in controlling the cladding effective refractive index as shown in Fig. 4.17d [120]. The platform not only supports a 1D guided mode but 2D slab mode [124]. This flexibility entails in-plane spectral-to-spatial mapping that yields a practical frequency-division multiplexer, as shown in Fig. 4.17e [121]. Further to that, the platform can interface directly with free space via different types of antennas [122, 125, 126]. An all-silicon horn antenna with radiation gain above 10 dBi across the 220–330 GHz band is shown in Fig. 4.17f. Connecting these and other peripheral components together on this monolithic platform will lead to operational systems to support various terahertz functions in broadband.

3.2 Topologically Protected Waveguides

Topological photonics is a rapidly emerging new field, which holds the promise of providing electromagnetic guidance on the edges of photonic lattices with no backscattering due to fabrication imperfections or other defects [127–129]. This field stems from the earlier discovery of novel phases of matter in condensed matter physics, in particular from the discovery of the quantum Hall effect [130–132] and topological insulators [133]. In photonics, the underlying idea is building structures, which guide light based on global properties of their dispersion bands in particular in the reciprocal k-vector space. These global properties, known as topological invariants, cannot be discerned locally and are therefore immune to local disorder such as sharp bends. The reflection-less propagation is achieved either by breaking time-reversal symmetry (with an external magnetic field or time modulation) [127, 128, 134] or by breaking the parity symmetry (does not require an external magnetic field) [129, 135, 136] similar to the terahertz photonic crystal structures shown in Fig. 4.18. These topological structures promise compact photonic waveguide devices that are not restricted to wavelength-scale smoothness [137].

Most of the topological photonic developments (delay lines, topological lasers, waveguides, and quantum circuits) have been demonstrated at microwave [128, 138, 139] and optical frequencies [127, 135, 137]. So far to the best of our knowledge, there has been two examples of topologically protected (TP) terahertz metawaveg-

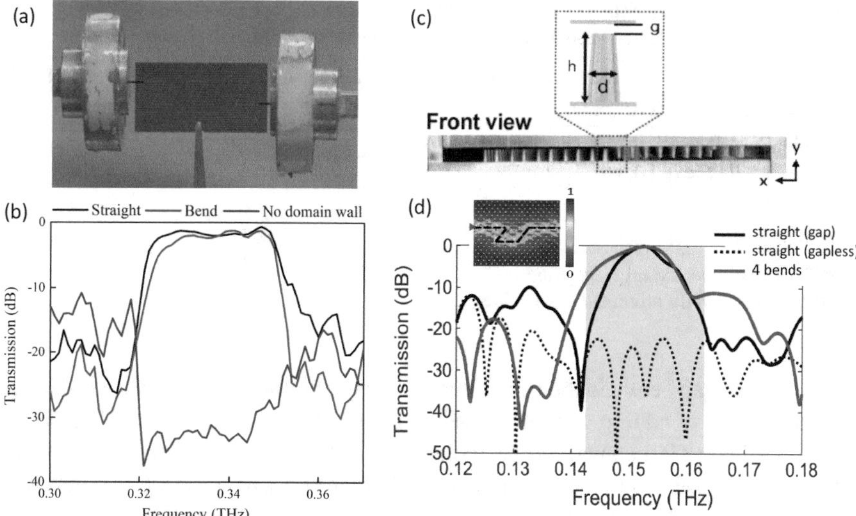

Fig. 4.18 Topologically protected terahertz waveguides: (**a**) An optical image of the fabricated twisted waveguide in the system. (**b**) Measured transmission curves for a straight path with and without domain wall and a twisted path with ten corners. Figures courtesy of Ranjan Singh [140]. (**c**) Schematic of the front view of metallic photonic crystal waveguide. The insert defines the geometric parameters of the waveguide. (**d**) Measured normalized field amplitude curves for similar length waveguides including a straight path with and without gaps and a twisted path with four corners as shown in the insert. The shaded region represents the bandgap. Copyright 2020 Wiley. Used with permission from [141]

uides demonstrated for terahertz guidance: all-silicon TP metawaveguide [140] and all-metallic TP metawaveguide [141]. In these metawaveguides, the time-reversal symmetry is achieved by breaking the parity symmetry of the crystal lattice, which emulates valley Hall photonic topological insulators (also known as valley Hall photonics crystals) [142] and spin-Hall photonic topological insulators (also known as spin-Hall photonics crystals) [143]. It is worth noting that the design procedure is the same for any operational frequency band, and it only requires scaling of the unit cell.

The first step is to design a photonic crystal that mimics the well-known graphene-like lattice. This means a photonic crystal that exhibits a pair of degenerate Dirac points at the K and K' symmetry point in the band diagram. This can be achieved by designing a silicon unit cell comprising a pair of inverted equilateral triangular holes when the side lengths of the triangles are identical [144] or by designing a hexagonal array of metallic cylinders sandwiched between two metallic parallel plates [143].

The next step is to break the symmetry of the lattice. This opens a bandgap at the vicinity of Dirac frequency (around the K symmetry point in the band diagram), which is a topologically protected bandgap. This is achieved by changing the length of triangles in each unit cell in all-silicon TP metawaveguide [140, 144] and by intro-

Table 4.1 Key characteristics of the Si-, metallic-, and hybrid-photonic crystal waveguides and metawaveguides for THz communication reported in the literature

Waveguide type	Central frequency (THz)	Loss (dB/mm)	Relative bandwidth % ($\Delta f / f_c$)
Si PCWG[a] [145]	0.328	<0.02	4
Metallic PCWG [146]	0.850	<0.25	58
Hybrid PCWG [147]	0.380	<0.1	18.5
Si TPWG[b] [140]	0.335	0.05	7.8
Metallic TPWG [141]	0.15	0.2	12.5

[a] PCWG: photonic crystal waveguide
[b] TPW: topologically protected waveguide

ducing an air gap between the pillars and the top or bottom metallic plate (which is also equivalent of adding washers) in all-metallic TP metawaveguide [141, 143].

In all-silicon TP metawaveguide, there are two topologically different valley Hall phases directly related to the sign of length difference ($\Delta\, l = l_1 - l_2$). If two opposite phases (opposite $\Delta\, l$ values) of valley Hall photonics crystals are assembled next to each other, the constructed domain wall will support two surface modes (known as "kink" states) within the bandgap: one in the K valley propagating forward, while the other one, in the K' valley propagating backward [144]. On the other hand, in all-metallic TP metawaveguide, the two topologically surface waves (one propagating forward and the other backward) will be supported when two spin-Hall photonic crystals with opposite spins (gap at top and bottom) are assembled next to each other [143].

The TP metawaveguides are promising for terahertz communication due to their unique properties. They are robust at present of defects and sharp bends, and they support single and linearly disperse guided mode indicating no mode coupling due to perturbation and negligible signal delay around the center of the bandgap [140, 141]. Similar to other terahertz waveguides, there is a trade-off between loss and operating bandwidth. It has been demonstrated that losses as low as 0.05 dB/mm with relatively narrow bandwidth (~7.8%) is achieved in all-silicon TP metawaveguide. This is while, a relatively larger bandwidth of ~12.5% with 0.2 dB/mm loss is achieved in all-metallic TP metawaveguide. It is expected that utilizing hybrid TP metawaveguides may lead to a middle ground with less loss compared to all-metallic counterparts while offering a larger bandwidth compared to all-silicon counterparts. This expectation is not far from reach as it has been demonstrated that using hybrid material in convectional photonic crystal waveguides leads to higher bandwidth with reduced lower losses compared to single material photonic crystal waveguides (Table 4.1).

4 Conclusion and Outlook

In this chapter, first the terahertz metasurfaces and the related achievable functionalities including amplitude, phase, and polarization control are reviewed. These metasurfaces overcome the limitations set by conventional bulk optics possessing large thickness, narrowband operation, and low efficiencies. Additionally, further research into reconfigurability, efficiency and bandwidth of metasurfaces would accelerate the integration of metasurfaces into various terahertz systems. A myriad of applications in terahertz sensing, imaging, and communications would benefit from the realization of such high-efficiency devices. These advancements in technology would then support development in areas such as public health, security, and defense.

Then terahertz metawaveguides and achievable improvements (loss, dispersion, flexibility, and single-mode operating bandwidth) due to utilization of metamaterials and topology were discussed. It is demonstrated that the unusual properties of anisotropic metamaterial-clad waveguides lead to characteristics not observed in convectional air-core waveguides such as subwavelength confinement, increasing single-mode operating bandwidth, and guidance due to magnetic and electric resonances. It is also demonstrated that effective-medium-clad waveguides and associated components can offer a large fractional bandwidth compared to their optical counterparts. Finally, it is shown that topologically protected metawaveguides are robust under the presence of defects and sharp bends and can support single and linearly disperse guided mode.

An ultimate goal for these meta-components is to deploy them in practical applications for effectively manipulating guided and free-space terahertz waves. To this end, a number of technical challenges unique to the terahertz band remain to be addressed. General requirements for free-space metasurfaces include high efficiency to preserve moderate terahertz power and broad bandwidth with fractional bandwidth beyond 40% to harness a large spectral resource. These devices should come with flexibility in controlling waves via different tuning mechanisms with high speed and high efficiency. Target applications include hyperspectral imaging and agile point-to-point communications. Metawaveguides for guided waves must overcome a similar set of challenges. Particularly, efficiency requirements become stricter with extended interaction length between the devices and waves. Wave confinement is a critical factor to suppress interference, crosstalk, and radiation loss. End applications for these waveguides include system integration, inter-chip connectivity, and inter-system fiber network. Active devices including terahertz emitters, detectors, and amplifiers can potentially be integrated onto these interconnecting waveguides and peripheral components that work collectively as a terahertz frontend. Transitions between these active devices and waveguides require attention to mode and impedance matching aiming for highest coupling efficiency. These research challenges could be engaged through either ingenious engineering solutions or radical physics approaches. In any case, we hope that significant outcomes will be impactful to society.

References

1. J.C. Wiltse, IEEE Trans. Microwave Theory Tech. **32**(9), 1118 (1984)
2. F. Vatansever, M.R. Hamblin, Photon. Lasers Med. **4**, 255 (2012). http://www.ncbi.nlm.nih.gov/pmc/articles/PMC3699878/

3. S. Matsuura, M. Tani, K. Sakai, Appl. Phys. Lett. **70**(5), 559 (1997)
4. A. Dreyhaupt, S. Winnerl, T. Dekorsy, M. Helm, Appl. Phys. Lett. **86**(12), 121114 (2005)
5. F.D. Brunner, O.P. Kwon, S.J. Kwon, M. Jazbinšek, A. Schneider, P. Günter, Optics express **16**(21), 16496 (2008)
6. X.C. Zhang, Y. Jin, X. Ma, Appl. Phys. Lett. **61**(23), 2764 (1992)
7. T.W. Crowe, W.L. Bishop, D.W. Porterfield, J.L. Hesler, R.M. Weikle, IEEE J. Solid State Circ. **40**(10), 2104 (2005)
8. D.R. Smith, W.J. Padilla, D. Vier, S.C. Nemat-Nasser, S. Schultz, Phys. Rev. Lett. **84**(18), 4184 (2000)
9. J. Valentine, S. Zhang, T. Zentgraf, E. Ulin-Avila, D.A. Genov, G. Bartal, X. Zhang, Nature **455**(7211), 376 (2008)
10. V.M. Shalaev, W. Cai, U.K. Chettiar, H.K. Yuan, A.K. Sarychev, V.P. Drachev, A.V. Kildishev, Optics Letters **30**(24), 3356 (2005)
11. R.A. Shelby, D.R. Smith, S. Schultz, Science **292**(5514), 77 (2001)
12. D. Schurig, J.J. Mock, B. Justice, S.A. Cummer, J.B. Pendry, A.F. Starr, D.R. Smith, Science **314**(5801), 977 (2006)
13. A. Alu, N. Engheta, J. Opt. A Pure Appl. Opt. **10**(9), 093002 (2008)
14. W. Cai, U.K. Chettiar, A.V. Kildishev, V.M. Shalaev, Nature Photonics **1**(4), 224 (2007)
15. I. Al-Naib, G. Sharma, M.M. Dignam, H. Hafez, A. Ibrahim, D.G. Cooke, T. Ozaki, R. Morandotti, Phys. Rev. B **88**(19), 195203 (2013)
16. M. Seo, H. Park, S. Koo, D. Park, J. Kang, O. Suwal, S. Choi, P. Planken, G. Park, N. Park, et al., Nature Photonics **3**(3), 152 (2009)
17. F. Baida, M. Boutria, R. Oussaid, D. Van Labeke, Phys. Rev. B **84**(3), 035107 (2011)
18. N.I. Zheludev, Y.S. Kivshar, Nature Materials **11**(11), 917 (2012)
19. J.B. Pendry, A.J. Holden, W.J. Stewart, I. Youngs, Phys. Rev. Lett. **76**, 4773 (1996)
20. C.R. Simovski, P.A. Belov, A.V. Atrashchenko, Y.S. Kivshar, Advanced Materials **24**(31), 4229 (2012)
21. J. Pendry, A. Holden, D. Robbins, W. Stewart, IEEE Trans. Microwave Theory Tech. **47**, 2075 (1999). https://doi.org/10.1109/22.798002
22. W. Withayachumnankul, D. Abbott, IEEE Photon. J. **1**(2), 99 (2009)
23. A. Bitzer, H. Merbold, A. Thoman, T. Feurer, H. Helm, M. Walther, Optics Express **17**(5), 3826 (2009)
24. L. Zou, W. Withayachumnankul, C.M. Shah, A. Mitchell, M. Klemm, M. Bhaskaran, S. Sriram, C. Fumeaux, IEEE Photon. J. **6**(4), 1 (2014)
25. W.S.L. Lee, R.T. Ako, M.X. Low, M. Bhaskaran, S. Sriram, C. Fumeaux, W. Withayachumnankul, Optics Express **26**(11), 14392 (2018)
26. S. Liu, M.B. Sinclair, T.S. Mahony, Y.C. Jun, S. Campione, J. Ginn, D.A. Bender, J.R. Wendt, J.F. Ihlefeld, P.G. Clem, J.B. Wright, I. Brener, Optica **1**(4), 250 (2014)
27. J. Huang, J.A. Encinar, *Reflectarray Antennas* (Wiley, 2008)
28. C.G. Ryan, M.R. Chaharmir, J. Shaker, J.R. Bray, Y.M. Antar, A. Ittipiboon, IEEE Trans. Antennas Propag. **58**(5), 1486 (2010)
29. I.E. Carranza, J.P. Grant, J. Gough, D. Cumming, IEEE J. Sel. Top. Quantum Electron. **23**(4), 1 (2016)
30. S.A. Kuznetsov, A.G. Paulish, A.V. Gelfand, P.A. Lazorskiy, V.N. Fedorinin, Prog. Electromagn. Res. **122**, 93 (2012)
31. L. Cong, S. Tan, R. Yahiaoui, F. Yan, W. Zhang, R. Singh, Appl. Phys. Lett. **106**(3), 031107 (2015)
32. Y.Z. Cheng, W. Withayachumnankul, A. Upadhyay, D. Headland, Y. Nie, R.Z. Gong, M. Bhaskaran, S. Sriram, D. Abbott, Adv. Opt. Mater. **3**(3), 376 (2015)
33. Y. Shang, Z. Shen, S. Xiao, IEEE Trans. Antennas Propag. **61**(12), 6022 (2013)
34. A. Ebrahimi, R.T. Ako, W.S. Lee, M. Bhaskaran, S. Sriram, W. Withayachumnankul, IEEE Trans. Terahertz Sci. Technol. **10**(2), 204 (2020)
35. W. Withayachumnankul, C.M. Shah, C. Fumeaux, B.S.Y. Ung, W.J. Padilla, M. Bhaskaran, D. Abbott, S. Sriram, Acs Photonics **1**(7), 625 (2014)

36. T. Niu, W. Withayachumnankul, A. Upadhyay, P. Gutruf, D. Abbott, M. Bhaskaran, S. Sriram, C. Fumeaux, Optics Express **22**(13), 16148 (2014)
37. Y.Z. Cheng, W. Withayachumnankul, A. Upadhyay, D. Headland, Y. Nie, R.Z. Gong, M. Bhaskaran, S. Sriram, D. Abbott, Appl. Phys. Lett. **105**(18), 181111 (2014)
38. W.S.L. Lee, S. Nirantar, D. Headland, M. Bhaskaran, S. Sriram, C. Fumeaux, W. Withayachumnankul, Adv. Opt. Mater. **6**(3), 1870010 (2018)
39. M. Wei, Q. Xu, Q. Wang, X. Zhang, Y. Li, J. Gu, Z. Tian, X. Zhang, J. Han, W. Zhang, Appl. Phys. Lett. **111**(7), 071101 (2017)
40. M. Khorasaninejad, W. Zhu, K.B. Crozier, Optica **2**(4), 376 (2015)
41. P.R. West, J.L. Stewart, A.V. Kildishev, V.M. Shalaev, V.V. Shkunov, F. Strohkendl, Y.A. Zakharenkov, R.K. Dodds, R. Byren, Optics Express **22**(21), 26212 (2014)
42. L. Liu, Y. Zarate, H.T. Hattori, D.N. Neshev, I.V. Shadrivov, D.A. Powell, Appl. Phys. Lett. **108**(3), 031106 (2016)
43. O. Paul, B. Reinhard, B. Krolla, R. Beigang, M. Rahm, Appl. Phys. Lett. **96**(24), 241110 (2010)
44. A. Arbabi, Y. Horie, A.J. Ball, M. Bagheri, A. Faraon, Nature Communications **6**, 7069 (2015)
45. L. Zou, W. Withayachumnankul, C.M. Shah, A. Mitchell, M. Bhaskaran, S. Sriram, C. Fumeaux, Optics Express **21**(1), 1344 (2013)
46. T. Niu, W. Withayachumnankul, B.S.Y. Ung, H. Menekse, M. Bhaskaran, S. Sriram, C. Fumeaux, Optics Express **21**(3), 2875 (2013)
47. I. Staude, A.E. Miroshnichenko, M. Decker, N.T. Fofang, S. Liu, E. Gonzales, J. Dominguez, T.S. Luk, D.N. Neshev, I. Brener, Y. Kivshar, ACS Nano **7**(9), 7824 (2013)
48. Z. Ma, S.M. Hanham, P. Albella, B. Ng, H.T. Lu, Y. Gong, S.A. Maier, M. Hong, ACS Photonics **3**(6), 1010 (2016)
49. K.E. Chong, I. Staude, A. James, J. Dominguez, S. Liu, S. Campione, G.S. Subramania, T.S. Luk, M. Decker, D.N. Neshev, I. Brener, Y.S. Kivshar, Nano Letters **15**(8), 5369 (2015)
50. Y. Yuan, X. Ding, K. Zhang, Q. Wu, IEEE Trans. Magn. **53**(6), 1 (2017)
51. F. Yue, D. Wen, J. Xin, B.D. Gerardot, J. Li, X. Chen, ACS Photonics **3**(9), 1558 (2016)
52. I. Al-Naib, W. Withayachumnankul, J. Infrared Millimeter Terahertz Waves **38**(9), 1067 (2017)
53. O. Bayraktar, O.A. Civi, T. Akin, IEEE Trans. Antennas Propag. **60**(2), 854 (2012)
54. P. Nayeri, M. Liang, R.A. Sabory-Garci´a, M. Tuo, F. Yang, M. Gehm, H. Xin, A.Z. Elsherbeni, IEEE Trans. Antennas Propag. **62**(4), 2000 (2014)
55. W. Hu, R. Cahill, J.A. Encinar, R. Dickie, H. Gamble, V. Fusco, N. Grant, IEEE Trans. Antennas Propag. **56**(10), 3112 (2008)
56. D.M. Pozar, S.D. Targonski, H.D. Syrigos, IEEE Trans. Antennas Propag. **45**(2), 287 (1997)
57. D. Headland, T. Niu, E. Carrasco, D. Abbott, S. Sriram, M. Bhaskaran, C. Fumeaux, W. Withayachumnankul, IEEE J. Sel. Top. Quantum Electron. **23**(4), 1 (2016)
58. S. Shrestha, A.C. Overvig, M. Lu, A. Stein, N. Yu, Light Sci. Appl. **7**(1), 1 (2018)
59. R. Jia, Y. Gao, Q. Xu, X. Feng, Q. Wang, J. Gu, Z. Tian, C. Ouyang, J. Han, W. Zhang, Adv. Opt. Mater. **9**(2), 2001403 (2021)
60. A. Saha, K. Bhattacharya, A.K. Chakraborty, Applied Optics **51**(12), 1976 (2012)
61. J.B. Masson, G. Gallot, Optics Letters **31**(2), 265 (2006)
62. C.F. Hsieh, R.P. Pan, T.T. Tang, H.L. Chen, C.L. Pan, Optics Letters **31**(8), 1112 (2006)
63. M. Reid, R. Fedosejevs, Applied Optics **45**(12), 2766 (2006)
64. B. Scherger, M. Scheller, N. Vieweg, S.T. Cundiff, M. Koch, Optics Express **19**(25), 24884 (2011)
65. X. You, R.T. Ako, W.S. Lee, M.X. Low, M. Bhaskaran, S. Sriram, C. Fumeaux, W. Withayachumnankul, Adv. Opt. Mater. **7**(20), 1900791 (2019)
66. R.T. Ako, W.S. Lee, M. Bhaskaran, S. Sriram, W. Withayachumnankul, APL Photonics **4**(9), 096104 (2019)
67. X. You, R.T. Ako, W.S. Lee, M. Bhaskaran, S. Sriram, C. Fumeaux, W. Withayachumnankul, APL Photonics **5**(9), 096108 (2020). Licensed under a Creative Commons Attribution (CC BY) license

68. R.T. Ako, W.S. Lee, S. Atakaramians, M. Bhaskaran, S. Sriram, W. Withayachumnankul, APL Photonics **5**(4), 046101 (2020)
69. W. Yu, A. Mizutani, H. Kikuta, T. Konishi, Applied Optics **45**(12), 2601 (2006)
70. T. Suzuki, M. Nagai, Y. Kishi, Optics Letters **41**(2), 325 (2016)
71. A. Lopez, H.G. Craighead, Optics Letters **23**(20), 1627 (1998)
72. N.K. Grady, J.E. Heyes, D.R. Chowdhury, Y. Zeng, M.T. Reiten, A.K. Azad, A.J. Taylor, D.A. Dalvit, H.T. Chen, Science **340**(6138), 1304 (2013)
73. G.P. Nordin, P.C. Deguzman, Optics Express **5**(8), 163 (1999)
74. R.C. Tyan, P.C. Sun, A. Scherer, Y. Fainman, Optics Letters **21**(10), 761 (1996)
75. D. Wang, Y. Gu, Y. Gong, C.W. Qiu, M. Hong, Optics Express **23**(9), 11114 (2015)
76. A.C. Strikwerda, K. Fan, H. Tao, D.V. Pilon, X. Zhang, R.D. Averitt, Optics Express **17**(1), 136 (2009)
77. C. Qu, S. Ma, J. Hao, M. Qiu, X. Li, S. Xiao, Z. Miao, N. Dai, Q. He, S. Sun, L. Zhou, Phys. Rev. Lett. **115**, 235503 (2015)
78. S.C. Saha, Y. Ma, J.P. Grant, A. Khalid, D.R. Cumming, Optics Express **18**(12), 12168 (2010)
79. J. Hao, Y. Yuan, L. Ran, T. Jiang, J.A. Kong, C.T. Chan, L. Zhou, Phys. Rev. Lett. **99**, 063908 (2007). https://doi.org/10.1103/PhysRevLett.99.063908. https://link.aps.org/doi/10.1103/PhysRevLett.99.063908
80. Y. Nakata, Y. Taira, T. Nakanishi, F. Miyamaru, Optics Express **25**(3), 2107 (2017)
81. W. Mo, X. Wei, K. Wang, Y. Li, J. Liu, Optics Express **24**(12), 13621 (2016)
82. L. Cong, N. Xu, J. Gu, R. Singh, J. Han, W. Zhang, Laser Photon. Rev. **8**(4), 626 (2014)
83. L. Cong, W. Cao, X. Zhang, Z. Tian, J. Gu, R. Singh, J. Han, W. Zhang, Appl. Phys. Lett. **103**(17), 171107 (2013)
84. Q.W. Song, M.C. Lee, P.J. Talbot, Applied Optics **31**(29), 6240 (1992)
85. L.B. Wolff, J. Opt. Soc. Am. A **11**(11), 2935 (1994)
86. R. Mendis, M. Nagai, W. Zhang, D.M. Mittleman, Scientific Reports **7**(1), 5909 (2017)
87. I. Yamada, K. Takano, M. Hangyo, M. Saito, W. Watanabe, Optics Letters **34**(3), 274 (2009)
88. C.W. Berry, M. Jarrahi, J. Infrared Millimeter Terahertz Waves **33**(2), 127 (2012)
89. A.G. Lopez, H.G. Craighead, Optics Letters **23**(20), 1627 (1998)
90. X.G. Peralta, E.I. Smirnova, A.K. Azad, H.T. Chen, A.J. Taylor, I. Brener, J.F. O'Hara, Optics Express **17**(2), 773 (2009)
91. J.S. Li, D. gang Xu, J. quan Yao, Applied Optics **49**(24), 4494 (2010)
92. T. Niu, A. Upadhyay, W. Withayachumnankul, D. Headland, D. Abbott, M. Bhaskaran, S. Sriram, C. Fumeaux, Appl. Phys. Lett. **107**(3), 031111 (2015)
93. M.V. Berry, Proc. R. Soc. Lond. A. Math. Phys. Sci. **392**(1802), 45 (1984)
94. S. Pancharatnam, in *Proceedings of the Indian Academy of Sciences-Section A*, vol. 44 (Springer, 1956), pp. 398–417
95. C. Menzel, C. Rockstuhl, F. Lederer, Phys. Rev. A **82**, 053811 (2010)
96. S. Atakaramians, S.A. V., T.M. Monro, D. Abbott, Adv. Opt. Photon. **5**(2), 169 (2013)
97. R. Ruppin, J. Phys. Condens. Matter **13**, 1811 (2001). https://doi.org/10.1088/0953-8984/13/9/304
98. R. Ruppin, J. Phys. Condens. Matter **16**, 599 (2004)
99. S. Atakaramians, A. Argyros, S.C. Fleming, B.T. Kuhlmey, J. Opt. Soc. Am. B **29**(9), 2462 (2012)
100. S. Atakaramians, A. Argyros, S.C. Fleming, B.T. Kuhlmey, J. Opt. Soc. Am. B **30**(4), 851 (2013)
101. S. Jahani, Z. Jacob, Optica **1**(2), 96 (2014). https://doi.org/10.1364/OPTICA.1.000096
102. J. Sultana, M.S. Islam, C.M.B. Cordeiro, A. Dinovitser, M. Kaushik, B. W.-H. Ng, D. Abbott, Fibers **8**(2), 14 (2020)
103. H. Li, S. Atakaramians, R. Lwin, X. Tang, Z. Yu, A. Argyros, B.T. Kuhlmey, Optica **3**(9), 941 (2016)
104. A. Tuniz, K.J. Kaltenecker, B.M. Fischer, M. Walther, S.C. Fleming, A. Argyros, B.T. Kuhlmey, Nature Communications **4**, (2013). https://doi.org/10.1038/ncomms3706

105. T.J. Huang, J. Zhao, L.Z. Yin, P.K. Liu, Optics Letters **46**(11), 2746 (2021). https://doi.org/10.1364/OL.421992. http://ol.osa.org/abstract.cfm?URI=ol-46-11-2746
106. S. Atakaramians, A. Stefani, H. Li, M.S. Habib, J.G. Hayashi, A. Tuniz, X. Tang, J. Anthony, R. Lwin, A. Argyros, et al., J. Infrared Millimeter Terahertz Waves **38**(9), 1162 (2017)
107. A. Stefani, R. Lwin, A. Argyros, in *2016 41ST International Conference on Infrared, Millimeter, and Terahertz Waves (IRMMW-THZ)* (2016)
108. N.M. Litchinitser, A.K. Abeeluck, C. Headley, B.J. Eggleton, Optics Letters **27**(18), 1592 (2002)
109. X. Tang, B.T. Kuhlmey, A. Stefani, A. Tuniz, S.C. Fleming, A. Argyros, J. Lightwave Technol. **34**(22), 5317 (2016)
110. Z. Wang, H. Jin, X. Sun, X. Li, Y. Xu, G. Liu, C. Gong, Optics Communications **474**, 126172 (2020)
111. Z.G. Wang, Y.Q. Zhou, L.M. Yang, C. Gong, J. Phys. D Appl. Phys. **50**(37), 375107 (2017)
112. S. Schwaiger, M. Bröll, A. Krohn, A. Stemmann, C. Heyn, Y. Stark, D. Stickler, D. Heitmann, S. Mendach, Phys. Rev. Lett. **102**, 163903 (2009)
113. E.J. Smith, Z. Liu, Y. Mei, O.G. Schmidt, Nano Letters **10**(1), 1 (2010)
114. T. Nagatsuma, G. Ducournau, C.C. Renaud, Nature Photonics **10**(6), 371 (2016)
115. IEEE Std 1785.1-2012, pp. 1–22 (2013)
116. T. Harter, C. Füllner, J.N. Kemal, S. Ummethala, J.L. Steinmann, M. Brosi, J.L. Hesler, E. Bründermann, A.S. Müller, W. Freude, et al., Nature Photonics **14**(10), 601 (2020)
117. H. Matsumoto, I. Watanabe, A. Kasamatsu, Y. Monnai, Nature Electronics **3**(2), 122 (2020)
118. W. Gao, W.S.L. Lee, X. Yu, M. Fujita, T. Nagatsuma, C. Fumeaux, W. Withayachumnankul, IEEE Trans. Terahertz Sci. Technol. **11**(1), 28 (2021)
119. D. Headland, W. Withayachumnankul, X. Yu, M. Fujita, T. Nagatsuma, J. Lightwave Tech. **38**(24), 6853 (2020)
120. W. Gao, W.S.L. Lee, C. Fumeaux, W. Withayachumnankul, APL Photonics **6**(7), 076105 (2021). Licensed under a Creative Commons Attribution (CC BY) license
121. D. Headland, W. Withayachumnankul, M. Fujita, T. Nagatsuma, Optica **8**(5), 621 (2021)
122. J. Liang, W. Gao, H. Lees, W. Withayachumnankul, IEEE Antennas Wireless Propag. Lett., 1–1 (2021)
123. W. Gao, X. Yu, M. Fujita, T. Nagatsuma, C. Fumeaux, W. Withayachumnankul, Optics Express **27**(26), 38721 (2019)
124. D. Headland, M. Fujita, T. Nagatsuma, Optics Express **28**(2), 2366 (2020)
125. D. Headland, W. Withayachumnankul, R. Yamada, M. Fujita, T. Nagatsuma, APL Photonics **3**(12), 126105 (2018)
126. W. Withayachumnankul, R. Yamada, M. Fujita, T. Nagatsuma, APL Photonics **3**(5), 051707 (2018)
127. F.D.M. Haldane, S. Raghu, Phys. Rev. Lett. **100**, 013904 (2008)
128. Z. Wang, Y. Chong, J.D. Joannopoulos, M. Soljačić, Nature **461**(7265), 772 (2009)
129. M. Hafezi, E.A. Demler, M.D. Lukin, J.M. Taylor, Nature Physics **7**(11), 907 (2011)
130. K.v. Klitzing, G. Dorda, M. Pepper, Phys. Rev. Lett. **45**, 494 (1980)
131. C.L. Kane, E.J. Mele, Phys. Rev. Lett. **95**, 226801 (2005)
132. B.A. Bernevig, S.C. Zhang, Phys. Rev. Lett. **96**, 106802 (2006)
133. M.Z. Hasan, C.L. Kane, Rev. Mod. Phys. **82**, 3045 (2010)
134. J. Koch, A.A. Houck, K.L. Hur, S.M. Girvin, Phys. Rev. A **82**, 043811 (2010)
135. A. Blanco-Redondo, I. Andonegui, M.J. Collins, G. Harari, Y. Lumer, M.C. Rechtsman, B.J. Eggleton, M. Segev, Phys. Rev. Lett. **116**, 163901 (2016)
136. A. Blanco-Redondo, B. Bell, D. Oren, B.J. Eggleton, M. Segev, Science **362**(6414), 568 (2018). https://doi.org/10.1126/science.aau4296
137. L. Lu, J.D. Joannopoulos, M. Soljačić, Nature Photonics **8**(11), 821 (2014)
138. A.P. Slobozhanyuk, A.B. Khanikaev, D.S. Filonov, D.A. Smirnova, A.E. Miroshnichenko, Y.S. Kivshar, Scientific Reports **6**(1), 1 (2016)
139. X. Cheng, C. Jouvaud, X. Ni, S.H. Mousavi, A.Z. Genack, A.B. Khanikaev, Nature Materials **15**(5), 542 (2016)

140. Y. Yang, Y. Yamagami, X. Yu, P. Pitchappa, J. Webber, B. Zhang, M. Fujita, T. Nagatsuma, R. Singh, Nature Photonics **14**(7), 446 (2020)
141. M.T.A. Khan, H. Li, N.N.M. Duong, A. Blanco-Redondo, S. Atakaramians, Adv. Mater. Tech., 2100252 (2020)
142. M.I. Shalaev, W. Walasik, A. Tsukernik, Y. Xu, N.M. Litchinitser, Nature Nanotechnology **14**(1), 31 (2019)
143. T. Ma, A.B. Khanikaev, S.H. Mousavi, G. Shvets, Phys. Rev. Lett. **114**, 127401 (2015)
144. A.B. Khanikaev, G. Shvets, Nature Photonics **11**(12), 763 (2017)
145. K. Tsuruda, M. Fujita, T. Nagatsuma, Optics Express **23**(25), 31977 (2015)
146. A.L. Bingham, D.R. Grischkowsky, IEEE Microwave Wireless Compon. Lett. **18**(7), 428 (2008)
147. H. Li, M.X. Low, R.T. Ako, M. Bhaskaran, S. Sriram, W. Withayachumnankul, B.T. Kuhlmey, S. Atakaramians, Adv. Mater. Tech. **5**(7), 2000117 (2020)

Chapter 5
Mechanical Robustness of Patterned Structures and Failure Mechanisms

Ehrenfried Zschech and Maria Reyes Elizalde

1 Reliability of Microelectronic Products and Failure Mechanisms

To boost performance and reduce power of microelectronic devices, chipmakers are moving to smaller and smaller transistors, now manufactured in sub-10 nm technology nodes. This trend goes along with a shrinking of the dimensions of on-chip interconnects and vertical structures used in advanced packaging. In addition to the geometrical shrinking, new integration schemes are applied for interconnects and packaging, particularly using heterogeneous 3D integration schemes, and new materials are used.

The increased complexity and new architectural solutions of microelectronic products; the scaling down of transistors, interconnects, and packaging structures; as well as the integration of new materials with changed properties have raised serious reliability concerns. In addition, the operation of microchips and chiplets in harsh environments, use cases that require lifetimes much longer than in the past, and safety-critical applications, where failures are potentially dangerous, are challenges to reliability engineers and physical failure analysis laboratories in the semiconductor industry. Particularly the high complexity of the 3D interconnect and packaging structures requires novel characterization techniques to assess their overall mechanical integrity and reliability.

E. Zschech (✉)
deepXscan GmbH, Dresden, Germany
e-mail: ehrenfried.zschech@deepxscan.com

M. R. Elizalde
CEIT-Basque Research and Technology Alliance, Donostia/San Sebastian, Spain

Universidad de Navarra, Tecnun, Donostia/San Sebastian, Spain
e-mail: relizalde@ceit.es

F. Iacopi, F. Balestra (eds.), *More-than-Moore Devices and Integration for Semiconductors*, https://doi.org/10.1007/978-3-031-21610-7_5

Reliability-limiting effects in microelectronic products can be categorized in electrical effects, which are enforced by mechanical stress, and in (thermo)mechanical effects. In this chapter, we will focus on mechanical degradation and failure mechanisms in backend-of-line (BEoL) structures of integrated circuits.

1.1 Electrical Effects

Electrical effects in multilevel interconnect stacks of integrated circuits that limit the lifetime of microelectronic products are:

- Electromigration (EM)
- Stress-induced voiding (SIV)
- Time-dependent dielectric breakdown (TDDB)

These reliability-limiting effects, and particularly the related degradation and failure mechanisms, have been widely studied and reported in literature [1–10]. Related tests are performed in industry to ensure the specified lifetime of the products.

These effects are particularly pronounced by several trends such as continuous scaling down of the on-chip interconnect structures, the introduction of advanced process steps in BEoL manufacturing, and the integration of thin film materials with low dielectric permittivity (k value) and metal interconnects with modified microstructure. In addition, changed stress states caused by advanced packaging approaches (3D TSV and micro-bump integration schemes, fan-out technology, hybrid bonding) have to be considered. Thermomechanical stress is accelerating the "conventional" degradation mechanism, i.e., the time to failure is reduced for the interconnect structures and consequently the lifetime of the products.

On the one hand, many "postmortem" studies were performed at failed EM, SIV, and TDDB structures, mainly based on scanning electron microscopy (SEM) images of focused ion beam (FIB) cross-sections through the region of interest where the catastrophic failure had occurred [1, 11]. On the other hand, several models were proposed to describe EM, SIV, and TDDB mechanisms [1, 12–15]. However, the experimental verification and validation of such models and the direct observation of the kinetics of degradation mechanisms in on-chip interconnect stacks require in situ imaging approaches using high-resolution microscopy [16–19]. The selection of the most appropriate microscopy technique depends on the geometry of the test structure and on the spatial resolution needed to image the degradation processes in metal/dielectrics structures, which are usually connected with directed atomic transport, with material inhomogeneities, and, in the case of metal interconnects, with the formation of voids. In particular, several in situ experiments directly revealed the degradation mechanisms in damascene Cu/SiO_2 and Cu/low-k interconnect stacks: SEM and transmission X-ray microscopy (TXM) studies of EM [16, 17], TXM study of SIV [18], and transmission electron microscopy (TEM) study of TDDB [19]. In situ TXM experiments allow the (indirect) visualization of directed material transport in fully embedded BEoL test structures by visualizing the evolution of voids in interconnects during the degradation process. Based on multi-modal and multi-scale microscopy studies, it was found that interfaces and grain boundaries are the most pronounced pathways of material transport in copper on-chip interconnects, and consequently, design, geometrical dimensions, and microstructure of the interconnects are playing an essential role for the degradation kinetics [20].

1.2 Stress-Driven Effects

The mechanical robustness of microchips is an increasing challenge because of the integration of new materials into advanced microchips, increased thermomechanical stress caused by new packaging technologies, and operation of microchips at harsh environments [21].

Thermomechanical effects in BEoL stacks can cause interface delamination or microcrack growth, which can eventually result in fracture failures, limiting the lifetime of the microelectronic products. Specific types of failures are:

- Failures in thinned silicon
- Failures in backend-of-line (BEoL) stacks
- Failures in redistribution layers (RDLs)
- Failures in 3D structures, e.g., through-silicon vias (TSVs) and micro-solder bumps

Microcrack formation and growth in microelectronic products is a reliability-limiting effect of increasing interest, in academia to study fracture mechanics in small dimensions and in industry to ensure the requested mechanical robustness of microchips [22, 23]. These effects are pronounced by new manufacturing technologies (thinned silicon, advanced packaging) and – since these are mainly thermomechanical effects – by the materials used having different coefficients of thermal expansion (CTE). The integration of materials with different CTE values results in mechanical stress caused by thermal processes during the microchip manufacturing – both BEoL manufacturing and (advanced) packaging – and during operation.

Advanced packaging and heterogeneous system integration solutions are essential boosters for the performance and functionality of advanced microelectronic products and miniaturized smart systems. However, despite this positive effect, product degradation caused by package-related stress and reliability-limiting effects in 3D IC structures including material integrity (e.g., failure modes like interface delamination, cohesive cracking, and metallurgical degradation at joints) have to be seriously considered. As an example of new integration schemes, 3D IC TSV stacking technologies for wafers or dies mainly use Cu vias and die-to-die interconnections like micro-solder bumps (e.g., SnAg) and Cu pillars. Since the

CTE values of copper and solder materials differ significantly from the CTE values of silicon as well as of intra-metal and interlayer dielectrics, all the new approaches for 3D heterogeneous system integration become increasingly challenging for the mechanical robustness of chips and chiplets and consequently for the product reliability.

The reliability-limiting effects of thermomechanical stress in the BEoL stack, originated from (advanced) packaging, are called chip-package interaction. This package-induced stress is accelerating the propagation of microcracks in the BEoL stack, and it increases the risk of failure caused by delamination along metal/dielectrics interfaces (adhesive failure) or fracture in dielectrics with low fracture toughness (cohesive failure). That means catastrophic failure of the microelectronic product will occur earlier.

The role of reliability engineering is increasing, specifically considering the requirements to stress management in modern microelectronic products and the broadening of use cases (operation of smart miniaturized systems in harsh environments and for longer lifetimes, safety-critical applications), with the goal to reduce package-induced thermomechanical stress and to mitigate reliability-limiting effects in 3D IC structures. The understanding of stress-driven mechanical effects in on-chip interconnect stacks, such as delamination and growth of microcracks, requires a multi-scale modeling considering the hierarchical structure of microelectronic products. Model-based numerical simulations and model validation require the experimental determination of accurate material properties, including Young's modulus (E) and CTE. Particularly for sub-100 nm structures, material properties change depending on the size of the structure [24]. For polycrystalline materials, their microstructure has to be considered [25].

Due to the relatively simple sample preparation, data analysis, and data interpretation, blanket thin films are the most widely used systems for the determination of mechanical data such as Young's modulus and fracture toughness [26, 27]. Nanoindentation experiments are usually used to measure Young's modulus of thin films [28]. However, recent years have shown increasing interests from the semiconductor industry in adopting more realistic 3D patterned structures for testing. Multi-scale finite element analysis studies were performed on nanopatterned structures [29, 30], and experimental techniques such as nanoindentation, four-point bending (FPB), and double cantilever beam (DCB) tests were applied [31–34]. A specially designed experiment to determine the CTE of Cu/low-k BEoL stacks in the SEM using two free-standing BEoL cantilevers was published in [35]. Recently, the determination of the energy release rate in patterned Cu/low-k BEoL stacks with 100 nm resolution was reported, based on a miniaturized DCB test, integrated into a TXM [36].

The experimental verification and validation of models describing the degradation kinetics and the direct observation of these processes and failure in on-chip interconnect stacks require in situ imaging approaches using high-resolution microscopy. Since transmission X-ray microscopy (TXM) and nano-X-ray computed tomography (nano-XCT) are nondestructive techniques for high-resolution 3D imaging of structures and defects, these techniques have been used for in situ studies of the propagation of microcracks in BEoL stacks [37]. The high-resolution X-ray imaging was performed, while a mechanical force is applied to the investigated sample by a miniaturized mechanical test setup [38]. The experimental study of controlled microcrack steering into regions with high fracture toughness provides knowledge for the design of so-called guard ring structures in microchips to stop the propagation of microcracks, e.g., generated during the wafer dicing process [36].

2 Risks of Microcrack Propagation and Design

As stated above, interfacial cracking (adhesive failure) is one of the most important failure modes in integrated circuits. The main contributions to the interfacial toughness are given by decohesion resistance (work of fracture) of the interface, the plastic deformation, and friction along contacting crack faces [39]. In the case of Cu/ILD interfaces, plastic deformation, either associated with the development of the crack tip plasticity or when one of the adjoining materials has an elastic-plastic behavior, constitutes the main contribution to interfacial toughness [40]. Continuum mechanics models have been proposed to account for the plasticity contribution. Among them, the most widely used is the embedded process zone (EPZ) model [41–44], where a traction-separation law ($\sigma - \delta$) is embedded in the continuum description of the interface to characterize the fracture process. The two more important parameters in this model are the work of fracture per unit area of interface, Γ_0, and the peak stress needed to cause interface separation, $\hat{\sigma}$ (Fig. 5.1). Molecular dynamics simulations have also been used to calculate Γ_0 in terms of mode mixity and to derive expressions for the traction-separation law [45, 46]. However, the shape of the $\sigma - \delta$ law has been demonstrated to be of secondary importance [47–49]. Finite element modeling with damage at the crack tip described using the EPZ approach has been used to simulate fracture propagation in patterned structures by inserting cohesive elements along potential crack paths [50].

All these models result in complicated functional relationships for the interfacial toughness, G_{iC}, of the form

$$G_{\text{iC}}/\Gamma_0 = F\left(\hat{\sigma}/\sigma_{\text{y}}, n, E_2/E_1, h/R_0, \psi\right) \tag{5.1}$$

where σ_{y} is the yield stress, n is the strain hardening exponent, h is a characteristic length of the metallic layer in the structure, and R_0 is a characteristic length representative of the size of the plastic zone at the crack tip. Only in two limiting cases the interfacial toughness is "thickness-independent": when the plastic zone is sufficiently small compared with the film thickness or in the limit of very thin metal films. All these theoretical and experimental analyses have revealed a strong mixed mode effect due to plasticity, which results in an increase in toughness with increasing proportions of mode II [51–53]. This is not surprising as, for a given

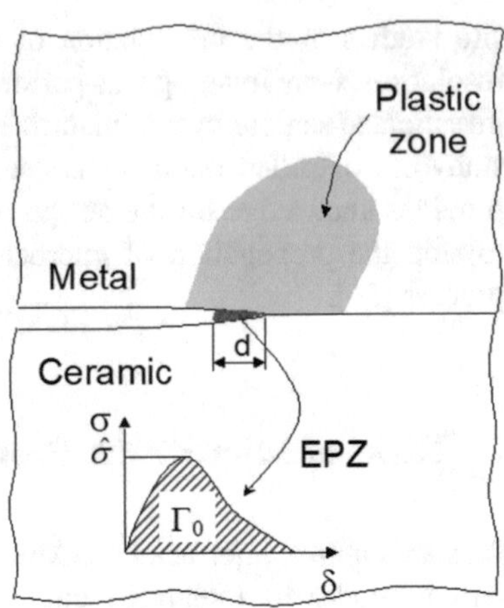

Fig. 5.1 Schematic of the EPZ model for the interface crack

value of G, the plastic zone in mode II loading is approximately twice as large as that for mode I.

In view of the relevance of the plasticity contribution to interfacial toughness in bi-material systems involving metals and the continuous reduction of the dimensions of the Cu lines [31], it is key to characterize the plastic behavior of these lines as a function of their actual microstructure and geometry. The expected size effect consists of an increase of the flow stress and a decrease of the toughness as the thin film thickness is reduced. An important role of crystal plasticity and even strain gradient plasticity is also expected [54–56].

The other relevant failure mode in the BEoL stack is the cohesive failure of interlayer dielectrics. High-performance microprocessors require thin films made of dense low-dielectric-permittivity (low-k) or porous ultralow-dielectric-permittivity (ultralow-k) materials. Unfortunately, reducing k usually produces an impairment of the mechanical properties. This is related to the introduction of "defects" (nanopores, C-doped, etc.) in the material's structure to reduce the dielectric permittivity. As for any brittle material, the behavior will depend on the distribution and size of the defects and the volume under stress [57].

One option to prevent the mechanical damage of microchips manufactured in leading-edge CMOS technology nodes is the integration of metallic guard ring (GR) structures at the rim of the microchip [21]. These specially designed metal structures are integrated into BEoL stacks to dissipate energy in such a way that crack propagation is efficiently slowed down and eventually stopped [58].

For the design of optimized, mechanically robust GR structures, a quantitative determination of the critical energy release rate G_c for crack propagation in patterned structures of fully integrated multilevel interconnect structures of a microchip is needed. Based on these quantitative values and the understanding of the kinetics of crack propagation, the risk of mechanical failure can be evaluated.

3 Characterization of Microcrack Behavior

3.1 Fracture Test at Microscale

Fracture mechanical properties of materials in micro- and nanoscale dimensions have become an important area of fundamental research, including the development and introduction of new techniques for micro- and nanomechanical testing. At the same time, there is an increasing need of industry to evaluate the risk of microcrack evolution at small length scales that can cause catastrophic failure in 3D structured systems and materials such as leading-edge integrated circuits.

Several tests have been developed to characterize blanket thin films. In the "channel cracking" [59], a crack is initiated from a scratch and propagated by bending the sample. The fracture energy of brittle films is calculated from the stress needed for crack propagation. This technique is extremely sensitive to the operator. Fracture toughness of brittle thin films has been also measured using nanoindentation techniques [60]. A sharp tip is pushed on the top surface until radial cracking occurs. The main drawback is that the crack patterns obtained depend on the system tested (thin film thickness, substrate properties, residual stresses, interfacial properties) which reduces the reproducibility and makes quantification very challenging. However, shallow and controlled cracking can be provoked using a particular geometry of the indentation tip, a dual tip [57]. The bulge test [61] consists of applying uniform differential pressure to a freestanding membrane while measuring the resultant bulge height. Membranes of different geometries, both brittle and ductile, are fabricated using standard micromachining techniques. Thin films with thicknesses below 100 nm have been tested. It has the advantage of being able to characterize a wide range of materials and magnitudes, i.e., residual stresses, elastic modulus, yield strength, and fracture toughness [61–63]. Another alternative widely used to characterize small volumes is testing microsamples [56, 64–66]. Beams with different geometries are machined from brittle or ductile blanket films and tested using a nanoindenter as a testing tool. The microsamples are usually produced using focused ion beam milling (FIB), but cantilever beams have also been machined using etching processes [64]. To calculate toughness, the specimens are pre-notched with FIB. Different experimental setups have been developed to conduct these tests in situ in a SEM or TEM. An effort is needed to rationalize the recorded data taking into account the experimental setup and the relevant material properties, including the effect of the FIB milling on the stresses and the material (Ga implantation, amorphization).

Regarding adhesion measurements, a set of techniques are available for blanket thin films [67]. In stub-pull tests, delamination is produced by bonding an actuator to the film surface by means of an adhesive followed by loading until the interface fails [68]. The drawbacks of this technique are the influence of the misalignment between the stub and the interface and the variability introduced by the strength of the adhesive. However, these methods are easy to apply and have been used as quick turn monitors. Peeling tests have been adapted for thin films by depositing a supra layer with a large intrinsic stress to produce spontaneous peeling [69]. The main limitation in this case is related to the complex fabrication of test specimens, which moreover can affect the structure of the interface of interest. Blistering tests are similar to bulge tests but with a crack propagating at the interface [70, 71]. The validity of the test is not clear as the measured interfacial energy release rate depends on the crack length, showing an R-curve behavior that could be induced by the test itself. The four-point bending test has been the reference technique used in the semiconductor industry [72]. A silicon beam with the same thickness as the sample substrate is adhered on top of the thin film structure by Cu diffusion bonding or using polymer adhesives. A notch machined on the silicon beam generates on loading a sharp pre-crack that deflects toward the interface of interest. The critical energy release rate can be calculated from the load plateau recorded during the bending test performed under displacement control. Some practical issues are the friction at the loading points and the lack of reproducibility when polymer adhesives are used. Cu diffusion is avoided due to the high temperatures needed. This technique has also been used to measure the interfacial adhesion of patterned arrays containing low-k or ultralow-k materials and Cu metal lines [31]. Nanoindentation on the top surface of thin films can produce blisters [73, 74] which allow for adhesion measurement. The driving force for delamination is obtained from the combined effect of intrinsic residual stresses and those produced by indentation. The driving force can be increased using overlayers. The main practical problem of this technique is that the cracking pattern is not reproduced for thinner or more brittle films. This problem arises also for scratching tests [75, 76].

An alternative method also used in industry to measure adhesion is the cross-sectional nanoindentation (CSN) [77]. In CSN, an indentation is performed in the silicon of a cross-sectioned IC structure close to the area of interest. Microcracks appear and a Si wedge is formed. This silicon wedge pushes on the structure, while the crack propagates and interacts with the stack [78, 79]. The cracks generated are imaged using optical or scanning electron microscopy. Interfacial toughness is measured using analytical models for elastic/elastic systems [77] and finite elements when an elasto-plastic layer is involved [40] (interlayer dielectrics/etch stop layer ILD/ES and copper/etch stop layer Cu/ES interfaces). The advantages of this technique are (i) the throughput time is short, sample preparation is straightforward, and no pre-cracking or adhesives are needed, (ii) actual stacks can be used, and (iii) spatial resolution as the indenter can be positioned in the area of interest and the surface delaminated is of the order of square μm not mm. The technique was modified (modified CSN) to be applied to on-chip interconnect structures [50]. In this case, FIB milling is used to prepare the cross-section obtained by cleavage and to mill a trench parallel to the cross-section. This produces a crack that grows in one direction interacting with the patterned structure and simplifying the FEM model combined with EPZ to extract adhesive and cohesive properties of the BEoL stack. However, if the goal is to study the interaction between cracks and BEoL, these experiments provide a two-dimensional (2D) information of a three-dimensional (3D) nanopatterned interconnect system.

3.2 Mode Mixity Dependence of Crack Path and Controlled Steering

As stated above, interfacial cracking is one of the most important failure modes found in patterned structures. Although the loading mode can be different for each specific structure, a common issue is the basic fracture mechanics problem of a crack interacting with an interface. For such a configuration, the singular stress field may consist of two modes, usually of unequal exponents, either a pair of complex conjugates or two unequal real numbers. For the case of a crack meeting an interface, the dominant stress singularity is of the type r^{-s}, with $0 < s < 1$, where s depends on the elastic properties of the materials bonded together and r is the distance to the crack tip [78–81]. For a crack crossing the interface, in addition to the typical $r^{-1/2}$ singularity at the crack tips, a different kind of singularity appears at the point at which the crack faces cross the interface [82, 83]. Stress singularities also appear at the junction of various dissimilar materials [84].

In the case of a crack lying along an interface between dissimilar materials, the elastic analysis predicts the interpenetration of the crack faces, which is reflected mathematically by an oscillatory stress distribution with singularity depending on the mismatch of the elastic properties [85–87]. This is, of course, unsatisfactory from the physical point of view. A number of models have then been developed to re-examine the problem [88–94] as well as to predict the tendency of the crack to propagate along the interface or to kink into one of the adjoining materials [95–100].

The elastic mismatch between the two materials can be rationalized in terms of two non-dimensional parameters defined by Dundurs [101] as

$$\alpha = \frac{(\kappa_2 + 1) - \gamma\,(\kappa_1 + 1)}{(\kappa_2 + 1) + \gamma\,(\kappa_1 + 1)} \equiv \frac{\overline{E}_1 - \overline{E}_2}{\overline{E}_1 + \overline{E}_2} \quad \text{and} \quad \beta = \frac{(\kappa_2 - 1) - \gamma\,(\kappa_1 - 1)}{(\kappa_2 + 1) + \gamma\,(\kappa_1 + 1)} \tag{5.2}$$

where subscripts 1 and 2 refer to the two materials, $\kappa = 3 - 4\nu$ for plane strain and $\kappa = (3 - \nu)/(1 + \nu)$ for plane stress, $\gamma = \mu_2/\mu_1$ with $\mu = E/[2(1 + \nu)]$, E is Young's modulus, ν is Poisson's ratio, $\overline{E} = E/\left(1 - \nu^2\right)$ in plane strain, and $\overline{E} = E$ in plane stress. Note that $\alpha = \beta = 0$ for homogeneous materials. The admissible values of α and β, assuming $\nu \geq 0$, lie within a parallelogram enclosed by $\alpha = \pm 1$ and $\alpha - 4\beta = \pm 1$ in the (α, β) plane. Most combinations of materials of practical interest give small values of β (falling between 0 and $\alpha/4$).

The anomalous oscillatory stress behavior at interface cracks does not appear in terms of the energy release rate, which is given by

$$G = \frac{1 - \beta^2}{E^*}\left(K_{\mathrm{I}}^2 + K_{\mathrm{II}}^2\right), \quad \text{with} \quad \frac{1}{E^*} = \frac{1}{2}\left(\frac{1}{\overline{E}_1} + \frac{1}{\overline{E}_2}\right) \tag{5.3}$$

where K_{I} and K_{II} are the stress intensity factors in modes I and II, respectively. The relative amount of mode II to mode I loading (mode mixity) is usually measured through the phase angle, ψ, defined as $\psi = \tan^{-1}(K_{\mathrm{II}}/K_{\mathrm{I}})$ for $\beta = 0$. When $\beta \neq 0$, the definition of ψ is slightly more complicated involving a length scale; but note that the effect of nonzero β, for small values of β, is of secondary importance; see Eq. (5.3). In the case of interface cracks, ψ is affected not only by the external loading but also by the elastic mismatch.

Crack advance can be characterized by a critical value of the energy release rate, G_{C}. In the case of homogeneous materials, the different criteria proposed to determine the direction of crack advance give very similar predictions [102, 103]. For instance, the criterion of maximum circumferential stress (which is approximately equivalent to locally opening mode I conditions and to the maximum G criterion) predicts an angle ω for the propagation direction given by

$$\tan\omega = -\frac{8\tan\psi}{3+\sqrt{1+8\tan^2\psi}} \tag{5.4}$$

This gives an angle of $\arctan\left(-\sqrt{8}\right) \approx -71^{0}$ for pure (positive) mode II. In the case of interface cracks, the critical value for crack advance (the interfacial fracture energy) is not a single material parameter but a function of the mode mixity, $G_{\mathrm{C}}(\psi)$, that has to be determined experimentally. A phenomenological relation has been proposed for this dependence in the form [104]

$$G_{\mathrm{C}} = G_{\mathrm{IC}}\left[1+\tan^2(\eta\psi)\right] \tag{5.5}$$

where G_{IC} is the mode I toughness and η is a coefficient that reflects the effect of interface roughness and plasticity in the adjoining materials. There is no mixed mode effect if $\eta = 0$, but a strong dependence exists when η approaches 1. This type of relationship has been validated by combined experimental and theoretical analyses of, for instance, organic electronic structures using Brazilian disc specimens [105] and other configurations [106, 107]. This strong dependence of G_{C} on ψ highlights the importance of measuring accurately the mode mixity level in the different tests designed to determine the interfacial toughness [67, 108, 109].

The problem of crack path, i.e., the preference of an interface crack to continue along the interface or to kink into one of the adjoining materials, is closely related to that of the mode mixity dependence of interfacial toughness. Residual stresses and the T-stress also affect the energy release rate of both the interfacial and the kinked crack. Relatively small values of residual compression substantially enhance debonding in preference to penetration [110]. The non-singular T-stress, acting parallel to the original crack before kinking, can also have a strong influence on the energy release rate at the tip of the kinked crack. The stress intensity factor and the

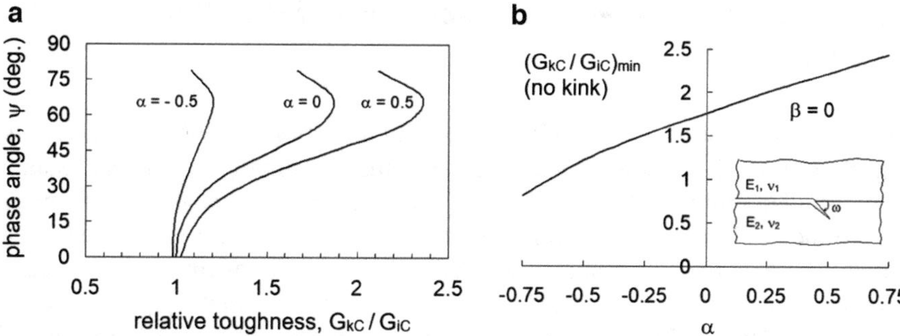

Fig. 5.2 Diagram of crack path prediction for interface cracks with $\beta = 0$: (**a**) phase angle below which the crack stays at the interface; (**b**) minimum material toughness for crack propagation along the interface for all phase angles of loading. (Adapted from He and Hutchinson [97])

T-stress have been accepted as a two-parameter fracture criterion in predicting the crack propagation direction and the shape and size of small-scale yield zones [111]. The general conclusion is that the crack path depends on the relative energy release rates for continued extension of the interfacial crack, G_i, and for crack kinking to one of the materials, G_k. The ratio G_i/G_k is a function of the elastic mismatch; the phase angle, ψ; and the kink angle, ω. Stress fields under realistic conditions are commonly non-uniform and difficult to capture by analytic expressions. Therefore, finite element calculations are required to compute the stress intensity factors and energy release rates of an interfacial crack accurately for arbitrary geometries and loading conditions.

As a result of the analysis, G_i/G_k^{max} becomes a function of α, β, and ψ. The fracture pattern at this scale is inherently dependent not only on the adhesive interface properties but also on the cohesive properties of the bulk materials surrounding the interface, resulting in a competition between adhesive and cohesive crack propagation. The crack path can be predicted by comparing the ratio G_i/G_k^{max} to the ratio G_{iC}/G_{kC}, where G_{kC} is the toughness of the material at the kinked crack tip [112, 113]. If $G_{iC}/G_{kC} < G_i/G_k^{max}$, the interface crack will tend to keep growing at the interface, while if the inequality is reversed, then crack kinking will be favored (Fig. 5.2a). The analysis also gives the minimum value of the toughness ratio, G_{kC}/G_{iC}, needed to ensure that the crack will not leave the interface for all combinations of loading (Fig. 5.2b). For smaller values of G_{kC}/G_{iC}, there is a range, $0 \leq \psi \leq \psi_{max}$, such that the crack stays at the interface, while for $\psi > \psi_{max}$, the interface crack will kink into one of the adjoining materials. Wang [100] extended the interface crack kinking problem to orthotropic and anisotropic bi-materials.

A closely related problem is that of a crack approaching or meeting a material interface in patterned structures and the competition between crack deflection along the interface and penetration across the interface [77]. This can be modeled using either EPZ models or enriched finite element formulations for the displacement fields [114, 115].

3.3 In Situ Monitoring of Microcracking with an X-Ray Microscope

The 3D visualization and in situ monitoring of crack evolution in Cu/low-k interconnect stacks require an experimental setup that allows mechanical loading of the studied sample within an X-ray microscope. Such a combination of miniaturized mechanical test and high-resolution imaging enables a precise control and monitoring of force and displacements in materials at the micro- and nanoscale. The application of a micro-double cantilever beam (micro-DCB) test in an X-ray microscope provides a unique capability for high-resolution 3D imaging of the on-chip interconnect stack of a microchip and of the microcrack evolution while a mechanical force is applied. In [37], a miniaturized piezo-driven DCB test positioned in the beam path of a laboratory transmission X-ray microscopy (TXM) tool (Xradia nanoXCT-100) [116, 117] was used to force a displacement-controlled crack propagation through the on-chip interconnect stack of an integrated circuit and at the same time to image the pathways of microcracks in fully integrated, nanopatterned multilevel interconnect structures with sub-100 nm resolution. Since the maximum photon energy used in state-of-the art laboratory TXM tools is 8 keV (Cu-Kα radiation), the sample size has to be <100 μm, at least in one direction parallel to the crack front, to enable the transmission of photons of this energy range [118]. The samples used in the micro-DCB test are much smaller than in the standard DCB test, and therefore, the sample mounting has to be carried out differently. In contrast to the hinged supports in the standard DCB test (flexible joints), clamps are rigidly connected to the cantilevers in the case of the micro-DCB test. The principle of the micro-DCB test and the experimental setup was described in detail in [38].

In the DCB test, the tensile stress normal to the crack plane is the dominating stress component, i.e., the mode I condition of crack propagation (opening or tension mode) is supposed to be the predominant fracture mode for the sample [119]. In contrast to the standard DCB test, asymmetric bending of the cantilevers occurs in the micro-DCB test because of a shear component in the crack plane, as a result of deviations from the "symmetric sample" geometry and of the contribution from the clamps. In addition, the thickness of the patterned on-chip interconnect stack in relation to the total beam height is not negligible, and the beams do not have a constant and equal height because of limitations of the sample preparation process. In addition, 3D effects at the crack tip have to be considered [120]. The typical double cantilever test geometry, characterized by hinged supports and free cantilever ends, is schematically shown in Fig. 5.3 [36]. This scheme includes a partially delaminated sandwich sample, and it indicates the force introduction to open the crack and to cause crack propagation. During the DCB test, the test beam is separated into two cantilevers.

Micro-DCB samples are characterized by a sandwich structure, consisting of two beams of usually similar dimension that are glued together with a thin layer of epoxy and the layer stack to be studied (region of interest, ROI) in the center. A piece

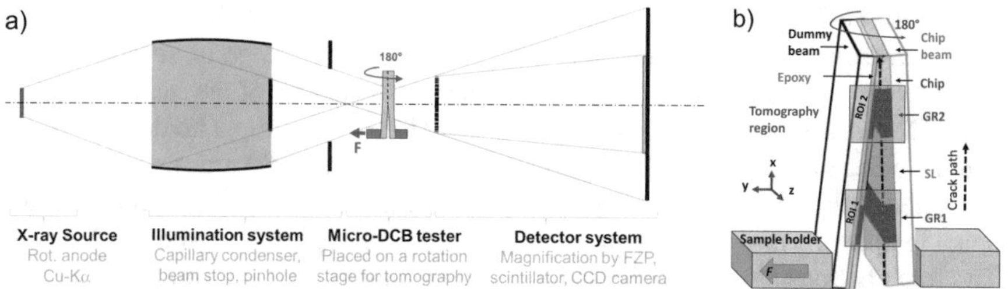

Fig. 5.3 (**a**) Scheme of the experimental setup: Micro-DCB test in the laboratory full-field transmission X-ray microscope. (**b**) Scheme of the micro-DCB test using a sandwich sample with the region of interest (ROI), the on-chip interconnect stack of a microchip featuring two copper guard ring structures (GR1 and GR2) separated by a scribe line (SL) [36]

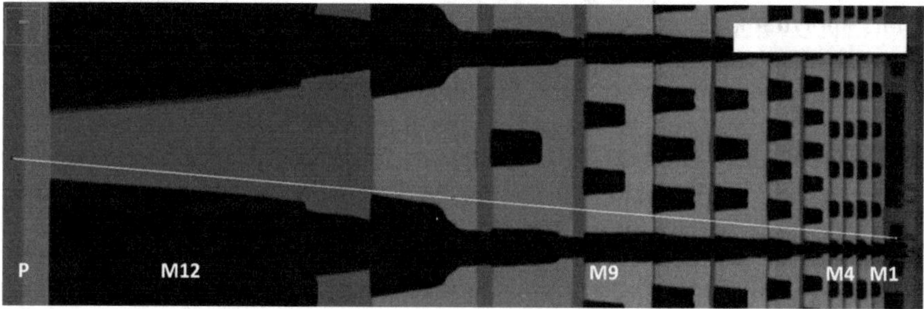

Fig. 5.4 TEM image of a part of a guard ring (GR) structure with 12 metallization of copper (M1 to M12) with different dimensions and a post-passivation layer, consisting of two sub-layers with different composition of the dielectrics. The different gray values indicate the copper structures (dark) and different dielectrics used to isolate the metal structures and for the passivation. Scale bar is 1 micron [121]

of a thinned wafer, containing on top of the silicon the ROI with two guard ring structures (GR1 and GR2) separated by a scribe line (SL) and a part of the BEoL stack, containing on top of the silicon the ROI with the BEoL stack, was glued to a dummy sample (silicon) of similar dimensions in length and width, but with varying heights. The integrated circuit was manufactured in 14 nm CMOS technology node, with a BEoL stack consisting of 12 layers of copper (M1 to M12) with different dimensions, insulated by low-k materials, and a post-passivation layer on top. The transmission electron microcopy (TEM) image of a cross-section through a part of a GR structure is shown in Fig. 5.4 [121]. The samples were grinded, polished, and sawed up to a length of 1 mm and a cross-section of approximately 50 μm × 50 μm, to fulfil the geometrical requirements for a micro-DCB test in a transmission X-ray microscope [38, 118]. The notch was cut in the center of one end face of the test sample applying a razor blade. This notch (the pre-crack) serves as a defined starting point to drive the microcrack within the on-chip interconnect stack toward the GR structures.

The micro-DCB tests that drive the crack into the on-chip interconnect stack were performed displacement-controlled. The load in the range of several tens of millinewtons was applied perpendicular to the rotational axis of the sample stage of the TXM, with a loading speed of 1 μm/s. One side of the sample holder was fixed, and the other side was moved horizontally, typically in 50–100 nm steps. The force is applied at the notched end of the test sample perpendicular to the direction of the notch to open the microcrack and to ensure stable crack growth in a controlled way. The force is transmitted into the sample through clamps that are rigidly connected to the cantilevers, while one clamp is fixed and the other is moving. The propagating microcrack causes progressing delamination, and it divides an increasing part of the sample into two individual cantilevers, while the sample holder acts as a support on the other end [118].

To visualize crack opening and propagation in the on-chip interconnect stack while applying a mechanical load, the micro-DCB test setup is positioned in the X-ray microscope (Xradia nanoXCT-100) on a rotation stage for tomography, and the notched sample with the ROI is mounted on this rotation stage. This geometrical arrangement allows to align the crack front parallel to the optical axis of the X-ray microscope, which is the preferred geometry to obtain free access to collect 2D radiographs without shadowing the ROI during the data acquisition, and, consequently, to achieve 3D tomography data of the ROI.

4 Understanding of Microcrack Behavior

Microcracks, introduced into microchips during the manufacturing process, e.g., sawing of wafers, are a serious reliability concern for microelectronic products since their stress-induced propagation can cause catastrophic microchip failure [21]. The risk of fracture is increased for BEoL stacks if metal interconnects are insulated with low-k or ultralow-k materials. These dense or porous CVD-deposited organosilicate glass materials are characterized by not only a low dielectric permittivity but also low Young's modulus and cohesive strength and consequently low fracture toughness [122].

In addition, the microchip manufacturing and particularly the subsequent advanced packaging technologies are using a large variety of materials, including hard lead-free solder materials, to form micro-bumps or copper pillars. Because of different coefficients of thermal expansion (CTE) of the used materials, temperature variations cause thermomechanical stress in the BEoL stack that can result in crack propagation [123]. The requirement to a mechanically robust on-chip interconnect structure against wafer processing and packaging stress is that the fracture driving force for pre-existing defects – e.g., microcracks – is smaller than the fracture resistance of the nanopatterned BEoL stack [21].

4.1 Local Energy Release Rate and Fracture Resistance (Size Effect)

A nondestructive 3D high-resolution imaging of the crack evolution in the on-chip interconnect stack, preferably using a micro-DCB test setup in an X-ray microscope, allows a 3D visualization of crack opening and propagation, i.e., without modifying the local stress state by sample preparation, and based on the measured crack geometry a quantitative determination of the critical energy release rate for crack propagation in patterned structures of fully integrated multilevel interconnect structures of a microchip. Using the measured geometry of the crack at several loading steps during the micro-DCB test together with a data analysis based on the linear elastic fracture mechanics and Euler-Bernoulli beam model allows the determination of the critical energy release rate for crack propagation in different regions of a processed silicon wafer [38].

The combination of a displacement-controlled crack propagation through different regions of the thinned wafer piece, particularly the on-chip interconnect stack, and at the same time imaging of the crack allows to determine the critical energy release rate G_c for crack propagation in these regions quantitatively. Conventional mechanical tests for the determination of the critical energy release rate G_c for crack propagation in macroscopic bulk samples or in unpatterned layer stacks, e.g., the double cantilever beam (DCB) test [72, 124], are performed at much larger samples. Therefore, a miniaturized mechanical test setup for samples that are transparent for photons with 8 keV energy had to be built and integrated into the beam path of an X-ray microscope [38]. This approach has the advantage that X-rays can be used to image the crack evolution in materials nondestructively.

Figure 5.5a shows two radiographs of the sample at two different stages of the micro-DCB experiment, for the crack tip toward the GR and after passing the GR, representing two different loading steps. It can be clearly seen from Fig. 5.5b that the crack length is growing with increasing crack opening in the SL. The crack growth rate is progressively reduced when the crack approaches the guard ring, and it stops propagating when it reaches the GR2. Only when the applied force, and consequently the crack opening (and beam displacement), is large enough, the crack propagates through the GR2 structure.

The uniqueness of the (ideal) DCB test is that only one stress component exists, i.e., the tensile stress normal to the crack plane. There is no shear stress in the crack plane. That means the sample is tested in mode I condition of crack propagation (opening or tension mode) [119]. Compared to the conventional macroscopic DCB test, the micro-DCB test is characterized by some specifics that have to be considered in the data analysis. Without momentums in the sample mounts, an important particularity of the micro-DCB test is that moments in the sample mounts caused by gluing change the cantilever deformation. That means the boundary conditions for the linear elastic model used are between the two extreme cases "fixed beam ends" (stiff clamping) and "free beam ends" (freely hinged). The contribution from the clamps causes a shear component in the crack plane, and consequently,

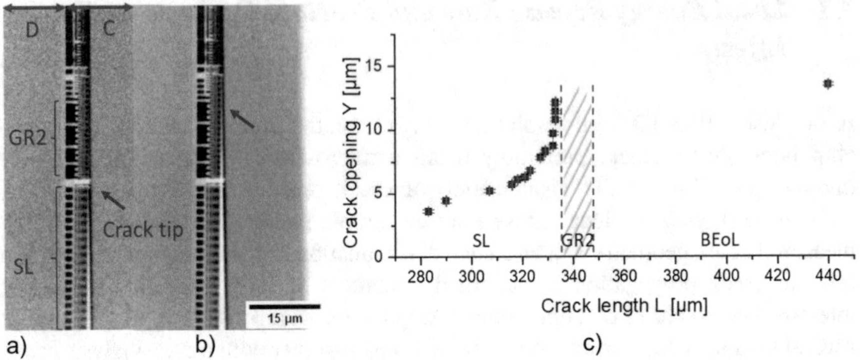

Fig. 5.5 (**a**, **b**) Radiographs of the micro-DCB sample with acquisition time per each of 15 seconds (D, dummy; C, chip; GR2, guard ring 2; SL, scribe line): crack tip (**a**) toward the GR2 and (**b**) after passing the GR2. (**c**) Crack opening Y vs. crack length L at several loading steps during the micro-DCB test. (Adapted from Kutukova et al. [36])

asymmetric bending of the cantilevers occurs. In addition, the assumption that the cantilevers have a constant and equal thickness, as it is fulfilled for ideal samples in the standard DCB test, is not true for micro-DCB samples because of micro-DCB sample geometry peculiarities like not negligible thickness of the patterned on-chip interconnect stack in relation to the total beam height as well as because of accuracy limitations of the sample preparation process. In addition, 3D effects at the crack tip have to be considered [125].

For an asymmetrical DCB sample configuration, i.e., for different cantilever widths h_n $\{n = 1, 2\}$, mode II (sliding mode or in-plane shear mode) condition for crack propagation occurs additionally to the mode I condition. That means the real test conditions reflect a combination of loading modes, i.e., mode mixity [21, 34].

The standard DCB test is the most common experimental technique for determination of the critical energy release rate G_c of macroscopic specimens with layers or layer stacks. A typical DCB test geometry is schematically shown in Fig. 5.6a [36]. This scheme includes geometric sizes of a partially delaminated sandwich sample, and it indicates the force introduction. In both, the macroscopic and the microscopic geometry, a force F is applied perpendicular to the interfacial plane to open the interfacial crack by a displacement _Y at the sample mount position ($x = L$, $x = 0$ at crack tip), causing a crack length L. With respect to the micro-DCB test, the situation of the two extreme cases "fixed beam ends" and "free beam ends" is shown schematically in Fig. 5.6a, right side. The geometric sizes in a real micro-DCB sample are indicated in a stitched radiograph in Fig. 5.6b [36]. Geometric inaccuracies of the sample, i.e., deviations from an ideal micro-DCB sample, result in asymmetric bending of the beams caused by an additional shear loading component.

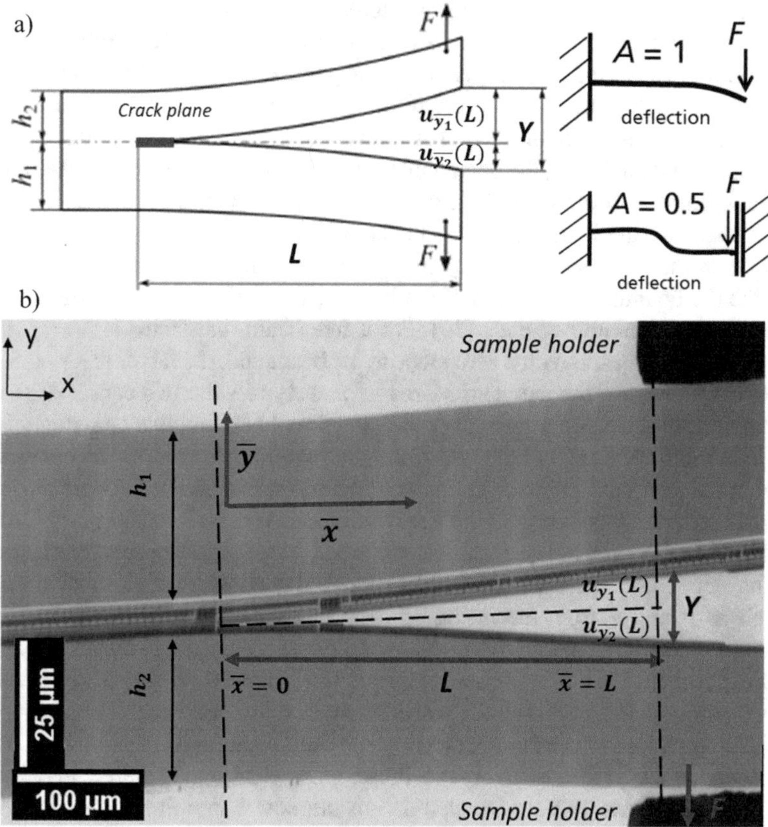

Fig. 5.6 Illustration of the (pre-cracked) DCB geometry: (**a**) in the standard DCB test with free beam ends (beam with central crack) and scheme for two extreme cases, fixed cantilever ends and free cantilever ends (F, applied force; h_n, cantilever height; $u_{\overline{y}}(L)$, cantilever deflections at the end of the beam in $\overline{y}$ $\{n = 1, 2\}$; $Y = u_{\overline{y_1}}(L) + u_{\overline{y_2}}(L)$, maximum crack opening; $\overline{x}$, coordinate in crack plane direction; L, crack length), and (**b**) stitched radiograph of a micro-DCB sample (rotated by 90° compared to Fig. 5.3b). The stitching array of 3 × 12 radiographs is compressed along the x coordinate for better visualization of the geometric sizes [36]

During the double cantilever beam test, the test beam is separated into two cantilevers. The energy release rate G for crack propagation is well described for the standard DCB test, based on the Euler-Bernoulli beam model [124, 126]. The critical energy release rate G_c that is needed for crack growth can be determined by the condition that the external energy is equal to the stored elastic energy. That means crack growth is initiated when the energy release rate is larger than a critical value G_c, i.e., $G > G_c$. The Euler-Bernoulli beam model [127], which is valid for beams under the assumptions that the beam cross-section dimensions are small compared to the beam length (equals crack length here) and that the constrained beam ends at the support side (crack tip here) are horizontally oriented, is applied to the two cantilevers for the determination of the stored elastic energy and ultimately of the energy release rate for crack propagation. Solving the Euler-Bernoulli differential equation for beam bending [128] provides the generalized result for the relationship between the applied force F and the resulting beam deflections $u_{\overline{y_n}}$ for a uniform static beam, here for each of the two cantilevers n, $\{n = 1{,}2\}$. That means, using the linear elastic beam theory, the beam deflection $u_{\overline{y_n}}$ can be expressed by the force F, the crack length L, beam width b_n, beam height h_n, and Young's modulus E (the last three parameters are included in the moment of inertia I_n).

The parameter A (see Fig. 5.6a) is a geometry parameter that is introduced to describe the boundary conditions of the beams (first derivative of the bending line) at the displaced cantilever ends [36]. For a free beam, the factor is $A = 1$. For a fully constrained beam, with a zero slope at the beam end, the factor is $A = 0.5$. The numerical values for the geometric sizes A_1 and A_2 (for the two cantilevers) were determined from the experimentally determined real bending line applying a least-square fitting procedure based on the Euler-Bernoulli beam model. The coordinates of the measured data points that characterize the crack geometry were extracted from radiographs acquired with the X-ray microscope at each loading step [36].

The critical energy release rate G_c for crack propagation is calculated at each loading step, considering geometrical and material parameters of the studied sample as well as F and A from the fitting procedure described above (for each beam). Figure 5.7 visualizes quantitative G values at several loading steps during the micro-DCB experiment [36]. The values for scribe line, guard ring, and BEoL stack are about 9 J/m^2, 34 J/m^2, and 6 J/m^2, respectively.

All three regions of the thinned wafer piece were manufactured using the Cu dual damascene process [129], with organosilicate glass as insulating dielectrics between the Cu interconnects. For unpatterned porous ultralow-k thin film materials as used in 14 nm CMOS technology node, G_c values in the range of 2–5 J/m^2 were reported [122, 130, 131], i.e., the fracture toughness of these materials is low. Compared to this G_c value for unpatterned OSG thin films, the G_c value is increased for patterned Cu/low-k structures. It depends on Young's modulus – and consequently on the critical energy release rate G_c as a measure for the fracture toughness – of the dielectrics as well as on the design of copper structures in the SL, GR, and BEoL stack regions of the studied wafer piece. Since the dielectric properties are changing only slightly for different levels of the (patterned) layer stack, the differences of the G_c values of different regions (SL, GR, BEoL) have to be explained with the design of the Cu structures.

As expected, the specially designed GR structures have the highest G_c value. It is several times higher than the G_c value of BEoL structures of the integrated circuit. As experimentally shown, the crack is stopped at the guard ring. This effect can be explained by the fact that the critical energy release rate G_c includes not only the breaking of chemical bonds across the interface (interface debond energy) but also the plasticity of adjacent ductile structures, such as copper guard rings [72]. The energy dissipation process that results in the increase of the G_c value close to the GR and a high G_c value of the GR structure itself can be explained with the geometry of the GR (GR length) and the size-dependent plasticity of

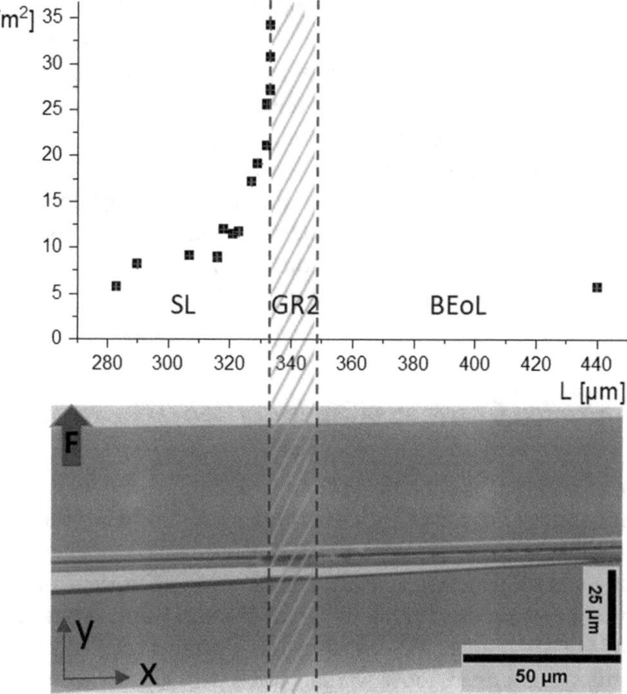

Fig. 5.7 Calculated G values at several loading steps (and respective crack lengths L) and stitched of the final loading stage in the BEoL region [36]

copper. Compared to the dimensions of the Cu interconnects in the BEoL stack (<1 μm, except for M12), the GR length in crack direction is 16 μm. For Cu structures with a dimension >1 μm, the dislocation confinement is relaxed, and the plasticity zone is enlarged when the crack is approaching the copper structure, thus dissipating more mechanical energy. This effect of copper plasticity is exploited in the functionality of the guard rings. Lane and Dauskardt [132] analyzed unpatterned layer stacks and reported an increase of G_c from 5 J/m^2 for Cu films with a thickness < 0.5 μm to 10 J/m^2 for 1.0 μm films and 40 J/m^2 for 5 μm films. A direct comparison between these data for unpatterned films with dense dielectrics and data for patterned Cu/low-k structures with porous dielectrics is not possible. However, considering Cu structures with dimensions up to 2 μm (M12) in the BEoL stack and GR lengths of about 16 μm, the G_c values determined in the micro-DCB test in this study are very reasonable.

To sum up, the implementation of a micro-DCB test in an X-ray microscope allows a 3D imaging of the pathways of microcracks in fully integrated multilevel on-chip interconnect structures with a spatial resolution of about 100 nm. Based on measured geometric shape of the crack and cantilever bending lines at several loading steps during the micro-DCB test as input data and the data analysis based on the Euler-Bernoulli beam model, the determination of the critical energy release rate G_c for crack propagation is possible for different patterned regions of the wafer manufactured in leading-edge technology node with a patterned Cu/low-k interconnect stack. The critical energy release rate G_c for crack propagation of a guard ring structure of >30 J/m^2 is significantly larger than the respective values in patterned surrounding regions with G_c values <10 J/m^2 and about one order of magnitude higher than the G_c values of the respective unpatterned dielectric thin films. These results show that, in addition to the material properties of the dielectric materials, the geometry of the metal structures is playing an essential role for the fracture behavior of Cu/low-k interconnect stacks.

4.2 Controlled Crack Steering into High-Toughness Regions

An option of energy dissipation is to slow down and eventually stop the crack propagation by steering the microcrack into regions of the BEoL stack with relatively high fracture toughness [58]. For a better understanding of fracture mechanics at small scales and for avoiding material cracking and interface delamination, as well as for providing guidelines for the design of GR and BEoL structures as well as the location of GR structures [30], the effect of a combination of loading modes, the mode mixity as described above, on the crack path has to be studied. In addition, fundamental questions of the role of plasticity of 3D micro- and nanoscaled metal structures [132] and of the mismatch of the elastic properties between two dielectric layers (so-called delta-E effect) [26] on the fracture mechanics of nanopatterned structures have to be considered.

An experimental approach that allows to steer the microcrack in a controlled way by tuning the fracture mode mixity locally at the crack tip and to acquire simultaneously the 3D image information of a region of interest (ROI) that includes the on-chip interconnect stack of an advanced microchip was reported recently [36]. The steering of microcracks in 3D nanopatterned structures and its simultaneous nondestructive imaging with high spatial resolution was performed by positioning a miniaturized piezo-driven DCB test setup in the beam path of a laboratory transmission X-ray microscopy (TXM) tool [38]. This experimental approach allows to force a displacement-controlled crack propagation through the on-chip interconnect stack of an integrated circuit and at the same time to image the pathways of microcracks in fully integrated, nanopatterned multilevel interconnect structures with sub-100 nm resolution [37].

The microcrack is either propagating along one level of the BEoL stack and then moving to a higher level or to a lower level, depending on the mode mixity, or – even more often – the crack front is running in several levels simultaneously. In the latter case, the propagation of the microcrack from one level to another level is often not abrupt for the whole crack front, i.e., usually for parts of the crack only, step by step. That means the crack front at a certain crack length is including not only one level of the stack. In some cases, it was observed that cracks moved from

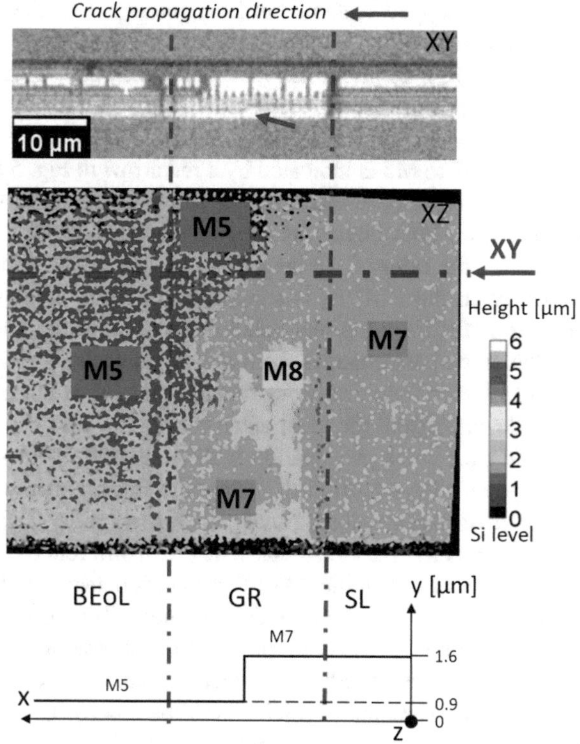

Fig. 5.8 Virtual cross-section (XY view) with indicated crack path change from M7 to M5 (top); XZ height map of the crack in the on-chip interconnect stack (SL, GR, Cu/low-k BEoL), heights above the Si substrate indicated by colors (middle); scheme of the cross-section XY at the indicated Z location (red line) with crack pathway, showing crack path change from M7 to M5, heights above the Si substrate in μm (bottom) [36]

one level to a neighbored level and subsequently back to the original one before the crack is moving irreversibly to a higher level or to a lower level (see, e.g., some regions in Fig. 5.8: M7 – M8 – M7). These observations underline the importance of a nondestructive 3D imaging technique.

A virtual cross-section through a piece of wafer (XY plane, perpendicular to the metallization layers) at the final loading stage of a micro-DCB test, visualizing the microcrack in the on-chip interconnect stack (SL, GR, Cu/low-k BEoL), is provided in Fig. 5.8 (top). This cross-section image is based on a reconstruction of X-ray computed tomography (XCT) data. The microcrack propagated from the SL within interlayer dielectrics or along weak interfaces toward the GR structure and eventually into the BEoL stack of the microchip. The nano-XCT data allow to determine where in the metal stack, i.e., in which layer or along which interface, the microcrack propagated. To visualize the propagating position in the stack, the heights of the crack above the silicon substrate (XZ horizontal map with height indicated by colors) are mapped as shown in false colors in Fig. 5.8 (middle). The color scale bar corresponds to the heights from zero (Si substrate) to about 6 μm (post-passivation layer P). In addition, the schematic crack path in an interconnect stack for one particular position along the *Z*-axis is provided in Fig. 5.8 (bottom) that shows the crack path along M7 and then transferring to M5. The change of the crack path down from M7 to M5 is indicated by a red arrow in Fig. 5.8 (top) [36].

The micro-DCB test results presented in Fig. 5.6 were achieved from a nearly "symmetric sample," i.e., both beams had the same target heights. That means, according to Fig. 5.6b, the ratio of the heights $e = h_1/h_2$ of the dummy beam h_1 and the chip beam (with the ROI) h_2 is supposed to be $e = 1$. However, due to sample geometry imperfections, the real value is $e = 1.3$ in the case provided in Fig. 5.8. Hence, the crack path changes from metallization layer M7 to the lower metallization layer M5, even for a nearly "symmetric sample" can be explained with the e value, which deviated from the ideal value $e = 1$, i.e., with the superposition of the fracture mode I for crack propagation with additional fracture modes. This example expresses already the opportunity to steer the crack by mode mixity, and it proves the importance of a thorough data analysis [36].

In [36], defined asymmetrical DCB samples, i.e., with different cantilever heights h_n $\{n = 1, 2\}$, are prepared and studied. For these asymmetric sample geometries, mode II (sliding mode or in-plane shear mode) and mode III (tearing mode or anti-plane shear mode) conditions for crack propagation occur in addition to the mode I condition. That means these test conditions reflect a combination of loading modes, the mode mixity [21, 34]. The chosen thickness ratios $e = h_1/h_2$ of dummy beam ($h1$) and chip beam ($h2$) were 2.0 and 0.5, respectively. These thickness ratios are related to the global mode angles 0° (symmetric case) and $\pm 22.5^\circ$ (both asymmetric cases). The real beam thickness ratios e were calculated from these image data. For each geometrical configuration (symmetric case and two asymmetric cases), a set of micro-DCB experiments in the X-ray microscope were performed, and nano-XCT 3D data sets were generated as described above. The crack pathways were analyzed based on the 3D tomography data set for each sample. The resulting data for each geometrical configuration are presented in Table 5.1:

Table 5.1 shows that microcracks are moving predominantly to higher metallization layers for $e < 1$ and predominantly to lower metallization layers for $e > 1$. That means microcracks can be steered into a particular level of the layer stack as a result of the fracture modes for crack propagation.

The steering of the crack path to higher metallization levels, as characteristic for beam thickness ratios $e < 1.0$, is the preferred option since the fracture resistance or the critical energy release rate for crack propagation is increased the closer the microcrack will be to M12. The reason is the thickness of the M12 Cu metallization

Table 5.1 Dominant metal layers of crack propagation for several beam height ratios e, based on 7 samples for $e \approx 0.5$, 9 samples for ≈ 1, and 12 samples for $e \approx 2$

Mode mixity	Asymmetric; $e = 0.5$	Symmetric; $e = 1.0$	Asymmetric; $e = 2.0$
Metal layer n	11 ± 2	8 ± 2	4 ± 2

of about 2 μm, in contrast to the M1 to M11 metallization layers with thicknesses lower than 1 μm. In addition to an enhanced near-tip sliding and friction caused by mode mixity [104], the plasticity of the copper in M12 increases the energy dissipation. This effect can be explained by the fact that the critical energy release rate G_c includes not only the breaking of chemical bonds across the interface (interface debond energy) but also the plasticity of adjacent ductile structures, such as relatively thick Cu lines [72]. The energy dissipation process that results in the increase of the G_c value close to the thick Cu lines can be explained with the size-dependent plasticity of copper. For Cu structures with a dimension >1 μm, the dislocation confinement is relaxed, and the plasticity zone is enlarged when the microcrack is approaching the Cu structure, thus dissipating more mechanical energy. This effect of copper plasticity is exploited in the M12 Cu lines. Lane et al. [132, 133] analyzed unpatterned layer stacks and reported an increase of G_c by >400% as the Cu layer thickness was increased from 0.3 to 3.3 μm. From a figure published in [132], G_c values of 5 J/m^2, 5–10 J/m^2, and 14 J/m^2 were extracted for Cu thickness values of $\leq$0.20 μm (M1–M6), 0.35–0.72 μm (M7–M11), and 2.02 μm (M12). A quantitative comparison between these data for unpatterned films with SiO_2 dielectrics and data for patterned Cu/low-k structures with porous dielectrics is not possible. In addition, chemical composition of the electroplated copper and process parameters for deposition and thermal treatment are certainly different. However, considering Cu structures with dimensions of 2 μm (M12) in the BEoL stack, the crack propagation is expected to be slowed down, and in the best case, the microcrack will be stopped. Since the plastic zone size exceeds the Cu layer thickness [132], this effect occurs also in the dielectrics next to the Cu structure. This toughening effect – energy dissipation caused by plasticity – is consistent with model predictions [134, 135].

The approach demonstrated and the experimental results have significant implications for the design of on-chip interconnects (including guard ring structures) of leading-edge integrated circuits and for the fundamental understanding of the fracture behavior of materials, e.g., composites, at the sub-micron and nanoscale. The micro-DCB experiment in an X-ray microscope allows to control fracture and to steer crack paths to regions in the on-chip interconnect stack with relatively high fracture toughness. With this technique, it is possible to study the complex failure modes in realistic BEoL stacks and to discuss the effects of process-induced thermomechanical stress and CPI on chip reliability. Based on the knowledge of how position of the microcrack initiation and mode mixity modulate the crack propagation, it is possible to study the fracture mechanics of small structures and draw conclusions for the dielectrics material selection to control the crack path and to ensure the required fracture resistance of BEoL structures for future advanced technology nodes that are often accompanied by advanced packaging solutions.

5 Summary and Outlook

5.1 Future Technologies and Mechanical Reliability Challenges

The future development of the microchip technology, for both transistors and interconnects, will be characterized by a further shrinking of feature sizes – the physical gate length of the transistors (at least up to 10 nm) and the dimensions of interconnects (on-chip and chip-to-chip/package), new 3D architectures for devices, and advanced packaging as well as the integration of new materials and material stacks. These technological developments require novel risk mitigation strategies for tightly packed 3D structures, and they force innovations to improve conventional or to develop completely new characterization techniques for 3D patterned systems.

In addition, the semiconductor industry – both chip design and technology – will be more and more driven by challenging use cases such as operation at harsh environments (space research, automotive industry), applications that require lifetimes much longer than in the past, and safety-critical applications (autonomous driving, drones, medicine), but also by requirements of data centers and of mobile gadgets. Some of the most advanced devices are expected to work for longer periods than in the past, with parts that are consistently in the fully "on" state expected to last decades rather than a couple years of intermittent service. All these customer requests challenge reliability engineering in the semiconductor industry to ensure the mechanical robustness of the microelectronic products for the specific use cases.

In particular, the mechanical robustness of increasingly complex interconnects, both on-chip and chip-to-chip/package, has become a very challenging aspect of More-than-Moore integration schemes. Mechanical failures of on-chip interconnects in the form of interface delamination and material cracking can pose a critical challenge for integrating ultralow-dielectric-constant (ultralow-k) materials in advanced integrated circuits. Subject to thermomechanical stress, mechanical failure of on-chip interconnects can take on various forms. The effective fracture resistance of the BEoL stack of microchips is governed by the ultralow-k dielectrics and the embedded metallic materials therein. To design robust interconnects against mechanical failures, it is desirable to characterize the fracture properties of the component materials using realistic integrated structures rather than blanket thin films, whose properties may not always truthfully reflect the impact due to patterning and processing.

With the methodology described in this chapter, i.e., the controlled steering of microcracks into regions with high fracture toughness while considering nanoscale mechanical properties of fully integrated 3D interconnect stacks (particularly the local critical energy release rate G_c), conclusions for the robustness of backend-of-line stacks can be drawn, and input for the design of BEoL and guard ring structures can be provided. The knowledge about the fracture behavior in microchips and particularly of how fracture mode mixity modulates the crack propagation in nanopatterned structures provides also a better understanding to the fracture mechanics of small structures. As a result, conclusions for interconnect design, materials, and processes can be drawn, with the goal to ensure the needed mechanical robustness of the BEoL stack, a 3D nanostructured system of materials. This new approach has important implications for the on-chip interconnect technology development and for the nondestructive study of the fracture behavior of materials in the nanometer scale. It opens a way for the fundamental study of the mechanical behavior of nanoscale structures and materials, and it provides the opportunity to establish appropriate risk mitigation strategies to avoid catastrophic failure of the microchip.

Considering the intrinsic advantages of nondestructive nano-X-ray computed tomography (nano-XCT) for high-resolution 3D imaging of opaque objects such as microchips, this technique is a promising future option for use in product development and physical failure analysis in the semiconductor industry, particularly for imaging of 3D advanced packaging, including wafer-level chip-scale packaging (WLCSP) and, e.g., hybrid bonding, and on-chip interconnect structures and defects such as microcracks. The combination of micromechanical testing and high-resolution X-ray imaging opens the way for the development of design concepts for novel engineered materials systems based on their local mechanical properties. Summarizing, the nondestructive 3D X-ray imaging of structures and defects and the kinetics of degradation processes in microelectronic products provide valuable information for reliability engineering and design-for-reliability (DFR) in the semiconductor industry. This approach also allows to evaluate process-induced material changes, and it provides a path to study the scaling effect on the fracture behavior of BEoL stacks for future on-chip interconnect design and technology development.

Laboratory sub-micro- and nano-X-ray computed tomography (XCT) at high photon energies (>10 keV) will allow a really nondestructive imaging of structures and localization of defects such as microcracks in advanced packaging and in BEoL stacks. Therefore, these techniques will have not only to be applied in reliability engineering to mitigate reliability-limiting effects as described in this chapter; however, they have the potential to be introduced in the semiconductor industry in physical failure analysis laboratories as well as for X-ray-based 3D metrology and diagnostics to support novel technologies for advanced packaging and for on-chip interconnects [136].

The experimental approach and the test samples used are not limited to samples from microelectronic products. The established concept for a controlled crack propagation can be adopted to study other materials systems as well. It opens new approaches for fundamental studies of the fracture behavior of constrained materials on a small scale.

5.2 Biomimetics: What Can We Learn from Nature?

Most of the engineered materials and systems are still being outperformed by natural materials with exceptional mechanical properties, e.g., simultaneous high stiffness and toughness, that have been "designed" during a long-term evolution process, with the goal for the living object to survive and to adapt to its surrounding environment. Hierarchically structured biocomposites are tailored according to their functionality [137–139]. As an example, protective mollusk shells consist of high-strength and high-stiffness building blocks (mineral phase) and ductile components (amorphous organic phase) with "weak" interfaces that allow to trap cracks in toughened regions. Firstly, microcracks propagate along interfaces between inorganic crystallites, with low resistance against crack growth, and subsequently, these microcracks are steered into the biopolymer regions where the cracks are stopped [140]. Targeting on engineered damage-tolerant 3D nanopatterned structures, e.g., microchips, as discussed here, the nature's design principles for damage-tolerant materials that result in a controlled crack steering into regions with high fracture toughness [137] can provide valuable information for the design of engineered materials systems such as backend-of-line stacks of integrated circuits or guard ring structures.

Studies at biological objects, combining experimental data and modeling, help to develop fracture mechanics at small scales and to understand particularly microcrack propagation in hierarchically structured materials systems. Since such studies provide valuable information for damage-tolerant design, fracture mechanics of materials in micro- and nanoscale dimensions will become an increasingly important field of fundamental research, including the development and introduction of new techniques for micro- and nanomechanical testing [22, 141]. The combination of miniaturized mechanical tests with high-resolution imaging enables the ability of a controlled steering of microcracks into regions with high fracture toughness and the visualization of the microcrack evolution [37].

On the other hand, bio-inspired, hierarchically 3D structured engineered materials and systems have a growing industrial importance. Examples are microelectronic products and MEMS systems as well as advanced battery electrodes and new downsized devices for medical applications. There is an increasing need of industry to evaluate the risk of microcrack evolution at small length scales that can cause catastrophic failure in 3D structured systems and materials. As presented in this chapter, the combination of micromechanical testing and in situ high-resolution X-ray computed tomography to image crack propagation in microchips provides implications to interconnect design concepts, BEoL technology, and material integration.

Acknowledgments The authors thank Kristina Kutukova (deepXscan GmbH, Dresden, Germany); Martin Gall and Kong Boon Yeap (both with GLOBALFOUNDRIES, Malta/NY, USA); Andre Clausner, Juergen Gluch, Matthias Kraatz, Christoph Sander, Zhongquan Liao, and Yvonne Standke (all with Fraunhofer IKTS, Dresden, Germany); and Martyna Strag (Institute of Metallurgy and Materials Science of the Polish Academy of Sciences, Krakow, Poland) for performing experiments and sample preparation, as well as José M. Martínez-Esnaola, Ceit, San

Sebastián, Spain; Paul S. Ho, University of Texas, Austin/TX, USA; and Reinhard Dauskardt, Stanford University, Palo Alto/CA, USA, for helpful discussions.

References

1. E.T. Ogawa, J.W. McPherson, J.A. Rosal, K.J. Dickerson, T.C. Chiu, L.Y. Tsung, M.K. Jain, T.D. Bonifield, J.C. Ondrusek, W.R. McKee, Stress-induced voiding under vias connected to wide Cu metal leads. Proc. IEEE IPRS **2002**, 312 (2002)
2. E.T. Ogawa, K.D. Lee, V.A. Blaschke, P.S. Ho, Electromigration reliability issues in dual-damascene Cu interconnections. Proc. IEEE IRPS **2002**, 403 (2002)
3. P.S. Ho, K.D. Lee, E.T. Ogawa, S. Yoon, X. Lu, Impact of low-k dielectrics on electromigration reliability for Cu interconnects. Proc. Charact. Metrol. ULSI Technol, AIP Proc. **683**, 533 (2003)
4. J.R. Lloyd, E. Liniger, T.M. Shaw, Simple model for time-dependent dielectric breakdown in inter- and intralevel low-k dielectrics. J. Appl. Phys. **98**, 084109 (2005)
5. R.C.J. Wang, C.C. Lee, L.D. Chen, K. Wu, K.S. Chang-Liao, A study of Cu/low-k stress-induced voiding at via bottom and its microstructure effect. Microelectron. Reliab. **46**, 1673 (2006)
6. R.S. Achanta, J.L. Plawsky, W.N. Gill, A time dependent dielectric breakdown model for field accelerated low- k breakdown due to copper ions. Appl. Phys. Lett. **91**, 234106 (2007)
7. V. Sukharev, E. Zschech, W.D. Nix, A model for electromigration-induced degradation mechanisms in dual-inlaid copper interconnects: Effect of microstructure. J. Appl. Phys. **102**, 053505 (2007)
8. E. Zschech, P.S. Ho, D. Schmeisser, M.A. Meyer, A.V. Vairagar, G. Schneider, M. Hauschildt, M. Kraatz, V. Sukharev, Geometry and microstructure effect on EM-induced copper interconnect degradation. IEEE Trans. Device Mater. Reliab. **9**, 20–30 (2009)
9. H. Ceric, S. Selberherr, H. Zahedmanesh, R.L. de Orio, K. Croes, Review – Modeling methods for analysis of electromigration degradation in nano-interconnects. ECS J. Solid State Sci. Technol. **10**, 035003 (2021)
10. P. S. Ho, C. K. Hu, M. Gall, V. Sukharev (eds.), *Electromigration in Metals: Fundamentals to Nano-Interconnects* (Cambridge University Press, Cambridge, 2022)
11. T. Oshima, K. Hinode, H. Yamaguchi, H. Aoki, K. Torii, T. Saito, K. Ishikawa, J. Noguchi, M. Fukui, T. Nakamura, S. Uno, K. Tsugane, J. Murata, K. Kikushima, H. Sekisaka, Suppression of stress-induced voiding in copper interconnects. Proc. IEEE IEDM **2002**, 136 (2002)
12. T.C. Huang, C.H. Yao, W.K. Wan, H.H. Lin, C.C. Hsia, M.S. Liang, Numerical modeling and characterization of the stress migration behavior upon v 90 nanometer Cu/low-k interconnects. Proc. IEEE IITC **2003**, 207 (2003)
13. Z. Suo, Reliability of interconnect structures, in *Reliability of Interconnect Structures, Interfacial and Nanoscale Failure*, Comprehensive Structural Integrity, ed. by W. Gerberich, W. Yang, vol. 8, (Elsevier, Amsterdam, 2003), pp. 265–324
14. V. Sukharev, E. Zschech, A model for electromigration-induced degradation mechanisms in dual-inlaid copper interconnects: Effect of interface bonding strength. J. Appl. Phys. **96**, 6337–6343 (2004)
15. T.L. Tan, N. Hwang, C.L. Gan, Bimodal dielectric breakdown failure mechanisms in Cu–SiOC low-k interconnect system. IEEE Trans. Device Mater. Reliab. **7**, 373–378 (2007)
16. M.A. Meyer, M. Herrmann, E. Langer, E. Zschech, In-situ SEM observation of electromigration phenomena in fully embedded copper interconnect structures. Microelectron. Eng. **64**, 375–382 (2002)
17. G. Schneider, G. Denbeaux, E.H. Anderson, B. Bates, A. Pearson, M.A. Meyer, E. Zschech, E.A. Stach, Dynamical X-ray microscopy investigation of electromigration in passivated inlaid Cu interconnect structures. Appl. Phys. Lett. **81**, 2535–2537 (2002)

18. E. Zschech, R. Huebner, D. Chumakov, O. Aubel, D. Friedrich, P. Guttmann, S. Heim, G. Schneider, Stress-induced phenomena in nanosized copper interconnect structures studied by X-ray and electron microscopy. J. Appl. Phys. **106**, 093711 (2009)
19. K.B. Yeap, M. Gall, Z. Liao, C. Sander, U. Muehle, K.Y. Yiang, P. Justison, O. Aubel, M. Hauschildt, A. Beyer, N. Vogel, E. Zschech, In-situ study on low-k interconnect time-dependent-dielectric-breakdown mechanisms. J. Appl. Phys. **115**, 124101-1–124101-7 (2014)
20. V. Sukharev, A. Kteyan, E. Zschech, W.D. Nix, Microstructure effect on EM-induced degradation in dual-inlaid copper interconnects. IEEE Trans. Device Mater. Reliab. **9**, 87–97 (2009)
21. H. Li, M. Kuhn, Controlled fracture and mode-mixity dependence of nanoscale interconnects. IEEE Trans. Device Mater. Reliab. **17**, 636–642 (2017)
22. R. Pippan, S. Wurster, D. Kiener, Fracture mechanics of micro samples: Fundamental considerations. Mater. Des. **159**, 252–267 (2018)
23. J. Ast, M. Ghidelli, K. Durst, M. Goeken, M. Sebastiani, A.M. Korsunsky, A review of experimental approaches to fracture toughness evaluation at the microscale. Mater. Des. **173**, 107762 (2019)
24. E. Zschech, R. Radojcic, V. Sukharev, L. Smith, Stress management for 3D ICs using through silicon vias, in *AIP Conf. Proc. 1378* (Melville, New York, 2011)
25. L. Zhang, M. Kraatz, O. Aubel. C. Hennesthal, E. Zschech, P. S. Ho, Grain size and cap layer effects on electromigration reliability of Cu interconnects: Experiments and simulation, in *Int. Workshop "Stress-Induced Phenomena in Metallization", Bad Schandau 2010, AIP Proc.*, vol. 1300 (2010), p. 3–11
26. H. Li, T.M. Shaw, X.H. Liu, G. Bonilla, Delayed mechanical failure of the under-bump interconnects by bump shearing. J. Appl. Phys. **111**, 083503 (2012)
27. H. Li, Y. Lin, T.Y. Tsui, J.J. Vlassek, The effect of porogen loading on the stiffness and fracture energy of brittle organosilicates. J. Mater. Res. **24**, 107 (2009)
28. G. Pharr, W. Oliver, Measurement of thin film mechanical properties using nanoindentation. MRS Bull. **17**(7), 28–33 (1992)
29. G. Wang, P.S. Ho, S. Groothuis, Chip-packaging interaction: A critical concern for Cu/low k packaging. Microelectron. Reliab. **45**, 1079–1093 (2005)
30. X. H. Liu, T. M. Shaw, M. W. Lane, E. G. Liniger, B. W. Herbst, D. L. Questad, Chip-package interaction modeling of ultra low-k/copper backend of line, in *IEEE IITC* (2007), p. 13–15
31. C. Litteken, R. Dauskardt, T. Scherban, G. Xu, J. Leu, D. Gracias, B. Sun, Interfacial adhesion of thin-film patterned interconnect structures, in *Proc. IEEE IITC 2003* (2003), p. 168–170
32. A.V. Kearney, A.V. Vairagar, H. Geisler, E. Zschech, R. Dauskardt, Assessing the effect of die sealing in Cu/low-k structures, in *Proc. IEEE IITC 2007*, (2007), pp. 138–140
33. D. Chumakov, F. Lindert, M.U. Lehr, M. Grillberger, E. Zschech, Fracture toughness assessment of patterned Cu-interconnect stacks by Dual-Cantilever-Beam (DCB) technique. IEEE Trans. Semicond. Manuf. **22**, 592–595 (2009)
34. H. Li, M.J. Kobrinsky, A. Shariq, J. Richards, J. Liu, M. Kuhn, Controlled fracture of Cu/ultralow-k interconnects. Appl. Phys. Lett. **103**, 231901 (2013)
35. C. Sander, Y. Standke, S. Niese, R. Rosenkranz, A. Clausner, M. Gall, E. Zschech, Advanced methods for mechanical and structural characterization of nanoscale materials for 3D IC integration. Microelectron. Reliab. **54**, 1559–1562 (2014)
36. K. Kutukova, J. Gluch, M. Kraatz, A. Clausner, E. Zschech, In-situ X-ray tomographic imaging and controlled steering of microcracks in 3D nanopatterned structures. Mater. Des. **221**, 110946 (2022)
37. K. Kutukova, S. Niese, C. Sander, Y. Standke, J. Gluch, M. Gall, E. Zschech, A laboratory X-ray microscopy study of cracks in on-chip interconnect stacks of integrated circuits. Appl. Phys. Lett. **113**, 091901 (2018)
38. K. Kutukova, S. Niese, J. Gelb, R. Dauskardt, E. Zschech, A novel micro-Double Cantilever Beam (micro-DCB) test in an X-ray microscope to study crack propagation in materials and structures. Mater. Today Commun. **16**, 293–299 (2018)

39. A.G. Evans, J.W. Hutchinson, The thermomechanical integrity of thin-films and multilayers. Acta Metall. Mater. **43**, 2507–2530 (1995)
40. M.R. Elizalde, J.M. Sánchez, J.M. Martínez-Esnaola, D. Pantuso, T. Scherban, B. Sun, G. Xu, Interfacial fracture induced by cross-sectional nanoindentation in metal-ceramic thin film structures. Acta Mater. **51**, 4295–4305 (2003)
41. A. Needleman, A continuum model for void nucleation by inclusion debonding. J. Appl. Mech. **54**, 525–531 (1987)
42. A. Needleman, An analysis of tensile decohesion along an interface. J. Mech. Phys. Solids **38**, 289–324 (1990)
43. V. Tvergaard, J.W. Hutchinson, The relation between crack growth resistance and fracture process parameters in elastic-plastic solids. J. Mech. Phys. Solids **40**, 1377–1397 (1992)
44. V. Tvergaard, J.W. Hutchinson, The influence of plasticity on mixed mode interface toughness. J. Mech. Phys. Solids **41**, 1119–1135 (1993)
45. X. Zhou, J. Zimmerman, E. Reedy, N. Moody, Molecular dynamics simulation based cohesive surface representation of mixed mode fracture. Mech. Mater. **40**, 832–845 (2008)
46. S.P. Patil, Y. Heider, A review on brittle fracture nanomechanics by all-atom simulations. Nanomaterials **9**, 1050 (2019)
47. R.D.S.G. Campilho, M.D. Banea, J.A.B.P. Neto, L.F.M. da Silva, Modelling adhesive joints with cohesive zone models: Effect of the cohesive law shape of the adhesive layer. Int. J. Adhes. Adhes. **44**, 48–56 (2013)
48. A.V.M. Rocha, A. Akhavan-Safar, R. Carbas, E.A.S. Marques, R. Goyal, M. El-zein, L.F.M. da Silva, Numerical analysis of mixed-mode fatigue crack growth of adhesive joints using CZM. Theor. Appl. Fract. Mech. **106**, 102493 (2020)
49. X. Chang, T. Guo, S. Zhang, Cracking behaviours of layered specimen with an interface crack in Brazilian tests. Eng. Fract. Mech. **228**, 106904 (2020)
50. I. Ocaña, J.M. Molina-Aldareguia, D. González, M.R. Elizalde, J.M. Sánchez, J.M. Martínez-Esnaola, J. Gil Sevillano, T. Scherban, D. Pantuso, B. Sun, G. Xu, B. Miner, J. He, J. Maiz, Fracture characterization in patterned thin films by cross-sectional nanoindentation. Acta Mater. **54**, 3453–3462 (2006)
51. C.F. Shih, Cracks on bimaterial interfaces: Elasticity and plasticity aspects. Mater. Sci. Eng. A **143**, 77–90 (1991)
52. K. Bose, P. Ponte Castañeda, Stable crack growth under mixed-mode conditions. J. Mech. Phys. Solids **40**, 1053–1103 (1992)
53. M. Kenane, Z. Azari, S. Benmedakhene, M.L. Benzeggagh, Experimental development of fatigue delamination threshold criterion. Compos. B. Eng. **42**, 367–375 (2011)
54. N.A. Fleck, J.W. Hutchinson, Strain gradient plasticity. Adv. Appl. Mech. **33**, 295–361 (1997)
55. J.R. Greer, J. De Hosson, Plasticity in small-sized metallic systems: Intrinsic versus extrinsic size effect. Prog. Mater. Sci. **56**, 654–724 (2011)
56. M. Trueba, D. Gonzalez, M.R. Elizalde, J.M. Martinez-Esnaola, M.T. Hernandez, H. Li, D. Pantuso, I. Ocaña, Assessment of mechanical properties of metallic thin-films through micro-beam testing. Thin Solid Films **571**, 296–301 (2014)
57. M. Trueba, D. Gonzalez, J.M. Martinez-Esnaola, M.T. Hernandez, D. Pantuso, H. Li, M.R. Elizalde, I. Ocaña, Fracture characterization of thin-films by dual tip indentation. Acta Mater. **71**, 44–55 (2014)
58. X. Zhang, R.S. Smith, R. Huang, P.S. Ho, Chip-package interaction and crackstop study for Cu/ultra low-k interconnects. AIP Conf. Proc. **1143**, 197–203 (2009)
59. R. Huang, J.H. Prévost, Z.Y. Huang, Z. Suo, Channel-cracking of thin films with the extended finite element method. Eng. Fract. Mech. **70**, 2513–2526 (2003)
60. A. Volinsky, J. Vella, W.W. Gerberich, Fracture toughness, adhesion and mechanical properties of low-k dielectric thin films measured by nanoindentation. Solid Thin Films **429**, 201–210 (2003)
61. J.J. Vlassak, W.D. Nix, A new bulge test technique for the determination of Young's modulus and Poisson's ratio of thin films. J. Mater. Res. **7**, 3242–3249 (1992)

62. Y. Xiang, *Plasticity in Cu Thin Films: An Experimental Investigation of the Effect of Microstructure*, PhD Thesis (Harvard University, 2005)
63. B. Merle, M. Göken, Fracture toughness of silicon nitride thin films of different thicknesses as measured by bulge tests. Acta Mater. **59**, 1772–1779 (2011)
64. K. Matoy, H. Schönkerr, T. Detzel, T. Schöberl, R. Pippan, C. Motz, G. Dehm, A comparative micro-cantilever study of the mechanical behavior of silicon based passivation films. Thin Solid Films **518**, 247–256 (2009)
65. D. Kiener, W. Grosinger, G. Dehm, R. Pippan, A further step towards an understanding of size-dependent crystal plasticity: In situ tension experiments of miniaturized single-crystal copper samples. Acta Mater. **56**, 580–592 (2008)
66. E. Hintsala, D. Kiener, J. Jackson, W.W. Gerberich, In-situ measurement of free-standing, ultra-thin film cracking in bending. Exp. Mech. **55**, 1681–1690 (2015)
67. J.M. Martínez-Esnaola, J.M. Sánchez, M.R. Elizalde, A.M. Meizoso, Interfacial cracking in thin film structures, in *Fracture Mechanics: Applications and Challenges, ESIS 26*, ed. by M. Fuentes, M. Elices, A. Martín Meizoso, J. M. Martínez-Esnaola, (Elsevier Science Ltd, Oxford, 2000), pp. 47–71
68. A. Atkinson, R. Guppy, Measurement of adhesion of oxides grown on nickel-alloys. Mater. Sci. Technol. **7**, 1031–1041 (1991)
69. M.Y. He, F. Xu, D.R. Clarke, Q. Ma, H. Fujimoto, The energy release rate for decohesion in thin multilayered films on substrates, in *Materials Reliability in Microelectronics VII*, ed. by J. J. Clement, R. R. Keller, K. S. Krisch, J. E. Sanchez Jr., Z. Suo, (Materials Research Society, Pittsburgh, 1997), pp. 15–26
70. R.J. Hohlfelder, H. Luo, J.J. Vlassak, C.E.D. Chidsey, W.D. Nix, Measuring interfacial fracture toughness with the blister test, in *Thin Films: Stresses and Mechanical Properties VI*, ed. by W. W. Gerberich, H. Gao, J. E. Sundgren, S. P. Baker, (Materials Research Society, Pittsburgh, 1997), pp. 115–120
71. B. Cotterell, Z. Chen, The blister test – Transition from plate to membrane behaviour for an elastic material. Int. J. Fract. **86**, 191–198 (1997)
72. R.H. Dauskardt, M. Lane, Q. Ma, N. Krishna, Adhesion and debonding of multi-layer thin film structures. Eng. Fract. Mech. **61**, 141–162 (1998)
73. D.B. Marshall, A.G. Evans, Measurement of adherence of residually stressed thin-films by indentation. 1. Mechanics of interface delamination. J. Appl. Phys. **56**, 2632–2638 (1984)
74. M.J. Cordill, D.F. Bahr, N.R. Moody, W.W. Gerberich, Recent developments in thin film adhesion measurement. IEEE TDMR **4**, 163–168 (2004)
75. S.K. Venkataraman, J.C. Nelson, A.J. Hsieh, D.L. Kohlstedt, W.W. Gerberich, Continuous microscratch measurements of thin-film adhesion strengths. J. Adhes. Sci. Technol. **7**, 1279–1292 (1993)
76. M.D. Kriese, D.A. Boismier, N.R. Moody, W.W. Gerberich, Nanomechanical fracture-testing of thin films. Eng. Fract. Mech. **61**, 1–20 (1998)
77. J.M. Sánchez, S. El-Mansy, B. Sun, T. Scherban, N. Fang, D. Pantuso, W. Ford, M.R. Elizalde, J.M. Martínez-Esnaola, A. Martín Meizoso, J. Gil Sevillano, M. Fuentes, J. Maiz, Cross-sectional nanoindentation: A new technique for thin film interfacial adhesion characterization. Acta Mater. **47**, 4405–4413 (1999)
78. A.R. Zak, M.L. Williams, Crack point stress singularities at a bi-material interface. J. Appl. Mech. **30**, 142–143 (1963)
79. D.B. Bogy, Plane elastostatic problem of a loaded crack terminating at a material interface. J. Appl. Mech. **38**, 911–918 (1971)
80. T.C.T. Ting, P.H. Hoang, Singularities at the tip of a crack normal to the interface of an anisotropic layered composite. Int. J. Solids Struct. **20**, 439–454 (1984)
81. D.H. Chen, K. Harada, Stress singularities for crack normal to and terminating at bimaterial interface on orthotropic half-plates. Int. J. Fract. **81**, 147–162 (1996)
82. F. Erdogan, V. Biricikoglu, Two bonded half planes with a crack going through the interface. Int. J. Eng. Sci. **11**, 745–766 (1973)

83. A. Romeo, R. Ballarini, A crack very close to a bimaterial interface. J. Appl. Mech. **62**, 614–619 (1995)
84. J.M. Martínez-Esnaola, Stress singularity at the junction of three dissimilar materials. Int. J. Fract. **113**, L33–L37 (2002)
85. A.H. England, A crack between dissimilar media. J. Appl. Mech. **32**, 400–402 (1965)
86. F. Erdogan, Stress distribution in bonded dissimilar materials with cracks. J. Appl. Mech. **32**, 403–410 (1965)
87. B.M. Malyshev, R.L. Salganik, The strength of adhesive joints using the theory of cracks. Int. J. Fract. Mech. **1**, 114–128 (1965)
88. M. Comninou, The interface crack. J. Appl. Mech. **44**, 631–636 (1977)
89. M. Comninou, Interface crack with friction in the contact zone. J. Appl. Mech. **44**, 780–781 (1977)
90. C. Atkinson, The interface crack with a contact zone (an analytical treatment). Int. J. Fract. **18**, 161–177 (1982)
91. C. Atkinson, The interface crack with a contact zone (the crack of finite length). Int. J. Fract. **19**, 131–138 (1982)
92. J.R. Rice, Elastic fracture mechanics concepts for interfacial cracks. J. Appl. Mech. **55**, 98–103 (1988)
93. L. Ni, S. Nemat-Nasser, Interface cracks in anisotropic dissimilar materials: General case. Q. Appl. Math. **50**, 305–322 (1992)
94. S.R. Choi, C.H. Chong, Y.S. Chai, Interfacial edge crack in two bonded dissimilar orthotropic quarter planes under anti-plane shear. Int. J. Fract. **67**, 143–150 (1994)
95. K. Hayashi, S. Nemat-Nasser, Energy-release rate and crack kinking under combined loading. J. Appl. Mech. **48**, 520–524 (1981)
96. K. Hayashi, S. Nemat-Nasser, On branched, interface cracks. J. Appl. Mech. **48**, 529–533 (1981)
97. M.Y. He, J.W. Hutchinson, Kinking of a crack out of an interface. J. Appl. Mech. **56**, 270–278 (1989)
98. G.R. Miller, W.L. Stock, Analysis of branched interface cracks between dissimilar anisotropic media. J. Appl. Mech. **56**, 844–849 (1989)
99. M.Y. He, A. Bartlett, A.G. Evans, J.W. Hutchinson, Kinking of a crack out of an interface: Role of in-plane stress. J. Am. Ceram. Soc. **74**, 767–771 (1991)
100. T.C. Wang, Kinking of an interface crack between two dissimilar anisotropic elastic solids. Int. J. Solids Struct. **31**, 629–641 (1994)
101. J. Dundurs, Discussion: Edge-bonded dissimilar orthogonal elastic wedges under normal and shear loading. J. Appl. Mech. **36**, 650–652 (1969)
102. S.K. Maiti, R.A. Smith, Comparison of the criteria for mixed mode brittle fracture based on the preinstability stress-strain field. Part I: Slit and elliptical cracks under uniaxial tensile loading. Int. J. Fract. **23**, 281–295 (1983)
103. S.K. Maiti, R.A. Smith, Comparison of the criteria for mixed mode brittle fracture based on the preinstability stress-strain field. Part II: Pure shear and uniaxial compressive loading. Int. J. Fract. **24**, 5–22 (1984)
104. J.W. Hutchinson, Z. Suo, Mixed mode cracking in layered materials. Adv. Appl. Mech. **29**, 63–191 (1991)
105. T.M. Tong, T. Tan, N. Rahbar, W.O. Soboyejo, Mode mixity dependence of interfacial fracture toughness in organic electronic structures. IEEE Trans. Device Mater. Reliab. **14**, 291–299 (2014)
106. A. Agrawal, A.M. Karlsson, On the reference length and mode mixity for a bimaterial interface. J. Eng. Mater. Technol. **129**, 580–587 (2007)
107. Z. Chen, K. Zhou, X. Lu, Y.C. Lam, A review on the mechanical methods for evaluating coating adhesion. Acta Mech. **225**, 431–452 (2014)
108. T. Scherban, D. Pantuso, B. Sun, S. El-Mansy, J. Xu, M.R. Elizalde, J.M. Sánchez, J.M. Martínez-Esnaola, Characterization of interconnect interfacial adhesion by cross-sectional nanoindentation. Int. J. Fract. **119/120**, 421–429 (2003)

109. A. Ajdani, M.R. Ayatollahi, A. Akhavan-Safar, L.F.M. da Silva, Mixed mode fracture characterization of brittle and semi-brittle adhesives using the SCB specimen. Int. J. Adhes. Adhes. **101**, 102629 (2020)
110. T.J. Lu, Crack branching in all-oxide ceramic composites. J. Am. Ceram. Soc. **79**, 266–274 (1996)
111. X.F. Li, L.R. Xu, T-stresses across static crack kinking. J. Appl. Mech. **74**, 181–190 (2007)
112. Q. Yao, J. Qu, Interfacial versus cohesive failure on polymer-metal interfaces in electronic packaging – Effects of interface roughness. J. Electron. Packag. **124**, 127–134 (2002)
113. H.J. Kim-Lee, A. Carlson, D.S. Grierson, J.A. Rogers, K.T. Turner, Interface mechanics of adhesiveless microtransfer printing processes. J. Appl. Phys. **115**, 143513 (2014)
114. Z. Zhang, Z. Suo, Split singularities and the competition between crack penetration and debond at a bimaterial interface. Int. J. Solids Struct. **44**, 4559–4573 (2007)
115. A. Tambat, G. Subbarayan, Simulations of arbitrary crack path deflection at a material interface in layered structures. Eng. Fract. Mech. **141**, 124–139 (2015)
116. A. Tkachuk, F. Duewer, H. Cui, M. Feser, S. Wang, W. Yun, X-ray computed tomography in Zernike phase contrast mode at 8 keV with 50-nn resolution using Cu rotating anode X-ray source. Z. Krist. **222**, 650–655 (2007)
117. A.P. Merkle, J. Gelb, Ascent of 3D X-ray microscopy in the laboratory. Microsc. Today **21**(2), 10–15 (2013)
118. E. Zschech, M. Loeffler, P. Krueger, J. Gluch, K. Kutukova, I. Zglobicka, J. Silomon, R. Rosenkranz, Y. Standke, E. Topal, Laboratory computed X-ray tomography – A nondestructive technique for 3D microstructure analysis of materials. Pract. Metallogr. **55**, 539–555 (2018)
119. P.G. Charalambides, J. Lund, A.G. Evans, R.M. McMeeking, A test specimen for determining the fracture resistance of bimaterial interfaces. J. Appl. Mech. **56**, 77–82 (1989)
120. F. Berto, A. Campagnolo, L.P. Pook, Three-dimensional effects on cracked components under anti-plane loading. Frat. ed Integrità Strutt. **33**, 17–24 (2015)
121. K. Kutukova, S. Werner, P. Guttmann, G. Schneider, E. Zschech, High-resolution spectromicroscopy of silicon compounds in integrated circuits, in *18th International Conference on X-Ray Absorption Fine Structure XAFS 2022*, (Sydney, 2022)
122. A. Grill, S.M. Gates, T.E. Ryan, S.V. Nguyen, D. Priyadarshini, Progress in the development and understanding of advanced low k and ultralow k dielectrics for very large-scale integrated interconnects—State of the art. Appl. Phys. Rev. **1**, 011306 (2014)
123. X. Zhang, Y.W. Wang, J.H. Im, P.S. Ho, Chip-package interaction and reliability improvement by structure optimization for ultralow-k interconnects in flip-chip packages. IEEE Trans. Device Mater. Reliab. **12**, 462–469 (2012)
124. J. Benbow, F.C. Roessler, Experiments on controlled fractures. Proc. Phys. Soc. **B70**, 201–211 (1957)
125. F. Berto, A. Campagnolo, Three-dimensional effects on cracked components under anti-plane loading. Frat. ed Integrità Strut. **33**, 17–24 (2015)
126. O. A. Bauchau, J. I. Craig (eds.), *Euler-Bernoulli Beam Theory* (Springer, Dordrecht, 2009), pp. 173–221
127. C.H. Wang, *Introduction to Fracture Mechanics* (DSTO Aeronautical and Maritime Research Laboratory, Melbourne, 1996)
128. J. M. Gere, S. P. Timoshenko (eds.), *Mechanics of Materials* (PWS Publishing Company, Boston, 1997)
129. M. Baklanov, P. S. Ho, E. Zschech (eds.), *Advanced Interconnects for ULSI Technology* (John Wiley & Sons, Chichester, 2012)
130. F. Iacopi, G. Beyer, Y. Travaly, C. Waldfried, D.M. Gage, R.H. Dauskardt, K. Houthoofd, P. Jacobs, P. Adriaensens, K. Schulze, S.E. Schulz, S. List, G. Carlotti, Thermomechanical properties of thin organosilicate glass films treated with ultraviolet-assisted cure. Acta Mater. **55**, 1407–1414 (2007)
131. S.W. King, Dielectric barrier, etch stop, and metal capping materials for state of the art and beyond metal interconnects. ECS J. Solid State Sci. Technol. **4**, N3029–N3047 (2015)

132. M. Lane, R.H. Dauskardt, A. Vainchtein, H. Gao, Plasticity contributions to interface adhesion in thin-film interconnect structures. J. Mater. Res. **15**, 2758–2769 (2000)
133. M. Lane, R.H. Dauskardt, N. Krishna, I. Hashim, Adhesion and reliability of copper interconnects with Ta and TaN barrier layers. J. Mater. Res. **15**, 203–211 (2000)
134. Y. Wei, J.W. Hutchinson, Models of interface separation accompanied by plastic dissipation at multiple scales. Int. J. Fract. **95**, 1–17 (1999)
135. P.A. Klein, H. Gao, A. Vainchtein, H. Fujimoto, J. Lee, Q. Ma, Micromechanics-based modeling of interfacial debonding in multiplayer structures. Proc. MRS **594**, 371–376 (2000)
136. www.deepxscan.com
137. M. Mirkhalaf, A.K. Dastjerdi, F. Barthelat, Overcoming the brittleness of glass through bio-inspiration and micro-architecture. Nat. Commun. **5**, 3166 (2014)
138. F. Barthelat, Z. Yin, M.J. Buehler, Structure and mechanics of interfaces in biological materials. Nat. Rev. Mater. **1**, 16007 (2016)
139. Y. Wang, S.E. Naleway, B. Wang, Biological and bioinspired materials: Structure leading to functional and mechanical performance. Bioact. Mater. **5**, 745–757 (2020)
140. M. Strag, L. Maj, M. Bieda, P. Petrzak, A. Jarzebska, J. Gluch, E. Topal, K. Kutukova, A. Clausner, W. Heyn, K. Berent, K. Nalepka, E. Zschech, A.G. Checa, K. Sztwiertnia, Anisotropy of mechanical properties of Pinctada margaritifera mollusk shell. Nanomaterials **10**, 634 (2020)
141. J. Ast, M. Ghidelli, K. Durst, M. Goeken, M. Sebastiani, A.M. Korsunsky, A review of experimental approaches to fracture toughness evaluation at the micro-scale. Mater. Des. **173**, 107762 (2019)

Chapter 6
Neuromorphic Computing for Compact LiDAR Systems

Dennis Delic and Saeed Afshar

1 Introduction

The last half century has seen remarkable advances in complementary metal-oxide-semiconductor (CMOS) electro-optic (EO) imaging systems. Device scaling due to Moore's law has been a driving factor in improvements, such as low light performance, pixel density, and noise reduction. But as device scaling approaches the physical limit of semiconductors, the exponential pace of miniaturization and cost reduction has begun to slow, motivating the search for technology solutions that can bridge the gap by improving performance while adding more functionality through integration. More-than-Moore devices represent this new paradigm of functional diversification of technologies and innovation by combining performance, integration, and cost which are no longer limited to CMOS scaling. Solid-state LiDAR imaging sensors packing more pixels per unit size is achievable, for example, by using novel three-dimensional (3D) stacking semiconductor assembly methods, a level of integration just not possible with standard monolithic techniques, while advanced computation methods in design such as machine vision techniques allow for increased functionality and performance.

EO sensing or imaging systems can be categorized as using either passive or active techniques. The sensing of electromagnetic radiation, in the form of visible, light, UV, infrared, and X-ray, has enormous practical applications. The ultimate sensitivity is the detection of individual photons or single-photon imaging. A type of active EO sensing system is laser imaging RADAR also known as Laser Detection

D. Delic (✉)
Defence Science and Technology Group, Edinburgh, SA, Australia
e-mail: dennis.delic@dst.defence.gov.au

S. Afshar
Western Sydney University, Sydney, NSW, Australia
e-mail: s.afshar@westernsydney.edu.au

F. Iacopi, F. Balestra (eds.), *More-than-Moore Devices and Integration for Semiconductors*, https://doi.org/10.1007/978-3-031-21610-7_6

and Ranging (LADAR) or Light Detection and Ranging (LiDAR) – these are the optical equivalent of RADAR but using light waves. Use of these terms is non-standardized; however, a typical distinction involves the target type. LADAR is used for hard targets (solids) or surface scattering while LiDAR for soft targets (e.g., gases or aerosols) or volume scattering within the medium [1]. But in this chapter, the term LiDAR will be used in a generic sense and will refer to both soft and hard targets.

LiDAR is used extensively in many diverse fields both military, biomedical, security, and civilian applications. For the military, advanced sensing technologies give enhanced situational awareness and decision superiority in the battlespace which significantly improve defense personnel survivability and mission success. There are many sensing systems which can address these needs; these include scanning LiDAR, 3D LiDAR imagers, thermal imaging, and surveillance RADAR [2]. LiDAR for military and security applications include target recognition, target location, aim point selection, tracking, and weapon guidance [3]. Such uses include rangefinders, designators, weapon guidance, and ability to undertake rapid environmental surveying, which includes both topographic (terrain) and hydrographic mapping (laser-based bathymetry) [3]. Navy vessels are exposed to threats, such as anti-ship missiles (ASMs), unmanned aerial systems (UAS), submarines, and submerged mines, and need advanced surveillance LiDAR technologies. Amphibious craft require rapid environmental assessment and littoral mapping for surface mine detection and threat avoidance in contested environments. Land tactical intelligence needs structural assessment and through-tree-foliage visibility to improve situational awareness of potential threats and aid decision-making.

A type of LiDAR called single-photon LiDAR (SPL) is a promising class of technology that relies on single-photon detectors (SPDs) or sensors to achieve extremely sensitive detection levels and performance. These systems dramatically increase the point (data) density on a target through detection and processing of low photon numbers at the receiver or sensor. They attempt to operate at the photon limit or detectability of light using high quantum efficient sensors. As such, a major benefit of SPL technology lies in the fact that by operating at such low photon fluxes, SPL can reduce the chance of detection by threat systems, detect a target in low levels of illumination, or reveal partially concealed targets at extended range especially through obscuring mediums such as water, fog, haze, smoke, dense foliage, and camouflage nets. Single-photon sensors offer revolutionary advances in performance over conventional technologies. Exploiting quantum properties such as photon entanglement and/or superposition will only see further enhancement to the performance of single-photon sensors enabling emerging technologies such as quantum LiDAR [4] and quantum imaging [5]. Long-range depth profiling of camouflaged targets using single-photon detection has already been demonstrated [6]. This alone provides powerful new defense capabilities in sensing behind obscurants with low detectability by adversaries.

A type of solid-state SPD is the Geiger-mode avalanche photodiode (Gm-APD) also known as a single-photon avalanche diode (SPAD), which are designed and biased in such a way that they can detect a single photon or quantum of light.

When accompanied with electronic circuits, a SPAD can perform both precise photon counting and timing functions. Creating an array of SPADs is possible by miniaturizing SPAD devices so that they all can fit onto a single integrated circuit (IC) chip which then can be used to build a SPAD camera. The level of accuracy and information processing offered from this type of camera allows hyper-sensitive and hyper-temporal 2D imaging, and when coupled with high-precision pulsed laser illuminators as part of a LiDAR system, accurate time-of-flight (TOF) information can be measured from the returning photons in a scene at both close and long ranges. Such TOF information can then be used to construct high-resolution 3D images, which can aid in eliminating ambiguities and uncertainties found in trying to classify targets or objects as taken by traditional 2D cameras. SPAD sensors in SPLs can be considered the key component in advanced single-photon EO systems designed for target detection, identification, and tracking. These systems can complement the existing sensor suite of tools employed by the war fighter, delivering tactical impact in three key priority areas which include early warning missile defense, anti-submarine warfare (ASW) [7],[1] and advanced surveillance assessment capabilities [8].

The use of small, cheap SPD and SPAD smart sensors is only likely to proliferate with equal impact for LiDAR technologies outside of the defense domain, with advantage for time-gated applications, fluorescence lifetime imaging (FLIM), time-correlated single-photon counting (TCSPC), positron emission tomography (PET), and single-photon mission computed tomography (SPECT) [9–12]. For commercial applications, SPL is a key enabling technology for autonomous vehicles; this includes automotive LiDAR for advanced driver-assistance systems (ADAS) as well as self-driving cars in future assistance systems [13, 14]. 3D imaging systems have also been incorporated into smartphones from several different manufacturers and broader use applications, which includes augmented reality/virtual reality devices (AR/VR), drones, robots, and industrial safety systems [15, 16].

There are many SPL configurations for 3D imaging which use SPAD sensors and cameras. TOF LiDAR scanning methods are reviewed and are typically single-pixel SPD devices that take time to build a 3D image by progressively rastering a scene. The problem with scanning SPL systems is that they suffer from long execution times which preclude their application to real-time analysis of highly dynamic and cluttered scenes. Another type of SPL, called 3D flash LiDAR, relies on a focal-plane array (FPA) sensor which is an array or collection of SPAD detectors integrated with CMOS digital circuits onto the same silicon chip or bonded separately to a readout integrated circuit (ROIC). 3D imaging cameras constructed using SPAD arrays/FPA sensors for 3D flash LiDAR systems are also known as 3D TOF cameras. These types of cameras acquire a 3D depth-resolved image of the scene instantaneously, increasing the number of 3D points and enabling high-speed imaging and classification of moving targets. SPAD-based FPA sensors operating in 3D flash LiDAR systems have no moving mechanical parts and scanning optics

[1] https://www.navy.gov.au/media-room/publications/tac-talks-52

like scanning TOF LiDAR systems; offer an improved Size, Weight, and Power (SWaP) footprint; and allow faster image acquisition/reconstruction time especially important when objects are moving or when large areas need to be surveyed quickly; good examples include spacecraft docking vehicles [17]. Being an imaging system, they have advantages in the information you can ascertain from targets or in topographic/hydrographic mapping scenarios, which requires intelligent recognition processing.

It should also be noted that airborne laser scanning (ALS) systems by virtue of the flight direction of the aerial vehicle carrying the LiDAR equipment can form a 3D topographic image rather quickly – but only along its flight line. Commercially available ALS systems are categorized as either linear-mode LiDAR (LM-LiDAR), Geiger-mode LiDAR (GM-LiDAR),[2] or SPL.[3] The SPL system available under the brand name Leica SPL100, for instance, is a scanning system that splits the output laser beam using a diffractive optical element (DOE) into 10 × 10 beamlets (this technique provides wider area mapping); the receiving beamlet is detected by its corresponding detector which is comprised of an array of silicon photon multipliers (SiPM), each SiPM containing many SPAD cells [18]. SiPM sensors are not imaging devices per se although they can reach single-photon sensitivities; the capabilities of LiDAR systems using SiPM are explored by [19]. But for applications requiring long-range imaging and tracking of moving objects, 3D flash LiDAR systems represent a promising technology of choice. The limitations and advantages of these types of LiDARs are examined.

Improving the performance of 3D flash LiDAR or 3D TOF cameras is very much dictated by SPAD detector design as well as by the sensors' ROIC capabilities. Applications such as satellite-based surveillance and hypersonic and ballistic tracking space sensor (HBTSS) would benefit from ultra-sensitive sensing and fast ROICs to detect and track faint objects hidden in background clutter [20]. The SPAD detector's ability to detect individual photons allows the tagging of individual photon arrival times to be done with such accuracy that the spatial resolution of the target can be in the millimeters' at very long ranges. This also means that there is only a small amount of energy needed to illuminate the target, and aggregate photons from the target (even if partially concealed) can be used to reconstruct its shape. These benefits make the technology particularly attractive for applications requiring improved situational awareness.

A quick introduction is then given to neuromorphic engineering and neuromorphic vision cameras, which are biologically inspired information processing systems which are implemented directly in hardware and are largely acknowledged as the next step in powerful computer processing to help develop advanced machine learning and artificial intelligence (AI). Neuromorphic sensor-processor systems operate much like the synapses in the human brain, achieving a high level of optimized computational efficiency in how data is processed. Neuromorphic

[2] https://www.l3harris.com/all-capabilities/geiger-mode-lidar

[3] https://leica-geosystems.com/products/airborne-systems/topographic-lidar-sensors

processing methods are highly compatible with a class of "event-based" imaging sensors/cameras known as neuromorphic vision cameras. Unlike normal (frame-based) cameras, each pixel in these cameras operates independently and only outputs asynchronous data packets or "events" in response to a change in the visual scene.

Research has also started on the development of SPAD-based imaging devices inspired by these neuromorphic vision sensors in which the SPAD pixels themselves also work in an asynchronous fashion responding only to events in the visual field. Operating at high frame rates in 3D flash LiDAR systems, SPAD imagers typically generate large volumes of noisy and largely redundant spatiotemporal data. This results in communication bottlenecks and unnecessary data processing. These factors make traditional frame-based SPAD imagers ideally suited for optimization through neuromorphic approaches, and when coupled with machine learning image processing and AI techniques, this new paradigm represents unparalleled target detection and recognition performance. This new class of SPAD imaging system based on these neuromorphic principles are known as "event-based" SPAD imagers (and/or sensors) and offer advantages over conventional frame-based SPAD imagers in their ability to minimize susceptibility to noise and reduce detection time and data output rates.

In the context of achieving More-than-Moore, the utility and limitations of current SPAD FPA sensors and imagers operating in 3D flash LiDAR mode will be discussed. This chapter will focus on researching CMOS IC design and layout techniques that can be used in the successful performance enhancements of SPAD arrays or FPA sensors and address their limitations for 3D imaging and tracking applications. Going beyond Moore's law in terms of functionality and diversification, FPA sensors can have an array or collection of SPAD detectors integrated with CMOS digital circuits onto the same silicon chip or bonded separately to a ROIC with advanced manufacturing techniques. Improving the performance of FPA sensors is only part of the story because as these FPAs scale with pixel count, there is a high density of data gathered resulting in very large datasets. This presents new challenges in image processing and how to deal with the huge volume of data generated; this is where event-based SPAD imagers and neuromorphic processing techniques can play a role.

1.1 Background to Single-Photon LiDAR Systems

3D range sensor systems provide advantages over 2D imaging systems especially for the purposes of removing ambiguities in a scene, but typically lack in terms of image quality and rendering time compared to 2D counterparts. Removing ambiguities means 3D systems possess the ability to segment the object of interest from the background obscurants and clutter; this benefit increases target recognition performance [21]. Various methods for acquiring 3D range images include methods based on interferometry,[4] pattern projection,[5] triangulation, and TOF principles [22]. It is the latter TOF approaches that will be the focus of further discussion.

SPL is an advanced single-photon EO system for detection and identification using photon counting techniques. It is comprised of four main components: (i) an illumination source, (ii) a SPD or sensor, (iii) fast timing electronics, and if doing imaging (iv) an image retrieval algorithm. The most common choice of detector for SPL systems is the solid-state SPAD device, which consists of a reverse-biased photodiode biased above the breakdown voltage so that an individual photon incident on the SPAD can cause an avalanche of electrical charge carriers that is directly detectable as a digital signal. Unlike "thermal detectors" which detect changes in temperature (such as bolometers, etc.), "photon detectors" measure the number of incident photons (and measure changes in electron mobility as a result) on the sensor. In the "photon detector" category, a SPAD detector has single-photon sensitivity, while photodiodes (as used in CCDs and 3D CMOS imaging [23] and low light sensors[6]) rely on a large flux (in cases millions) of photons [15]. Other examples of photon detector technologies include PMT [24], EMCCD [25], etc. In recent years, single-photon counting LiDAR systems have been widely used in 3D imaging under weak light conditions because of its single-photon sensitivity and picosecond timing resolution.

TOF systems rely on a single sensor, and depth information is obtained by means of active light illumination (of a specific wavelength); here, the sensor can capture distances by measuring the travel time taken for an emitted light signal (light wave or pulse) to hit a target and then reflect back toward the receiver. There are various 3D imaging systems available based of TOF principles; see Fig. 6.1.

TOF systems are typically classified according to the type of light signals used, modulated if the signal is a continuous wave or pulsed TOF if light pulses are used. TOF techniques can be grouped into direct TOF (DTOF) and indirect TOF (ITOF) categories. DTOF methods directly measure the time delay by means of a very accurate timer. Depending on the accuracy or resolution of the timer, this method is typically used for long (kilometers) distance (or range) measurements and at very-high-precision (millimeter) depth resolutions. The ITOF method in contrast reconstructs the time delay (hence distance) from the measurement of the phase delay of the reflected signal when compared to the periodic emitted light signal. This technique is more suited to short or medium distances (tens of meters) and with depth resolutions of a few centimeters [28]. For the ITOF technique, two methods can be implemented: either (1) continuous-wave ITOF (CW-ITOF), whereby a sinusoid modulated light source illuminates the scene and the returned signal is sampled a few times during the modulation period to compute the phase delay, or (2) a pulsed-light ITOF (P-ITOF) method where the illuminator uses square pulses

[4] Stereo LiDAR Vision System Clarity: https://light.co/clarity

[5] 3D Imaging using Structured Light: https://www.photoneo.com/motioncam-3d/

[6] Low Light Camera with IMX428 CMOS imaging sensor: https://diffractionlimited.com/product/sbig-stc-428-p/

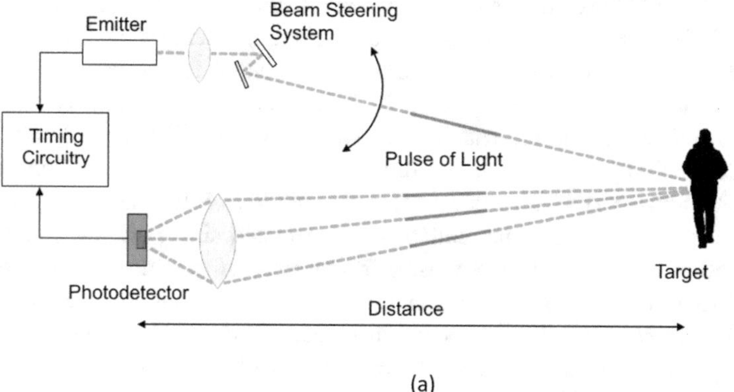

(a)

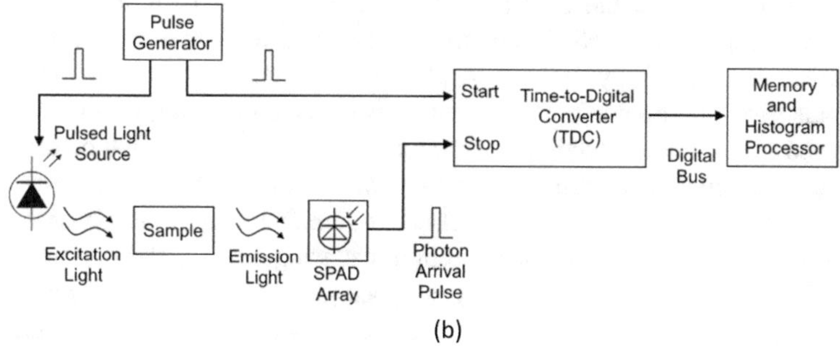

(b)

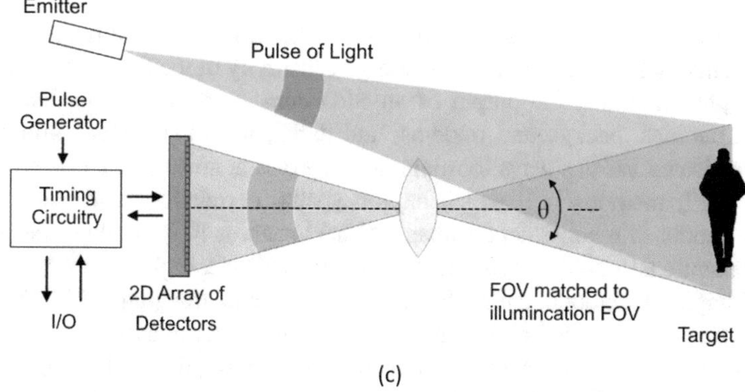

(c)

Fig. 6.1 Different LiDAR operating configurations. (**a**) Scanning LiDAR system, Ref. [26]. (**b**) TCSPC system for fluorescence lifetime imaging microscopy and 3D imaging, Ref. [27]. (**c**) Staring or 3D flash LiDAR system, Ref. [26]

of light [29]. This chapter will focus on using technology to implement specifically DTOF methods for LiDAR; this is inferred when referring to TOF techniques.

The first 3D TOF cameras were constructed as scanning laser systems comprising a laser range finder with a rotating or scanning element(s) to progressively scan the field of view (FOV). These scanning LiDAR systems are effectively single-pixel devices collecting TOF information in a single direction which build up a 3D image progressively by moving the pointing direction of the sensing element; see Fig. 6.1a. SPAD-based TOF devices are often referred to as SPL sensors [30]. SPL systems often employ a technique called time-correlated single-photon counting (TCSPC), and it is considered the most effective way to measure the temporal structure of weak optical signals. Originally developed for fluorescence lifetime imaging microscopy [3, 6], TCSPC began to be used in the field of laser ranging [31, 32] and gradually developed to the direction of laser long-distance 3D imaging based on TOF algorithms [33–35].

The TCSPC technique is essentially a statistical sampling method that records the arrival time of a photon with respect to a synchronization signal, with picosecond temporal resolution. The basic idea of TCSPC is like that of a stopwatch: the laser starts a timer with each illumination pulse, and the timer is stopped with the detection of a photon by the sensor. The time difference between the stop and start signals gives the photon's TOF; see Fig. 6.1b. Due to timing uncertainty and the presence of nuisance detections due to ambient light, the illumination signal is repeated to build up a histogram of photon detection times with the histogram peak typically indicating the true TOF. There has been significant research demonstrating 3D imaging using TCSPC techniques, through scattering media and obscurants such as fog, underwater, and at long range in free space [36–38]. Static targets have been imaged at ranges 45 km and then 200 km kilometers away; such long-range imaging is achievable with new noise suppression techniques to extract the weak TOF signal from high background noise, requiring as few as 0.44 signal photon counts per pixel [39, 40].

In general, for a TCSPC system, the image result is greatly affected by environmental noise because of the ultra-high sensitivity of the SPDs. At extremely low flux photon levels, the output of an SPD consists of a Poisson distribution of signal photons, background photons, and dark counts [41]. Many approaches have investigated various ways to minimize the Poisson noise, such as first-photon imaging (FPI) methods [42] which exploit spatial correlations and use the first detected photon to reconstruct an image. Others such as [43] employ first-photon and computational ghost imaging techniques to realize a high-efficiency photon-limited imaging technique, called fast first-photon ghost imaging (FFPGI), which uses less than 0.1 detected photon per pixel to retrieve an image. However, these approaches are limited to raster scanning with a SPD to build a 3D image; others have used an array of SPDs to realize highly efficient images with excellent noise rejection [44].

Further techniques used to decrease the impact of noise include range-gated technology [39, 45] and asynchronous acquisition methods [46]. Range-gated imaging or gate viewing (GV) is a well-known active imaging technique where a pulsed laser is synchronized to a gated sensor. Under the laser illumination, only target reflectance that arrives at the sensor within the right timing window is considered. The timing window is determined by the time delay between laser pulse and gate pulse as well as the gate time. This prevents background photons triggering the detectors too early in each frame. Various scanning TCSPC LiDAR systems implementing a range-gated approach have been used to identify targets that have been obscured by clutter, in other words having the ability to "see" behind or through various obscuring media such as camouflage nets [6]. At the expense of maximizing the range measurement, it is possible by intensity dividing a laser-return pulse into two SPDs (in this case SPADs) as demonstrated by [47], and comparing the arrival times using an AND gate function, this can drastically reduce false positives caused by environmental noise. Others, such as [48], propose a depth image denoising method (DIDM) based on photon TOF correlation and target spatial correlation to decrease the effect of environmental noise.

Advanced computational processing techniques are now increasingly being used to enhance SPL imaging capabilities. This includes methods to reconstruct images from sparse photon data [49] to non-line-of-sight imaging [50]. Among the various possible computational imaging algorithms, machine learning, and in particular deep learning, provides a statistical or data-driven model for enhanced single-photon 3D imaging retrieval. Good examples of using deep learning-based single-photon 3D imaging for processing multiple LiDAR returns are presented by [51, 52]. For imaging reconstruction using scanning SPL in [53, 54], and using a deep learning and sensor fusion approach to photon efficient 3D imaging is explored by [55]. Finally, imaging from temporal data via spiking convolutional neural networks (CNNs) is presented by [56]; here, arrival times of photons received by the SPD are in the form of spikes distributed over time. These promising machine learning methods have a major disadvantage in that they suffer from long execution times which precludes their application to real-time analysis of highly dynamic and cluttered scenes.

There are various non-scanning or scanner-less 3D imaging LiDAR system techniques available [57]. This includes direct-detection flash LiDAR which is based on DTOF ranging; these are also known as staring, 3D flash LiDAR systems, or recently 3D TOF cameras [58]. Here, the principle of operation is like single-pixel LiDAR scanning systems, except the whole FOV is illuminated at once using a wide-angle laser source; see Fig. 6.1c. A laser pulse irradiates a target with a short-duration laser pulse (i.e., a laser flash), and photons that are backscattered off objects produce an instantaneous image onto the sensor; the TOF measurement of returning photons gives a 3D (distance measured) image. The sensor in scanner-less (i.e., staring) 3D flash LiDAR systems is usually comprised of an array of detectors also known as a FPA and hence able to achieve more rapid scene capture than scanning systems. This technique can avoid problems associated with scanning/beam steering systems such as mechanical wear, vibration, and/or motion blur [32, 59]. 3D flash systems have some advantages over existing LiDAR scanning methods. For instance, they have no moving mechanical parts and scanning (beam steering) optics; hence, they offer an improved SWaP (compact) footprint which means they can fit easily on power-constrained and mobile platforms such as an unmanned aerial system (UAS) or helicopter [60]. They acquire a 3D depth-resolved image of a scene instantaneously and are especially useful when targets are moving or when large areas need to be surveyed quickly – allowing faster identification of targets obscured or hidden in background clutter. The trade-off is that 3D flash LiDAR system due to the number of detectors on the FPA is less sensitive than single-pixel scanning systems and any SWaP saving may be offset by the need for larger optics and laser just to compensate for this reduction in sensitivity.

1.2 Flash 3D Imaging Laser RADAR and Applications

LiDAR technology has been employed in defense active EO systems for target detection and identification and tactical applications requiring imaging and ranging information for many years. Of those, SPL systems which operate in 3D flash LiDAR or 3D TOF mode employ a FPA or "SPAD array" sensor as the main sensing component. The utility of these types of systems to wide area surveillance [61] and the long-range sensing (and tracking) of low-signature (hard-to-see) targets in highly dynamic (moving) scenes and clutter are explored in this section.

A FPA is an array of pixels (or detectors) coupled to a ROIC. The term "focal plane" comes from the placement of the device in the optical system. The type of detector can be either a photodetector (i.e., photodiode or photoconductive) or a thermal detector [62]. Avalanche photodiode (APD) is a type of photodetector which consists of an absorption region and a multiplication region. Two layers – a p-doped and n-doped semiconductor material which creates a pn-junction depletion (or multiplication) region (i.e., a diode). The multiplication region is designed to exhibit a high electric field to provide internal photo-current gain by impact ionization (or avalanche). The APD structure can operate in either linear mode (LM) or Geiger mode (GM); see Fig. 6.2.

The linear-mode APD (LM-APD) is biased slightly below breakdown of the pn-junction and provides a linear gain with photoelectrons detected; a flux of photons (many) is needed for a signal to be registered. A Geiger-mode avalanche photodiode (GM-APD), also called a SPAD, operates at a bias voltage above breakdown, so in principle only a single photon is needed to set off an avalanche, triggering the detector. SPAD sensors when coupled with electronic circuits can then count individual photon events and precisely record their time of arrival. The speed with which photons can be recorded means SPAD sensors are capable of capturing transient events at a trillion frames per second (FPS) [63] and also useful for passive single-photon 2D intensity imaging of low light and highly dynamic (fast motion) scenes, ideal for burst photography applications [64].

Both LM-APD and GM-APD sensor modalities underpin 3D flash LiDAR imaging technologies. LM-APDs have a noise floor significantly higher than a single photon (hence, more laser energy is required to image a given scene than SPADs), but there has been effort in developing higher-sensitivity LM-APDs [65]. The

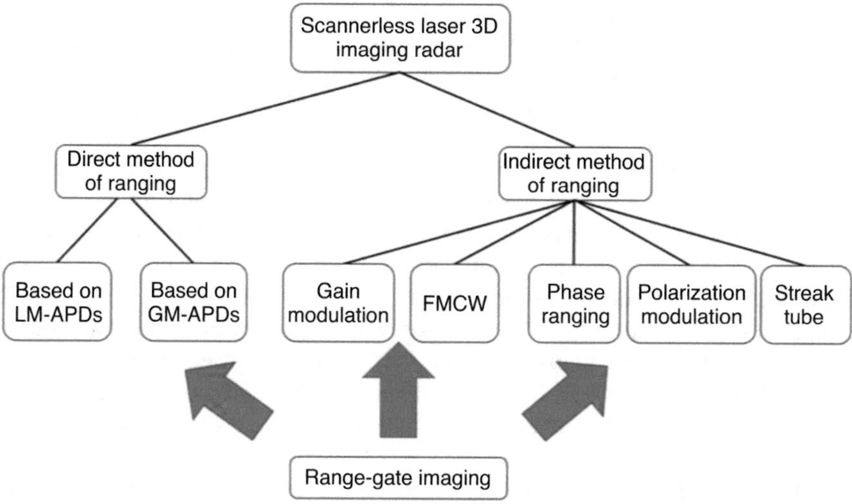

Fig. 6.2 Non-scanning or scanner-less laser 3D imaging radar classification, Ref. [57]

limitations of GM-APD/SPAD array's operation for direct-detection flash LiDAR are compared to LM-APD detector-based systems in [66–68]. SPAD performance is greatly hampered by its ability to detect more than one photon per dead time period. The dead time is the interval of time required to reset the SPAD detector, so it is ready for the next incoming photon event. Because a SPAD's dead time is long relative to the rate of incoming photons, later target photon arrivals at the SPAD sensor's optical aperture from the same pulse are not detected if an earlier photon event has been recorded. These shortcomings were shown to be a direct result of the "dead time" characteristic of SPAD sensors, which causes the SPAD sensor to behave non-linearly with changes in target range, laser power, detector efficiency, and the scattering properties of foreground objects. However, these issues can often be mitigated with range gating techniques, and with improved SPAD sensor design recovery times are getting less. Other disadvantages are that the detection efficiency of the SPAD sensor is diminished by the dark counts (triggering events in the absence of photons) generated within the detector, as well as by background passive photon events, or scene-reflected active laser returns. Due to these combined effects, a significant amount of target signal information present at the optical aperture of the LiDAR receiver is not collected by the SPAD sensor.

It is widely accepted that more information would be available from LM-APDs which preserve the amplitude information (intensity) as well as the timing (TOF) information. However, the practical downside is that each LM-APD requires a linear electronic amplifier chain (including biasing and automatic gain control) to bring its signal up to the level (above the noise floor) at which it can be applied to a threshold detector or analog-to-digital converter. These amplifiers needed by the LM-APD detectors require continuous biasing into their linear amplification region, so they are ready for a minimum of 50–100 photons to register a response. For single-channel ranging receivers, this may be acceptable, but for larger array implementations, this would present system design challenges as all the extra power and heat dissipation would need to be managed with dedicated cooling.

For these reasons, this chapter will focus exclusively on Geiger-mode APD, or SPAD arrays/FPA sensor implementations, specifically those based on silicon CMOS technology. The FPA material type is usually dependent on the spectral range required; intrinsic silicon is commonly used for visible wavelengths, although other materials such as indium gallium arsenide (InGaAs) can be used to construct SPAD image sensors targeting short-wave infrared (SWIR, 1.5 μm) 3D flash LiDAR system applications [69–71]. These longer wavelength systems may have better eye safety laser energy limits; however, they suffer decreased spatial resolution (as compared to visible wavelengths) when identifying objects with detail due to diffraction limitations (for practical aperture sizes). There are various assembly techniques used to construct FPAs such as via monolithic techniques, where the detector array and ROIC are implemented on the same planar chip, or via 3D stacking methods, such as flip chip bonded or through-silicon via (TSV) technology, but this will be explored in more detail in later sections.

Figure 6.3 illustrates a typical 3D flash LiDAR setup using a SPAD sensor/camera. The FPA or SPAD array chip is positioned at the correct focal plane of the accompanying objective in the camera and represents the key component in the sensing system.

In operation, a pulsed laser light source illuminates the scene by using an optical dispersive lens to achieve a wide FOV covering the target. The laser echo is reflected

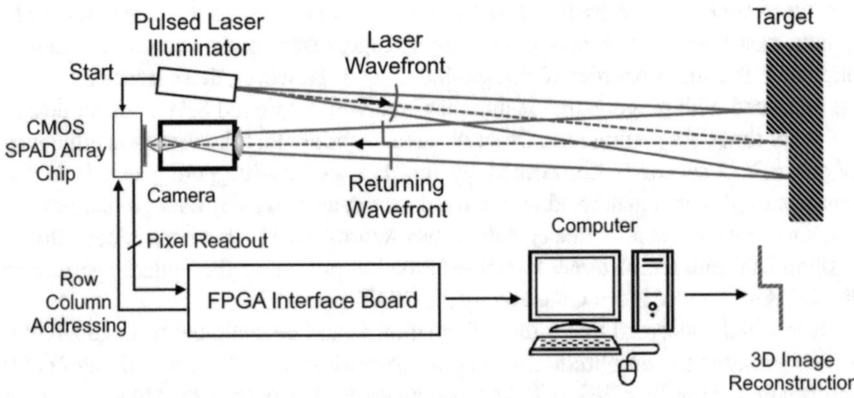

Fig. 6.3 How a SPAD array chip is typically used in a flash LiDAR setup for 3D imaging. The SPAD imager operates in DTOF mode using a pulsed laser illuminator. The TOF data is read from the SPAD chip by the FPGA interface board and then uploaded to a computer which then can perform image processing in real time, Ref. [72]

from the target and then concentrated evenly across the SPAD array sensor by means of a FOV receiving lens. An optical narrowband filter in series with the receiving lens is finely tuned to the wavelength of the laser which ensures high background noise rejection. Usually, there is only a paucity of photons being received by the image sensor which is also required to resolve the TOF of individual photons. Hence, the sensitivity of each SPAD to detect photons, a measure represented by its quantum efficiency (QE) and/or photon detection efficiency (PDE), is an important indicator of SPAD performance. By recording the TOF between light emission and reflected signal detection, it is then possible to compute the distance between an object and the sensor using the speed of light. To acquire a 3D depth-resolved image of a scene, it is possible to measure the TOF information pixel by pixel for an entire array of pixels. Each pixel contains the SPAD detector and associated timing electronics. The process of collecting TOF information across the SPAD array is repeated for every pulse generated by the laser, which is also called the repetition rate or frame rate.

Using Fig. 6.3 as a reference, the corresponding author's home organization at DST Group has developed innovations in SPAD array sensors/microchips, advanced PCB designs, embedded FPGA firmware, GUI development, and real-time processing algorithms to develop SPAD TOF imaging cameras for 3D flash LiDAR systems; see examples in Fig. 6.4a–e. DST Group has on-site underground tunnel testing facilities to evaluate and characterize the LiDAR systems under development, with the ability to control ambient lighting and introduce environmental effects (such as obscurants, e.g., fog and smoke); see Fig. 6.4f–j where an indoor experimental 3D flash LiDAR setup successfully imaged a target obscured by different mediums; see Ref. [73]. DST Group has also tested two concept demonstrator SPAD-based 3D flash LiDAR detection systems for tactical applications; these include (1) a long-range, high-powered ground-based SPL as a part of a collaboration with defense industry contractor BAE Systems Australia[7] (see Fig. 6.4k) and (2) a low SWaP version, compact enough to be attached to stabilized gimbal mount for agile deployment and/or power-starved applications (see Fig. 6.4l). This compact version of the 3D flash LiDAR imaging system was also mounted to the guardrails of a Navy vessel to demonstrate further maritime applications (see Fig. 6.4m) and carried by a remotely piloted heavy lift multirotor UAS platform to demonstrate the capability of underwater imaging and land-based surveillance of targets (see Fig. 6.4n).

The latter was part of a NATO Science for Peace and Security (SPS) project titled "Microelectronic 3D Imaging and Neuromorphic Recognition for Autonomous UAVs," an international collaboration led by DST Group and the Polytechnic University of Milan. The aim was to fly a low SWaP SPAD-based 3D flash LiDAR system on an unmanned aerial vehicle (UAV) to collect real-time imagery and perform neuromorphic processing for accurate target detection and classification, further details of which can be found here [74]. All the developed demonstrator

[7] https://news.defence.gov.au/technology/spotlight-highly-sensitive-light-sensor

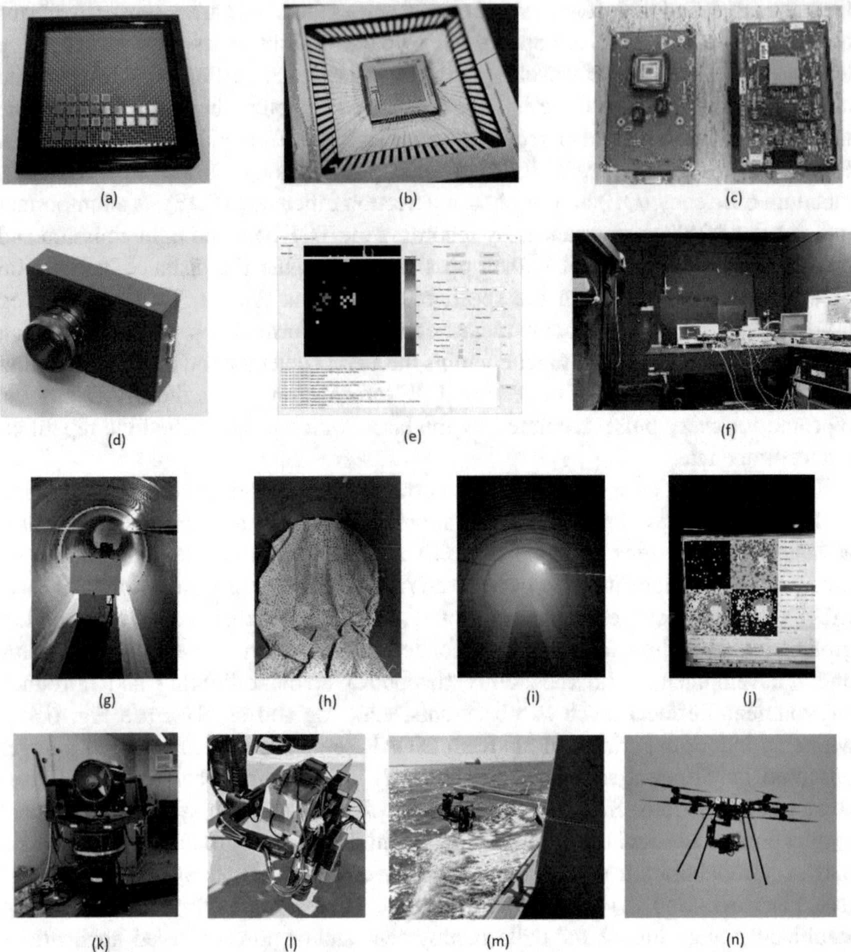

Fig. 6.4 (**a**) SPAD array (FPA) sensor bare dies. (**b**) 32 × 32 SPAD array microchip wire-bonded and packaged. (**c**) Interfaced with power and readout PCB electronics. (**d**) SPAD TOF camera housing with optics. (**e**) GUI configuration setup and readout. (**f**) Laboratory testing and evaluation of a 3D flash LiDAR system. (**g**) Static square target in underground testing facility. (**h**) Camouflage net placed in front of the target shown in (**g**). (**i**) Fog introduced to obscure the target shown in (**g**). (**j**) 3D flash LiDAR GUI output showing the target. (**k**) BAE Systems ground-based long-range SPAD LiDAR system for early warning surveillance applications. (**l**) Gimbal-mounted low SWaP 3D flash LiDAR system. (**m**) Gimbal stabilized and mounted on a Navy vessel for underwater sensing and imaging. (**n**) DST Group low SWaP flash LiDAR system mounted on a multirotor UAS system for ground-based surveillance, Ref. [74]

systems mentioned relied on TCSPC and range/time gating techniques to effectively image the target and remove background noise.

The BAE Systems 3D flash LiDAR system, shown in Fig. 6.4k, also known as a low observable platform detection (LOPD) system, used a 32 × 32 SPAD array sensor and had the harder task of both detecting and tracking incoming (moving) sea-level threats at long ranges. The intended application was for Navy surface combatants who are vulnerable to threats such as subsonic sea skimming anti-ship missiles (ASMs). Modern fielded passive ASMs are guided by either radiofrequency (RF), electro-optic, or a combination of both and are designed to defeat modern ship phased array radar and sensing systems by employing low radar cross-sections, low infrared signatures (stealth technologies), and ultra-low sea skimming ride heights. The LOPD was a successful concept demonstrator to evaluate the technology against such mock threats but had limitations such as the tracking system needed to be pointed in the right direction (i.e., cued) by a dedicated passive staring visible/infrared sensor and processing latency meant tracking targets at range proved difficult. The LOPD also demonstrated its ability to effectively detect and track UAS threats at long ranges; others have explored these applications for 3D flash LiDAR systems as well; see Refs. [75, 76].

As some of these examples have illustrated, the application space for SPAD-based 3D flash LiDAR is diverse and includes rapid environment assessment, space situational awareness and space sensing, hyper-temporal imaging, detecting stealth/camouflaged targets, MAWS (missile approach warner), and counter-UAS and hostile fire indicator systems. But 3D flash LiDAR systems do have limitations, especially for long-range sensing and tracking applications; this will be further explored next.

1.3 Challenges for Flash LiDAR Systems

In this section, the limitations of current SPAD FPA sensors operating in 3D flash LiDAR mode for long-range sensing are explored. Factors, such as limited FOV, need for cueing, and latency associated with the real-time processing and tracking of moving objects in real time, are covered.

Flash LiDAR based on the principle of TOF is the least computationally demanding technique among other ranging methods, such as triangulation, interferometry, and structured light, as the flood illumination eliminates the mechanical complexity of a scanning emitter [77]. It is worth noting that 3D flash LiDAR techniques can also be applied to conventional CCD/CMOS sensors; these are known as burst illumination gated viewing (GV) systems or 3D range-gated imaging (3DRGI) [78]. Although a GV camera has typically higher pixel resolution than SPAD arrays as used in 3D flash LiDAR, they are not suitable for range imaging in dynamic scenarios (moving scenes) as many laser pulses are required to calculate range. Recent work, by [79], attempt to improve the real-time performance of 3DRGI systems, by using a 3D super-resolution range-gated flash LiDAR (using a high-resolution CCD sensor) based on triangular algorithm of range-intensity correlation, and further present a coding method based on triangular algorithm for high depth-to-resolution ratio.

As range performance and FOV are linked for SPAD array-based 3D flash LiDAR systems, the SPAD sensor FOV is limited by the number of pixels in the array which impacts the system performance. Despite its advantages, solid-state "integrated" SPAD arrays currently have far fewer pixels than other types of image sensors such as CCD/CMOS imagers. This means that the SPAD image sensor would capture an image with much less detail. This is due to SPAD detectors requiring additional circuitry which needs to be incorporated within each pixel in the SPAD sensor. With conventional photon counting SPAD microchip design, each pixel or cell contains one or more SPADs and corresponding digital circuitry to count when photon events occur. For every pixel, the count data is collectively communicated off-chip in a synchronous manner at the frame rate of the laser. Hence, the pixel resolution is very poor and the FOV in the optical system limited. The challenge with SPAD sensors having more pixels (higher-resolution arrays) is the demand to communicate more data off the chip and how to process this data quickly; this adds latencies in processing the image. This also means these types of sensors typically struggle with background clutter and highly dynamic scenes making it hard to discriminate objects of interest. Unfortunately, many documented applications in the literature have been demonstrated in controlled indoor/laboratory environments with static (non-moving) objects in uniform backgrounds where the imaging signal-to-noise ratio (SNR) can be arbitrarily improved by capturing and post-processing a large-enough amount of scene data.

As sensor pixel count/resolution increases, this presents new challenges in terms of developing low-power TOF SPAD-based image sensors that can address the large data volumes being generated. This causes a major data processing bottleneck on the device when either the number of photons per pixel is large, the time resolution is fine, or the spatial resolution is high, as the space requirement, power consumption, and computational burden of the depth reconstruction algorithms scale with these parameters; in either case, the TCSPC data must be recorded, stored in memory, and transferred from the chip for each pixel capturing the scene. Various methods have attempted to tackle the trade-off between depth resolution and computational/space complexities associated with the TCSPC histogram. These methods will be covered more in Sect. 2.

Some of the constraints involved with long-range sensing are laser power, small FOV, and reduced sensitivity (as compared to scanning systems – the reason for this is that the returning laser energy is spread over many SPAD pixels in an array rather than just one as for scanning systems). A small FOV is usually addressed by having an accompanying cueing method. There are techniques that can address some of these system-level limitations, for example, employing block-based illumination techniques [80]. Here, only part of the system FOV is illuminated by each emitted laser pulse. The receiver would then consist of a SPAD array covering the whole system FOV, but since only a small part of the FOV is being illuminated per emitted laser pulse, only the corresponding set of receiving SPAD elements are activated and further processed.

Without having a cueing method in space and range, LiDAR data acquisition is typically performed by scanning several beam locations at each depth. This can lead to a high acquisition time, creating a bottleneck for deployment in real-world applications. Approaches to deal with this problem include non-local fusion-based image processing methods such as [81], which combines information from an optical image with sparse sampling of the single-photon array data, providing accurate depth information at low-signature regions of the target. Other approaches include rapid panoramic 3D flash imaging of larger scenes as investigated by [82]. For real-world tracking systems such as the LOPD system (see Fig. 6.4k), this means line of sight (LOS) needs to be cued very close to the target location for the tracker to initiate lock; hand-off is usually complicated by complex scenery and cloud cover. Small perturbations in target location can easily push the target out of the SPAD FOV faster than the control system can correct for and slew the LOS back to the target, particularly at shorter ranges where the LOS rates are higher. The SPAD FOV also increases the reliance on a highly responsive and accurate control system for the gimbal motion to maintain the target in the FOV. Another limitation of the demonstrated LOPD system is the laser emitter, having the following impacts on system performance:

- Use of a visible or near-infrared (NIR) laser wavelengths that are aligned with Fraunhofer absorption lines can decrease the amount of backscatter and hence increase the SNR and the processing algorithm's ability to maintain target identification.
- Higher laser pulse powers are required to improve SNR, but laser sources with high repetition rates and required peak pulse powers are difficult to source. In the current LOPD system, the update rate for target returns and track data are such that the LOS rates that the control system can accommodate are limited by the relatively slow pulse repetition rate of the laser rather than the SPAD sensor output data rates.
- The available laser pulse repetition rate of 20 Hz and 20 mJ pulse energy limited the processing algorithms that could be applied. Higher pulse repetition rates would enable use of more effective processing algorithms with improved object detection performance and noise rejection without slowing down the tracking performance. Multiple pulses in air technology may address some of these limitations [83–85].

For tracking at long ranges, the LOPD concept demonstrator project highlighted the need for larger format SPAD/FPA sensors coupled with both higher repetition and high-powered lasers tuned to specific wavelengths. However, achieving better FPA sensors is only part of the story because as these FPAs scale with pixel count, there is a high density of data gathered resulting in very large datasets. This presents new challenges in image processing and how to deal with the huge volume of data generated, especially as frame/laser repetition rates increase, so this is where neuromorphic processing and deep learning techniques can play a role in fast object tracking and classification. An introduction to neuromorphic vision systems is given in the next section.

1.4 What Is Neuromorphic Event-Based Vision?

The goal of neuromorphic engineering is to design and implement microelectronic systems that emulate the function and performance of the biological nervous systems. Neuromorphic systems are designed to operate in the same distributed, spike-based, low-power regime as biological brains but adapted to and optimized for silicon. By attempting to emulate biological information processing principles in silicon hardware, neuromorphic engineering seeks to better understand how such energy-constrained systems can operate under challenging real-world sensory environments. How does a dragonfly navigate its incredibly complex dynamic environment while evading predators and capturing ten mosquitoes a day on a power budget of ten mosquitoes a day? Despite remarkable recent progress, current state-of-the-art machine learning-based solutions can provide no solution that matches biology in this power, hardware, and bandwidth-constrained requirement space. Unsurprisingly, through seeking to emulate biology, neuromorphic systems tend to be most useful in sensing and processing in remote or challenging environments where power and bandwidth are limited making these systems particularly attractive in defense applications.

While neuromorphic sensing principles have been applied to a wide range of sensory modalities such as audio, tactile, and smell, the sensory modality with the greatest research focus has been that of neuromorphic vision and silicon retinas where a range of neuromorphic sensors and processors have been developed that emulate the human retina and the human vision system, respectively. Visual processing is an energy-intensive task and therefore must be used sparingly and efficiently. This is particularly true for biological eyes, which also require significant infrastructure to process and interpret the visual information. By contrast, conventional cameras used in computer vision are designed to be generic imaging devices and capture as much information as possible in the hopes that the recorded data contains the information relative to the task that is currently being undertaken.

Neuromorphic vision sensors attempt to emulate biological nervous/visual systems which have evolved via natural selection for optimized speed, power efficiency, and precision in noisy and dynamic environments. Each pixel in these devices operates in an asynchronous and independent manner, emitting events only in response to contrast changes sensed at each pixel. This greatly reduces the amount of redundant information generated and allows these devices to operate at low power levels and across a wide range of different scene conditions. These devices do not output frames, but rather a stream of events with a high temporal resolution using a protocol called address-event representation (AER) which transmit each event asynchronously [86]. The AER protocol is an efficient communication protocol for sparse event-based data that reports events as they occur removing the need for global frames [87].

In other words, instead of sampling the environment at the maximum sampling rate required as dictated by the speed of the fastest changing stimulus, the localized, independent change detectors in neuromorphic sensors trigger asynchronous sampling events only in response to local stimulus change. In the presence of the spatiotemporal sparsity that is present in essentially all real-world visual scenes, the amount of generated data and thus required bandwidth and computation is significantly reduced as compared to traditional CMOS sensors. Indeed, the spatiotemporal sparsities which neuromorphic event-based sensors exploit are essentially the same as those used in conventional video compression methods. However, whereas the latter methods are applied post-data capture using power-consuming computation, event-based sensors inherently perform this compression at the sensor by not generating data in response to unchanging stimuli.

By implementing in silicon hardware the architectures and principles present in biology, neuromorphic sensors are often able to generate extremely sparse yet highly salient data that are several orders of magnitude smaller than conventional systems offering imaging at a lower computational load than conventional cameras. Commercial examples of these sensors can be found here[8,9,10,11,12] and recently [88].

In summary, the major advantage of event-based sensors and event-based sensing is that it allows scene analysis to be performed in environments where conventional sensors inherently struggle such as heavily occluded or distorted environments. By implementing sparsely activated event-based networks in dedicated hardware, high-speed occlusion and distortion invariant scene analysis may be performed over extended periods using minimal power. The difference between an event-based camera and a normal 2D camera or even a TOF camera (frame based) is that the former only outputs information (i.e., an event) when there is a change in the visual scene; see Fig. 6.5 for an illustration of how this principle works.

These types of sensors then give you several advantages, which include:

- Capturing fast-moving transient objects in the field of view of the sensor
- Ability to filter out lots of redundant information in the scene
- Inherent data compression in the amount of information coming out of the sensor/camera
- Fast, lightweight, low-power event processing

Through the inherent data compression properties, event-based sensing enables a range of scene analysis algorithms to be performed in heavily occluded environments where conventional sensors inherently struggle such as heavily occluded or distorted environments. By implementing sparsely activated event-based networks in dedicated hardware, high-speed occlusion and distortion invariant scene analysis may be performed over extended periods using minimal power.

---

[8] The Dynamic Vision Sensor (DVS) from https://inivation.com/dvs/

[9] DAVIS camera (by iniLabs; see https://inilabs.com/products/dynamic-and-active-pixel-vision-sensor/)

[10] ATIS camera (see http://www.pixium-vision.com/en/technology-1/smart-neuromorphic-event-based-camera)

[11] Prophesee event-based cameras. https://www.prophesee.ai/

[12] Samsung Home security (https://www.samsung.com/au/smart-home/smartthings-vision-u999/

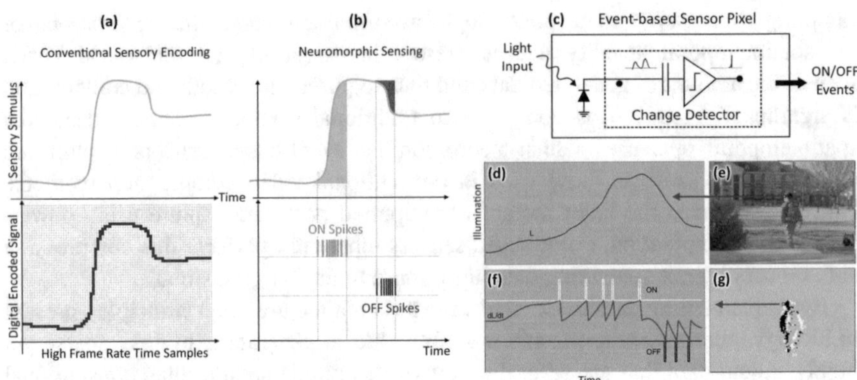

Fig. 6.5 Illustration of the difference between conventional sensing and neuromorphic sensing. The above figure shows a hypothetical real-world signal and how it would be represented through a conventional sampling-based sensing approach, see (**a**) top and (**a**) bottom, and a spike-based neuromorphic approach, see (**b**) top and (**b**) bottom, using event-based pixels as shown in (**c**). For a given scene (**e**), the conventional approach needs to quantise the sample space for the illumination signal at the indicated pixel plotted in blue in (**d**), resulting in both quantisation and redundant data. For the same scene the neuromorphic approach (**g**) only encodes the changes in the illumination value and emits these as spikes or events (**f**). Redundant information is inherently suppressed without affecting the temporal resolution of the sensor. This figure, therefore, serves to illustrate how neuromorphic approaches to sensing can break the relationship between fidelity (either spatial or temporal) and data rate. Ref. [89]

When the neuromorphic event-based paradigm used in neuromorphic sensors is extended to the area of deep neural networks (DNNs), the data reduction effects become even more dramatic since the total number of processing nodes in large deep CNNs can dwarf the number of input channels of any sensor. As shown in Fig. 6.6, event-based scene segregation continues the sparse regime of the sensor where the nodes of the event-based or spiking neural network (SNN) activate only in response to novel input at the lower levels dramatically reducing the amount of computation required and speeding up the operation of the system. Unlike conventional DNNs where computation occurs at every node of every layer at every frame, in an event-based or SNN, computation only occurs at nodes where a change has been detected at the lower layer. This sparse activation within the network results in orders of magnitude less node activation in an event-based network and a corresponding reduction in power consumption if the network is implemented directly in neuromorphic hardware. This sparse high-speed mode of operation allows the DNN to match features with its developed model at the same speed as the sensor and tracking subsystems. This preservation of event timing throughout the system enables seamless per-event scene segregation without the complex operations required to partition synchronous frames of data.

While neuromorphic processing using SNNs is often demonstrated for imaging applications in the visible band (using conventional CMOS/CCD detectors and sensors), the underlying principle of using event-based networks for hierarchical

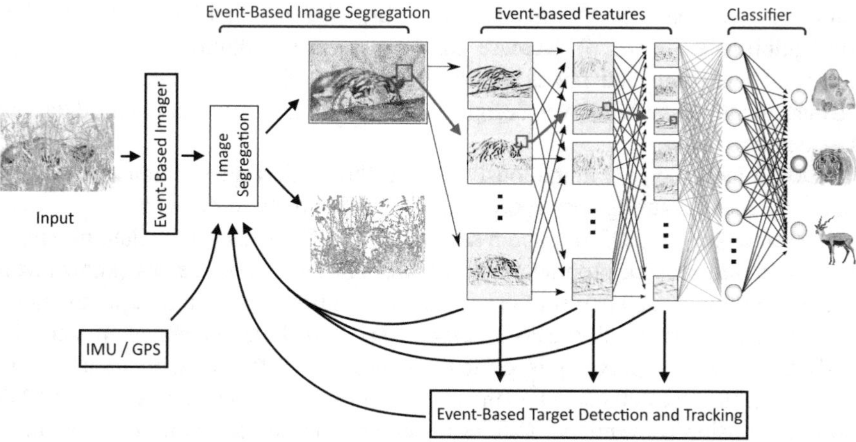

Fig. 6.6 Event-based detection classification and tracking in highly cluttered environments. Sparsely activated event-based feature extraction networks allow high-speed power-efficient processing of complex environments by only updating the internal model of the system in response to new and unexpected stimuli

data compression and preservation of temporal information applies to any data and is particularly well suited to high-precision temporally coded data that is generated via TOF LiDAR systems employing SPAD sensors. Such SPAD-based neuromorphic systems would have a diverse range of benefits for applications that include detection, tracking, and situational awareness in edge environments. In Sect. 3, we discuss examples of such systems.

2 Innovating SPAD Array Sensors for 3D Flash LiDAR

It has been seen how the key to SPL systems operating in 3D flash LiDAR mode are GM-APD or SPAD FPA sensors. In terms of sensor design, addressing the limitations of FOV, sensitivity, depth resolution, large data volumes being generated, and latency involved with object tracking is examined. This involves innovations to SPAD array sensors in the following areas: SPAD detector design, IC design, optical enhancements, and manufacturing techniques. These methods are all explored in this section.

There represents a huge community research effort in going beyond Moore's law in terms of functionality and diversification in overcoming the current limitations of SPAD array sensors in terms of high-resolution, high-sensitivity SPADs and real-time processing. Addressing this will deliver unparalleled functionality and high-quality imaging performance for future LiDAR systems. The current technical challenges needed to improve sensor performance include the following: 1) SPAD array microchips with the highest integrated SPADs (pixel density) while at the same time (for each SPAD) maintaining a high QE and/or PDE, 2) minimizing susceptibility to noise, 3) increase timing precision/resolution and 4) manage the data bandwidth and power consumption as pixel counts scale. This means advancements in system-level design, individual pixel electronics, and data bus architecture.

CMOS technology is ubiquitously used in the semiconductor industry because of its low manufacturing cost, speed, and the level of transistor integration possible. However, to get good QE or photon sensitivity, SPAD designs in the literature have often required special, proprietary, non-CMOS fabrication processes (known as a custom process) which prevent the integration of supporting circuitry onto the same silicon die. Whereby other SPAD designs that have been integrated in standard CMOS technology prove it is difficult to produce detectors with high photon sensitivity, the research challenge then is to produce a working high-density SPAD array chip using conventional CMOS manufacturing process while maintaining a high QE. This will not only deliver cheap sensors but also allow the easy integration of additional functionality opening new and exciting applications. The development of higher-density arrays improves the viability of using SPAD arrays in higher-resolution imaging applications.

New techniques to reduce the size of the SPAD detectors and novel supporting circuits to provide the photon timing information in each SPAD pixel are presented. Constant innovation in SPAD detector design, e.g., vertical avalanche photodiode (VAPD), (see Ref. [90]) makes them smaller in size. To miniaturize a high density SPAD array, much effort has focused on the individual SPAD pixel in the SPAD array, this includes the SPAD detector, corresponding front-end circuitry, i.e. detecting the avalanche event/ignition, quenching the avalanche current, and recharging/resetting the detector after any ignition event, and finally developing a digital counter mechanism in each pixel, which counts the TOF of the photon and stores this information in its local register.

By combining high pixel resolution with high-sensitivity SPADs will then deliver improved LiDAR systems with increased FOV. Trends in SPAD arrays and imagers have already been complied by others [91, 92]; see Fig. 6.7. As CMOS process technology nodes have become progressively smaller and manufacturing techniques moved from monolithic to 3D stacking methods, the number of SPAD pixels has increased from 32 to a recently reported 1 megapixel[13] [93], over a 17-year period, and this corresponds to roughly a doubling every year. To focus on the number of SPAD pixels alone does not tell the whole story as the SPAD arrays reviewed in Fig. 6.7 are a collection of different performance metrics and targeted applications, but they do indicate a level of technology advancement and scaling.

Figure 6.8a shows the functional diagram of typical SPAD pixel elements. Depending on system and application requirements in performing photon counting and timing functions, pixel circuit architectures vary. SPAD pixels typically contain

[13] Cannon Develops World's First 1-megapixel SPAD Sensor: https://global.canon/en/technology/spad-sensor-2021.html

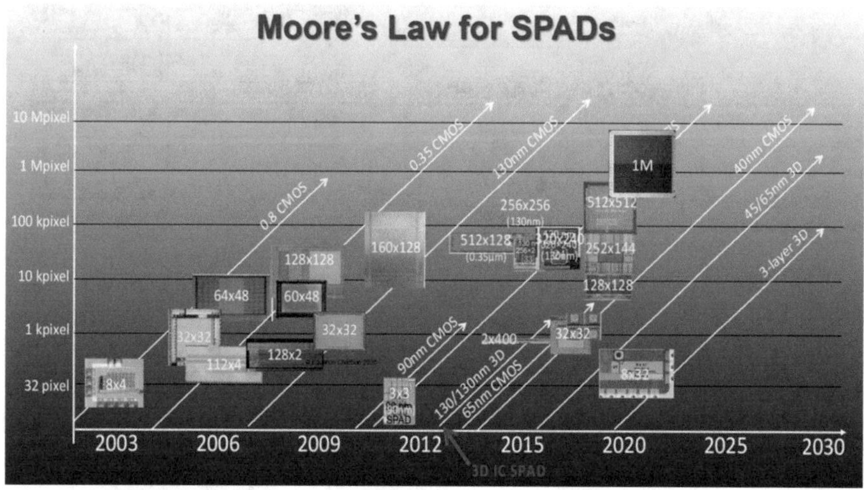

Fig. 6.7 Moore's law for SPADs, Refs. [91, 92]

the SPAD detector, which require specialized guard rings, quenching and recharge circuits, counters, memories, readout, gating circuitry, time-to-digital converters (TDCs), and digital processing and communication units (DPCUs) [92].

Each SPAD pixel can then be duplicated many times in a two-dimensional array to create an FPA. Novel circuit design and layout techniques can then be applied to reduce the footprint of the supporting electronics used to provide photon timing information in the SPAD pixels and associated back-end high-speed acquisition and readout (counting) functionality. The process of SPAD array miniaturization will need to address problems such as dead time, dark count rate (DCR), pixel area (or fill factor), PDE, after-pulsing, and crosstalk [94].

In operation, it is expected that at the beginning of a new pulse, all counters are reset and then photon timing begins; data can then be read out sequentially by selecting each pixel storage register by employing a row-by-column addressing scheme as similarly used in conventional memories. The overall system array architecture design is thus also a critical area of design as it can allow the detection of photons during the current pulse while reading out data held in the registers from the previous pulse. Achieving a high data readout rate capability is a particularly important objective especially for 3D flash LiDAR imaging applications.

With all this in each pixel, the fill factor (FF) of the sensor is reduced. The FF is a term used to denote the ratio of the pixel area that is sensitive to light to the total pixel area, and it can be often expressed as a percentage. Pixel FF is also another way of measuring how effective the pixel design utilizes layout area. Compared to other photodetector structures such as 2D CMOS imaging sensors, SPADs and SPAD pixels are larger; hence, the sensor arrays have a smaller pixel count. Having a relatively low FF impacts effective sensitivity and DCR (or background noise photons), as there is a tradeoff with timing/circuit performance and FF.

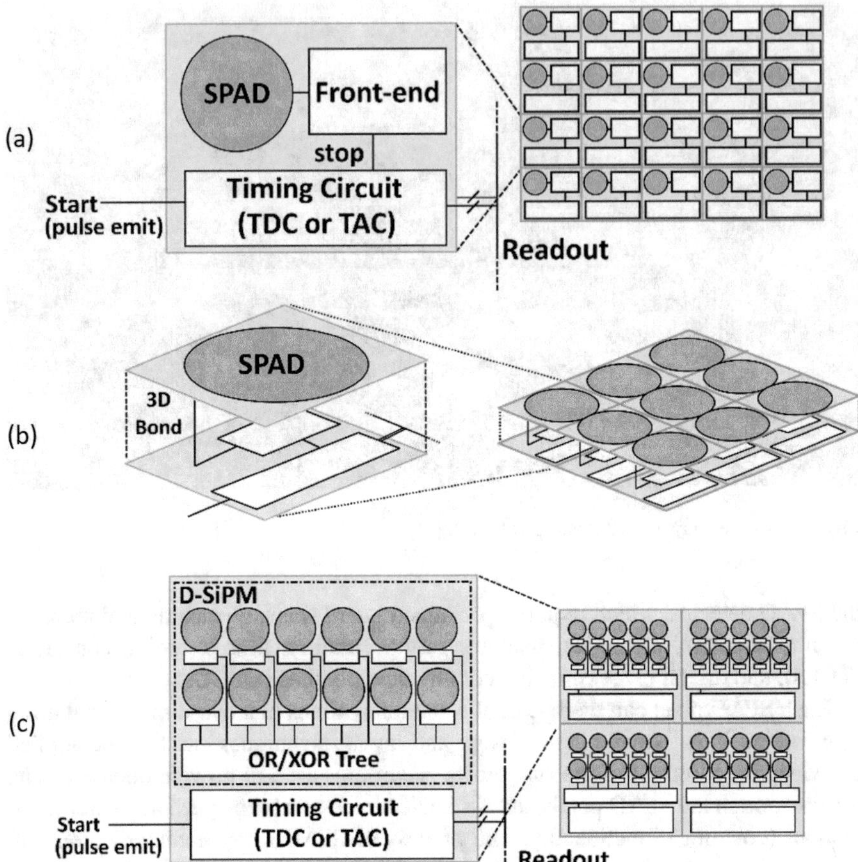

Fig. 6.8 Functional diagram of a typical SPAD pixel and array architectures. (**a**) Planar (in pixel) with column or serial readout processing, (**b**) 3D stacked, and (**c**) D-SiPM (macro pixel). Ref. [92]

Various strategies are available to implementing large pixel count SPAD-based DTOF depth sensors. The advantage of using an advanced CMOS technology will allow a level of miniaturization of smaller front-end (supporting) and back-end circuits; thus, the FF can be improved, and/or new functionality can be added to SPAD arrays, which in turn will allow performance enhancements useful for 3D flash LiDAR applications. Other methods to increase the FF include 3D stacking manufacturing techniques, which will be explored in later sections (see Fig. 6.8b), and use of macro pixels comprised of digital silicon photomultipliers (D-SiPMs) [95] (see Fig. 6.8c). In the latter, multiple SPADs are combined to form a single pixel increasing the total active area [92].

Designing smart and novel pixel array architectural CMOS ICs and layout design techniques for SPAD sensors are explored in the next section.

2.1 Smart SPAD Sensor IC Design

The goal for smart SPAD sensor IC design is to have a high PDE for the SPAD detector, minimize susceptibility to noise, increase timing precision/resolution, and manage both the data bandwidth and power consumption as pixel count scales.

Clearly developing high-density SPAD arrays is a challenging task because of the various technological and physical problems that need to be solved. It has been discussed already how there are two different design and manufacturing philosophies used, either employing standard CMOS process technology or using a custom SPAD process technology. There are advantages and disadvantages to both approaches: but a custom process is undesirable as it requires the quenching, recharging, and processing circuitry to be performed off-chip and a dedicated silicon foundry to be found all at the expense of designing and optimizing a high-performance SPAD structure. For 3D flash LiDAR applications, the high speed of the required timing and processing circuitry, and the large amount of information which must be processed in real time from many pixels on the SPAD array, a CMOS approach is needed. Work by [96] developed a figure of merit (FOM) to compare the performance of SPAD detectors for photon imaging applications and found custom SPADs present better performance than CMOS SPADs when few pixels are needed; conversely, when multi-pixel arrays are required, CMOS SPADs are the only choice to provide real time imaging at single photon levels.

A good comparison of CMOS SPAD DTOF sensor performance metrics developed over the last two decades is given by [92], and innovations can broadly be divided into sensor architecture, use of macro pixels, and on-chip histogramming, all in the attempt to address the following limitations:

- *Noise Reduction and Improving SNR*

There have been efforts to reduce circuit noise by developing novel front-end circuits for active quench and reset of SPADs, which eliminated after-pulsing effects and boasting the counting bandwidth of CMOS SPAD imagers [97]. Improving the SNR of LiDAR measurements can be achieved through multiple repetitions and by exploiting the TCSPC technique, but there are other approaches which include minimizing false alarms by using multiple SPADs per pixel or detector to filter out noise, see Refs [98, 99], using multi-echo detection techniques whereby each SPAD can detect 3 echoes (or multi-events) per laser cycle [100], adaptive confidence detection [101] and progressive time gating [102]. Automatic region-of-interest (ROI) selection of just those pixels illuminated by the laser spot is another way to improve SNR of the TCSPC histogram [103].

- *Data Rate Reduction/Data Compression*

It has already been stated how SPAD array sensors with large pixel formats that can deliver fine spatial resolution produce high volumes of data per frame, which must be transmitted off-chip and processed externally to generate the 3D image, consuming I/O power and external system resources. Various methods have been proposed that histogram TCSPC data at each pixel to pre-process photon time stamps and, in doing so, reduce the output data volume for each frame. Good examples of in-pixel or on-chip histogramming and data compression are [30, 104, 105]. A high-speed SPAD TOF camera is presented by [30] that can output direct depth maps, and [105] embed histogram processing in the design so the output for each pixel is the histogram peak location. Another approach used to reduce the data transfer of the information needed to reconstruct the LiDAR image is to compress the data on-chip, such as statistical distribution [106] or use compressed sensing methods [107]. Lastly, TDC sharing and event-driven readout architectures present more efficient and optimal ways of transmitting information off-chip; see Ref. [108].

- *Improving Timing Resolution and Counting*

A triple integration time-to-amplitude conversion (TAC) scheme for high-resolution SPADs to remove the dependence on process, voltage, and temperature (PVT), readout noise, and capacitive discharge effects is presented by [109, 110]. Such design techniques allow circuits to be miniaturized without causing a large timing non-uniformity across the sensor array, resulting in improved timing performance metrics such as resolution and jitter.

- *Low Power Design*

Despite data compression techniques, TOF systems still read out high spatial and temporal resolution data for every frame irrespective of scene activity. This continuous operation in TOF systems results in high power consumption due to the uninterrupted triggering of the laser emitter, the generation of high-frequency time gates in-pixel, and the continuous transfer of ranging data to an external processor, the highest power contributors in a TOF system. In this case, motion detection solutions have been demonstrated, allowing reduction in system power by avoiding readout and processing of high-resolution TOF frames with no motion activity; see Ref. [77]. Other approaches include new logic counting architectures such as proposed by [111] to improve the area and performance of on-chip counters, allowing larger-scale array designs.

- *Upscaling Depth Images (Resolution Enhancement) and Image Denoising*

Finally, methods known as super-resolution, hybrid-mode imaging and guided upscaling are techniques used to increase the native resolution of depth images from SPAD cameras without designing a high-resolution sensor, either by using a standalone SPAD camera operating in a dual mode such that it captures alternate low-resolution depth and high-resolution intensity images at high frame rates or by using another sensor to provide the high-resolution intensity images. These techniques use the intensity images and multiple features extracted from down-sampled histograms to guide the up-sampling of the depth maps; see Refs. [112–114].

Figure 6.9 shows different circuit architecture and layout examples of sensors developed by the corresponding author at DST Group that apply some of the strategies mentioned above. Figure 6.9a shows a conventional SPAD pixel design

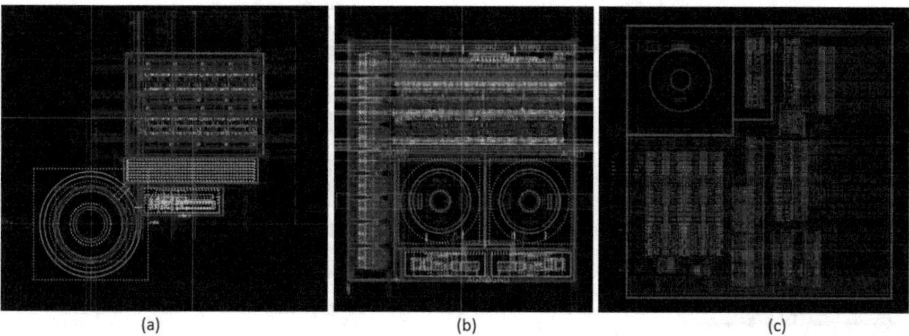

Fig. 6.9 Different SPAD pixel layout architectures. Left, single SPAD detector per pixel with passive quench and recharge circuit including the digital counter, Ref. [116]. Middle, two SPAD detectors per pixel with shared counters and memory, Ref. [99]. Right neuromorphic event-based SPAD pixel showing a "receptive field"; the term pixel no longer applies. The "receptive field" is effectively a macro cell that connects 16 neighboring single SPADs together in a 4 × 4 arrangement. Each "receptive field" contains its own circuitry that digitally "processes" a 4 × 4 array of neighboring SPADs such that it determines when four distinct shapes are detected in the "field," Ref. [72]

with a passive quench/recharge circuit and digital counter. Figure 6.9b shows the architecture for a novel counting design using two SPADs per pixel for correlated detection; see Ref. [99]. Figure 6.9c shows a neuromorphic event-based topology; see Ref. [115] for more details.

Besides these innovations in circuit design and architecture, other methods exist to improve the FF of SPAD arrays, such as microlensing and 3D stacking, and these will be covered in the following sections.

2.2 *Optical Enhancement Through Microlensing*

One disadvantage of SPADs implemented in standard CMOS technology is the limited FF due to the need for guard rings and the placement of in-pixel circuitry and electronics. The lower the FF, the less sensitive the sensor is, and the longer exposure times are required. Work by [117] examines various methods to compensate for low FFs. Suggested approaches range from simplifying circuit complexity at the expense of reduced functionality and decreasing the layout size of the electronic circuits by using a more advanced technology node (i.e., leveraging technology process scaling and using smaller transistor feature sizes), although this last option introduces impurities in the process resulting in higher DCRs and after-pulsing effects. Alternatively, simply increasing the SPAD detector size by using larger and rectangular-shaped SPADs will also have the effect of increasing the active area to pixel size, but again this will produce higher DCRs. Figure 6.10a–c shows how implementing these strategies affect the FF; the higher the percentage, the better.

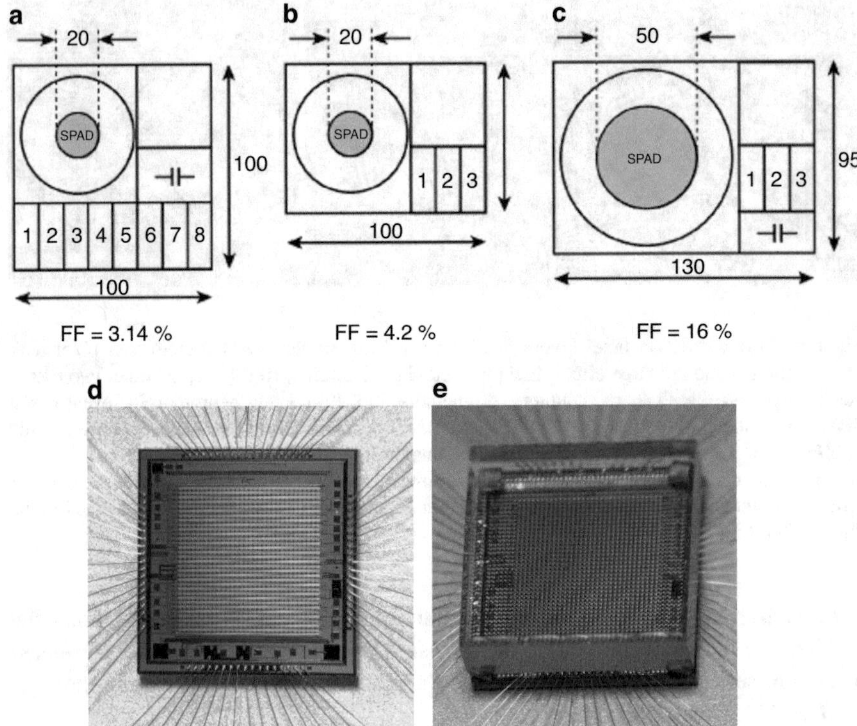

Fig. 6.10 (**a**) Pixel structure with SPAD and electronics. (**b**) Effect of reducing the number of counting bits. (**c**) Increasing the SPAD size, but higher DCR. Ref. [117]. (**d**) Developed at DST Group, a SPAD array microchip wire-bonded. (**e**) DST Group SPAD array microchip with a commercial off-the-shelf microlens fitted. Ref. [116]

Other options include moving the electronics altogether outside of the pixel or onto different chips/die completely and connecting to the SPAD detector via innovative manufacturing techniques such as 3D stacking methods (this will be examined in the next section). The final option is to use an array of microlenses or micro-optical concentrators post-chip fabrication, to collect photons that would otherwise hit non-sensitive areas of the pixel and focus onto the SPAD active areas, thereby improving collection efficiency. Figure 6.10e shows a typical microlens array (32 × 32) mounted and matched to a 32 × 32 SPAD array die underneath [116]. Critical are the alignment of each microlens of the array directly above each SPAD detector and the spacers which are used to hold the array at the correct distance above the die; methods to do this are explained in [118]. With the use of microlens arrays, the FF can be increased by as much 25% [119, 120]. Other more innovate approaches use non-uniform microlens arrays to mimic the large FOV feature of a compound eye [121]; these techniques show the FOV of a typical SPAD array based imaging system can be substantially increased.

The downside of using a microlensing solution is the increased fabrication complexity (including alignment precision) and cost, which becomes more evident as pixel counts scale especially for very large format sensors.

2.3 3D Stacked Assembly Methods for Imaging SPAD Sensors

As more advanced CMOS technologies and design techniques have become available, FF has been steadily increasing, allowing higher density of pixels and higher functionality per pixel. This trend will only increase with the use of 2.5D technologies and 3D stacking methods for sensor arrays, which represents a viable alternative to transistor scaling to sustain Moore's law.

It has been discussed how from a monolithic architectural point of view there are several approaches to increasing FF. For in-pixel-level processing results in the highest computation speed and functionality per pixel is possible but at the cost of FF, while column processing and micropixel processing allow for better FF [122]. 2.5D technologies and 3D stacking solutions offer heterogeneous integration of different devices and/or processing functions between dies; in other words, for a sensor array, all the supporting circuitry can be placed on a different layer or die than the sensors themselves. This offers the ability to integrate many SPADs on a dedicated die, maximizing FF and going beyond Moore's law. With a 2.5D IC structure, the dies are integrated side by side using a silicon interposer; however, this method is typically not used with imaging devices as there are larger interconnects (resulting in delay lines) between say the SPADs on the SPAD dedicated die and corresponding processing electronics on the digital die. 3D stacking methods are preferred as here dies are vertically stacked on top of each other. In this case, the SPAD arrays are implemented on the top-tier die, while all the pixel electronics are placed on the bottom-tier die; there is a one-to-one correspondence between each SPAD and electronic circuit positioned underneath; this reduces wirelengths compared to 2.5D and results in more uniform length interconnects across the SPAD array.

The 3D stack manufacturing process varies depending on whether the SPAD sensor is front-illuminated (refer to Fig. 6.11a) or back-illuminated (refer to Fig. 6.11b). Front-illuminated SPAD sensors rely on a through-silicon via (TSV) 3D stack method, whereby a post-process TSV is made for each individual device to vertically connect the SPAD output to the pixel circuit while back stacking two dice together. Here, only the SPAD detectors/array is implemented on the top tier (die/wafer), and all the pixel electronics are placed on the bottom tier, while back-illuminated 3D SPADs can be realized with the TSV-less face-to-face direct connection [123]. An advantage of 3D stacking means the top tier can be implemented in a technology optimized for SPADs and the bottom tier can be implemented in a state-of-the-art (CMOS) technology for lower power consumption and higher functionality, resulting in improvements in timing resolution and data processing. In general, front-illuminated 3D stacked SPADs with shallow junc-

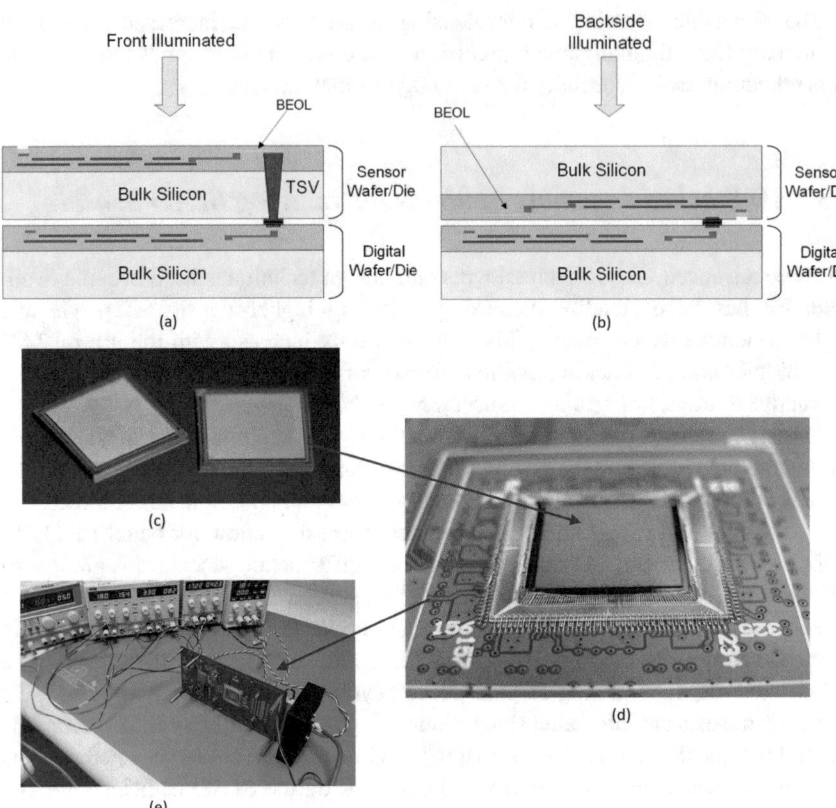

Fig. 6.11 (**a**) Frontside-illuminated TSV 3D die stacking technology. (**b**) Backside-illuminated 3D die stacking technology; see Refs. [122, 123, 126]. (**c**) Front-illuminated 256 × 256 3D stacked (TSV) SPAD array microchip designed by Milan Politechnic, Monash University, Manufactured by SilTerra and Fraunhofer ISM, 3D stacked TSV assembled by Fraunhofer IZM. Developed at DST Group, courtesy Fraunhofer IZM. (**d**) 256 × 256 3D TSV SPAD array sensor wire-bonded via interposer directly to a DST Group custom-designed PCB. (**e**) 256 × 256 3D IC PCB under test with heatsink attached

tions are more useful for applications in near-ultraviolet (NUV), blue, and green wavelength regions, while back-illuminated 3D stacked SPADs have a deepened junction in the face-to-face stacking and achieve higher red and NIR photon sensitivities. Recent examples in the literature of 3D stacked SPAD DTOF imagers include [30, 124, 125]. Figure 6.11c is the first and only reported case of a front-illuminated 3D stacked 256 × 256 SPAD imager developed by DST Group with collaborating partners. Here, a 180 nm high-voltage (HV) process was used for the top SPAD layout design and a 130 nm CMOS process for the digital bottom circuits. A specialized ceramic interposer was manufactured by DST Group and used to facilitate the wire-bonding of the 3D IC die directly to the pads on the PCB; see Fig. 6.11d. Each pad on the 3D IC die was individually wire-bonded to the interposer and then another wire bond made from the interposer to the PCB. For testing, a protective window was then fastened over the 3D IC sensor and heatsink mounted to the underside and routed offboard; see Fig. 6.11e. The 3D IC sensor had such a high FF (close to 40%) that a microlensing solution was not required.

2.4 Frame-Based Versus Event-Based SPAD Architectures

The most common approach to 3D flash LiDAR processing using SPAD sensors has been to encode the TOF of the arriving photons using high-precision counters for each SPAD cell and to transfer this timing data off-chip for processing [127]. Typical 3D flash LiDAR systems read out high spatial and temporal resolution data for every frame, where a frame is an interval of time synchronized to the laser repetition rate, usually irrespective of scene activity. We have seen how various methods have been proposed to reduce the output data volume from SPAD array sensors for each frame. This approach typically involves as a first step some form of averaging over several frames which would significantly increase the cost of on-chip processing. This transfer process also creates an information bottleneck which is currently one of the major limiting factors in the speed of operation of high frame rate SPAD cameras. In addition, the use of conventional CPUs or GPUs for processing this temporal data makes processing SPAD data computationally intensive using conventional signal processing techniques and results in significant power and hardware requirements. Figure 6.12a shows the conventional approach; all SPAD captured data is communicated off-chip for further image processing.

These data issues have motivated the development of novel hardware-based solutions such as in-pixel histogram [104, 128]. Yet the attributes that make SPAD data challenging for conventional processors, when combined with the significant level of temporal redundancy present in real-world visual data, make the SPAD cell activation patterns ideal for event-based and spiking neuromorphic processors that are designed to operate directly on noisy temporal data in a parallel fashion. GPU-based implementation such as [129] demonstrates the feasibility of realizing a high-speed hardware-efficient feature extraction and classifications system for noisy low-resolution SPAD imagers.

In [72, 115], the measurement of the TOF of the laser pulse is abandoned entirely in favor of a neuromorphic processor that operates directly in the time domain and on the inter-spike intervals within local regions of the SPAD array. The proposed approach illustrated in Fig. 6.12b motivates the development and hardware implementation of event-based feature extraction algorithms and circuits that generate sparse event-based local representations from the non-sparse event-based SPAD activation data and in this way drastically reduce the I/O requirements of the overall system. This design seeks to combine the inherently temporal nature of TOF SPAD spatiotemporal data with neuromorphic event-based feature extraction

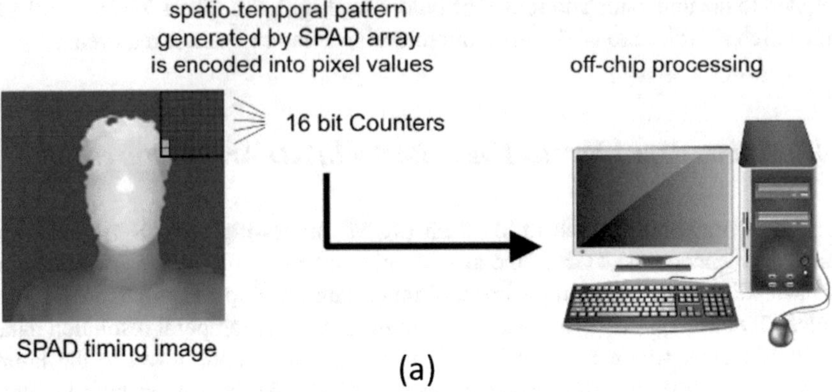

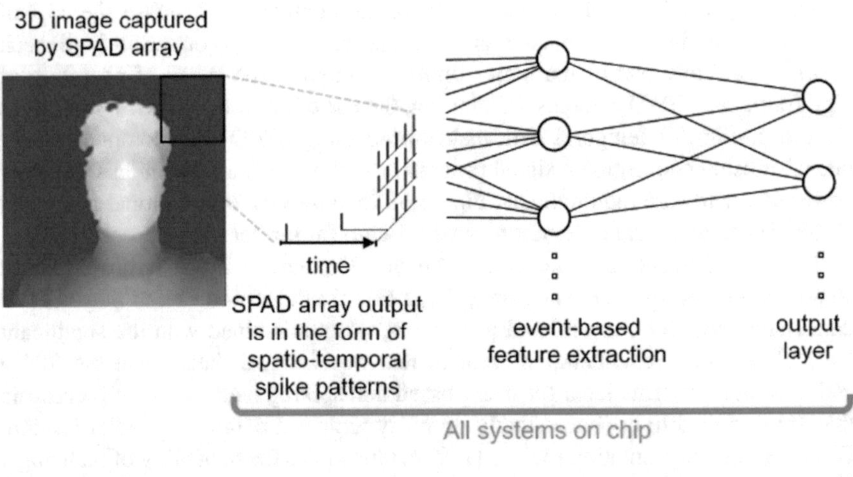

Fig. 6.12 Conventional and neuromorphic SPAD TOF data processing. (**a**) Standard approach to processing SPAD imaging data using (16 bit) on-chip counters and off-chip processing. (**b**) Proposed event-based approach to SPAD data processing. Ref. [72]

and processing. The new type of sensors presented is called "event-based" SPAD imagers.

These concepts and how deep learning techniques can play a role in fast object tracking and classification are explored further in the next section.

3 SPAD Neuromorphic Event-Based Sensing and Processing

We have already seen how deep learning techniques are prevalently used in single-photon 3D LiDAR for imaging retrieval, detection, and classification. Recently, neural networks were used in conjunction with TOF SPAD sensors operating in 3D flash LiDAR mode at short ranges to showcase high-speed object detection, having the ability to localize and classify objects in the FOV with low latency [130]. By combining 3D flash LiDAR systems based on SPAD detectors with the advantages of event-based sensing paradigms, we have the best of both worlds, enhanced imaging sensitivity and responsivity, fast 3D imaging, neuromorphic processing compatibility, and reduction of IO bandwidth, all with minimal computational overhead, which makes it ideal for compact and low SWaP LiDAR applications.

The utility of taking a more bio-inspired approach to SPAD processing is gathering more interest. In [131], a scalable 20×20 SPAD imaging array using asynchronous AER readout for low light (non-LiDAR-based) imaging applications was presented. The same approach is proposed by [132] for use in PET applications. In these works, the SPAD cells operate in photon counting mode where an analog photon-counting circuit counts incoming photon until the counter reaches a pre-set threshold causing the pixel to generate an event indicating a pre-set level of illumination. This mode of operation is like previously proposed non-SPAD event-based sensors [87] albeit with the advantage of the SPAD's high QE. Reference [129] shows how an event-based architecture can be used on SPAD sensors operating in TOF mode. The first 128×128 SPAD array TOF sensor with offboard event-based processing using a FPGA for LiDAR was implemented by [72] and recently a complete 5×5 SPAD array vision sensor with onboard neuromorphic processing for LiDAR presented by [133]. The techniques used in converting TOF SPAD information to a stream of events as presented by [72, 133] highlight the easy for scalability in design with the ability to accommodate larger format (pixel count) sensors.

The inherent temporal nature of TOF SPAD data, the relatively high level of information redundancy in high frame rate imaging, and the noisiness of the signal all make TOF SPAD data an ideal candidate for SNN and event-based processing. The case for processing temporal spike-like SPAD latch events with an event-based SNN is straightforward and would entail the following:

- Processing SPAD latching data in its inherent event-based form should result in a more efficient system since the conversion of millions of high-speed timing signals to high-precision digital representations via a high-speed clock and a time-to-digital converter at each pixel can be avoided along with their post-processing via an equally high-precision over-engineered processor.
- The vast majority of high-speed TOF SPAD imaging contains redundant information in practice. This is true both temporally, where sequences of frames attempt to image identical or near-identical scenes, and spatially, due to the natural redundancy and self-similarity in feature space that is present across the FOV in most imaging contexts. The inherent data compression and redundancy

suppression resulting from the inherent mode of operation of event-based systems, which only process unexpected changes at the input of each layer, greatly reduce the amount of data and thus calculations performed when an SNN operates on SPAD latching signals.
- The noise present in most SPAD imaging applications makes the application suitable for processing by biologically inspired SNNs which by virtue of being modelled on noise-robust biological systems tend to be designed and tested for noisy applications and typically over-perform in these contexts [89, 134, 135].

The following sections will briefly highlight the various methods used in developing SPAD sensors compatible with neuromorphic processing techniques.

3.1 First AND SPAD Feature Event Generation

In the method proposed in [72], instead of encoding, storing, and transferring the photon time-of-flight data off-chip for processing, a neuromorphic processor is used which operates directly in the time domain and on the inter-spike or inter-latch intervals within local regions of the SPAD array. Thus, only the *relative* timing of SPAD cell latching events is detected by the system. This approach illustrated in Fig. 6.12b generates a sparse event-based local representation from the non-sparse event-based SPAD activation data.

To simplify the sensor-processor design, the feature extraction operation at each receptive field was performed by competing AND gate neurons. These neurons may be wired in any configuration. In [72], they were used to encode a 4×4 edge detector circuit where at each laser pulse, the first AND gate whose receptive field was fully latched would prevent subsequent latching of any later gate or feature in the same receptive field via a recurrent enable connection that gates all AND gates. This temporal inhibitory feedback structure was introduced in the synaptodendritic kernel adapting neuron (SKAN) network [136]. Furthermore, each receptive field in this design contains a memory of the last detected feature if the same feature is detected at the same receptive field the sensor outputs no events. Thus, by only generating an event in response to a change in feature detections, the sensor processor reduces output data by approximately two orders of magnitude [72].

3.2 First SPAD Photon Counting Event-Based Design for Temporal Intensity Imaging

The receptive field-based design of the first AND system takes advantage of spatiotemporal structure of the SPAD sensor latch times in TOF mode to extract features from the illuminated scene. While this mode of operation is particularly suitable for processing via a spike-based system, neuromorphic time-based processing principles can also be utilized for SPAD imagers in photon counting mode.

One example of such an event-based temporally controlled solution is using a logarithmically varying decay process which treats each SPAD pixel count as the membrane potential m of a leaky integrate and fire (LIF) neuron whose leak, implemented as a decrement toward zero of the photon counts C, occurs at periodic intervals of

$$T = 2^{\tau} \tag{6.1}$$

multiple of the system clock and where N is adapted via a negative feedback path from the spike output of the same pixel. The photon count feedback mechanism is illustrated in Fig. 6.13b and operates as follows: the photon count C increases with photon detection(s) and is decremented at every

$$t = Tk \tag{6.2}$$

where k is an integer. If the photon count C passes an upper threshold θ^{+}, then the count is reset to zero, and a positive event or spike is generated indicating that the photon detection rate at the pixel is above the expected photon count rate or in other words that the current membrane leak is insufficient and the decrementation interval T is too long. In response to a positive event, decrementation interval T is halved via

$$\tau_{\text{new}} = \tau - 1 \tag{6.3}$$

If instead the photon count passes a lower threshold θ^{-}, the count is reset to zero, and a negative event or spike is generated at the pixel indicating the photon count rate is below expectation. In response to this event, the rate of decay is halved by doubling T.

In this way, by adapting the time interval between decrementation of the counter, a steady state is achieved where the photon count rate and the decrement rate approximately match each other, assuming the photon count rate is approximately constant. As shown in Fig. 6.13a when the increment and decrement rates match each other, the counter will not, or will rarely, reach either threshold and therefore does not generate new events or modify τ, with the pixel receiving photons without generating an output until a positive or negative change in the photon count rate is observed.

Thus, such a SPAD sensor can temporally modify and report a photon count signal like an event-based neuromorphic sensor, where only signal changes are reported, significantly reducing the data generated from the sensor while capturing the key transition events in the signal.

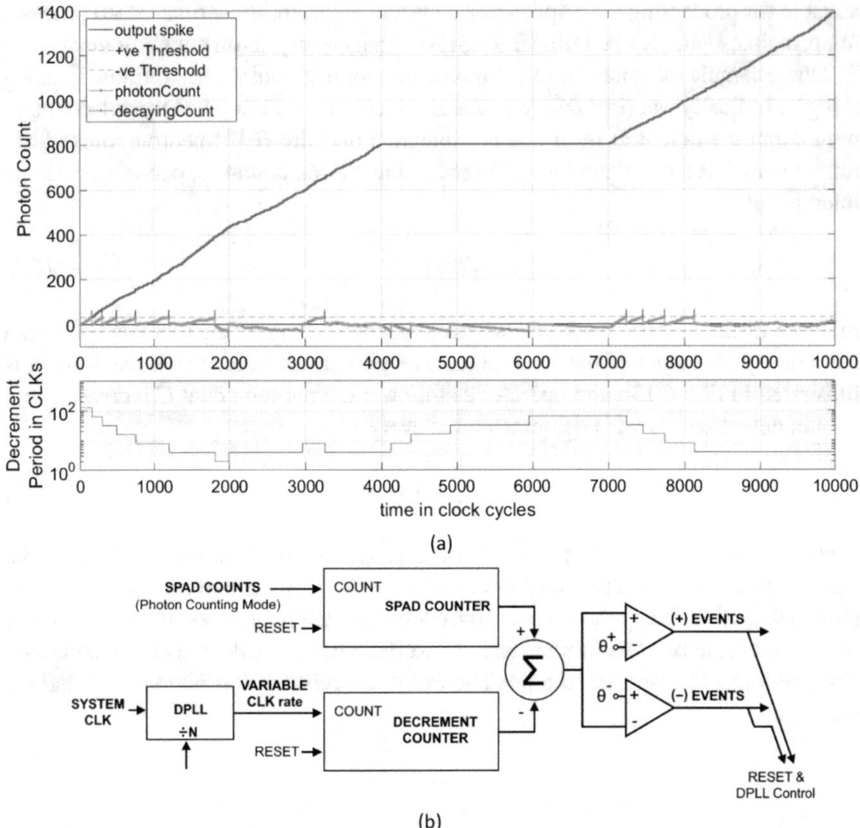

Fig. 6.13 (**a**) Diagram of a temporally modulated leaky event generator operating on a SPAD sensor in photon counting mode. (**b**) Compression of SPAD photon count into events

3.3 SPAD Data Processing Using Integrated Spiking Neural Networks (SNNs)

The conversion of SPAD sensor data to the event domain allows highly efficient event-based or SNNs to operate directly on the sensor output. SNNs differ from conventional artificial neural networks (ANNs) in that each node of an SNN only generates an event or spike in response to salient stimulation, whereas in a conventional neural network, every node of every layer performs its respective calculation's periodic sample regardless of the stimulus it receives.

Thus, the first approximation of every layer of such an event-based neural network can be viewed as operating in a similar regime as an event-based sensor or the retinal cells on which it is modelled. In such a network, each node or neuron at each layer detects changes at its input relative to a recent background context, and using localized competition among neurons detects changes in the local feature space. Like the event-based sensor itself, if this perceived change is above some threshold level, a spike is generated. In this way, each of the layers of an SNN encodes their environment in an ever-sparser spatiotemporal spike pattern representing higher-scale features with each layer silencing any activity from previous layers that is unchanging or predicted at the higher feature scale. This structure mirrors the function of the brain where every cortical layer acts as a mesh of dynamic filters blocking the activation of subsequent layers except at points in time and space where new higher-level features have been detected.

It is only through such a hierarchical event-based feature detecting and prediction architecture and the suppression of redundant activation at every stage of processing that biological brains can operate in challenging dynamic environments in real time on highly constrained power budgets.

In contrast, deep CNNs require high-precision numerical processing at every single node of every layer on every input frame regardless of whether the input carries highly salient information or none. This highly inefficient mode of operation would be analogous to every single neuron in an animal's brain firing at its maximum firing rate. When the event-based paradigm used in neuromorphic sensors is extended to deep neural networks, the data rate reduction becomes even more dramatic since the total number of processing nodes in deep CNNs typically dwarfs the number of input channels. The block diagram of such an end-to-end system is shown in Fig. 6.14.

3.4 Algorithm Testing and Temporally Constrained Dataset Generation

Unlike controlled image collection environments typically used in machine vision research, real-world imaging environments are unpredictable, dynamic, and noisy precluding many commonly used LiDAR image enhancement methods such as arbitrarily long frame averaging. To better evaluate the algorithms which aim to operate in such dynamic environments, datasets need to replicate the temporal imaging constraints present in such contexts.

An example of such a dataset was presented in [72] where the test targets in the dataset were imaged using a SPAD sensor in a photon timing mode at 100 kHz, providing photon TOF information at an extremely high frame rate. The experimental setup involved high-speed model airplane classification where model airplanes were dropped at high speed close to the sensor. The experiment involved the use of civilian airplane models in the background as a distractor which becomes increasingly prominent as the number of frames collected for an image is increased. The inclusion of the larger distractor with the free-moving high-speed target classes ensures that such a dataset can only be processed at extremely high speed ensuring a high noise floor that better represents real-world imaging environments. Such dynamic time high-speed experiment designs can encourage the development of algorithms that are more robust to noise and can more readily be applied to challenging real-world imaging environments.

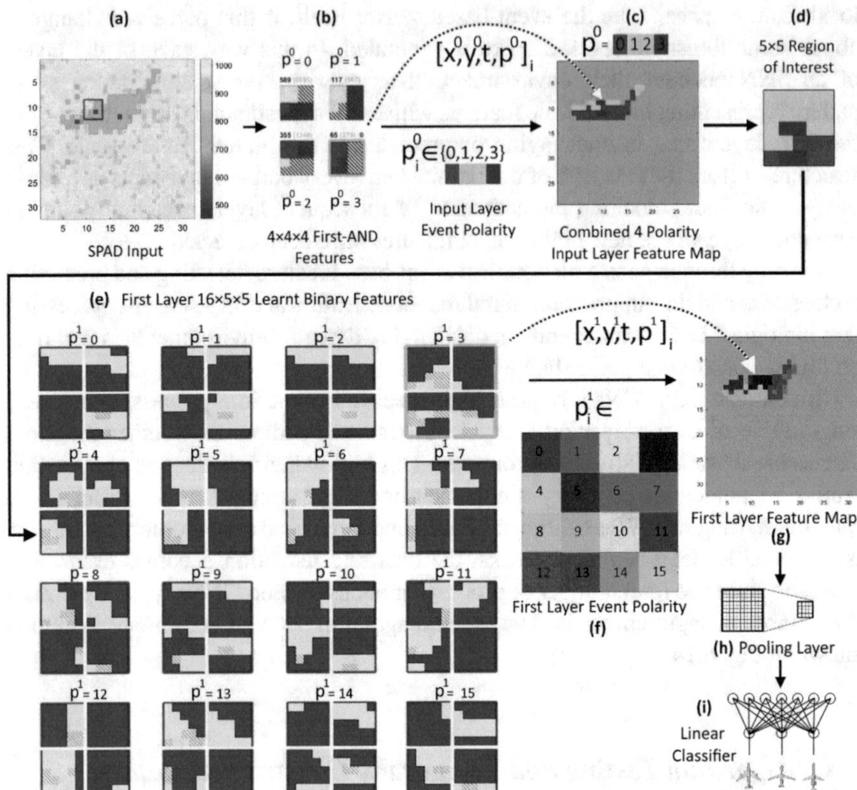

Fig. 6.14 Block diagram of the end-to-end first AND event-based processing system. (**a**) Shows the raw image generated by the SPAD sensor in time-of-flight mode as a B-2 airplane model enters the field of view. The red box indicates the receptive field of the current generated event. (**b**) Shows the four first AND features and their binary bar-shaped weights. Superimposed are the state of the latched SPAD pixels at the moment the first AND feature generates an event (diagonal black lines). The third feature is the first AND gate to latch disabling the others and passing its event to the next layer. (**c**) Shows S0 i, the binary-valued four-polarity time surface with activation over τ0 seconds. This surface serves as a feature map for the next layer of processing. (**d**) shows the 5 × 5 region of interest (ROI) extracted from Si0. (**e**) shows the 16 four-polarity binary event-based features which operates on Si0. (**f**) Shows the encoding of the 16 features. (**g**) Shows the 16-polarity binary-valued time surface S1i. Panels (**h**) and (**i**) show the pooling and classification layers, respectively. Ref. [72]

3.5 Spiking Neural Networks (SNNs): Trends and Challenges

In practice, however, there are two major challenges to the use of SNNs even in the context of an ideally suited application such as TOF SPAD imaging. These challenges, while not insurmountable, significantly handicap any SNN-based solution relative to a contemporary conventional ANN-based solution which the former must compete within real-world applications. These challenges to SNN systems can be broadly grouped into two categories:

1. The first category of challenges is due to the relative lack of theoretical progress in the SNN space; this is due in part to the nature of the spike signal used in SNNs. The spikes are typically modelled as delta functions and are inherently non-differentiable signals. This makes theoretical analysis of SNNs much more challenging than ANN. Thus, insights and advances within the SNN field are typically either heuristically developed through trial and error or simply adapted from concepts that are better understood from the more tractable ANN field and assumed to apply to SNNs without proof. Finally, unlike the ANN field which aims exclusively to maximize the performance of a system against objective and clear benchmarks, the aim of many researchers in the SNN field is muddied by the additional requirement of making a system "bio-plausible." The nebulous and often subjective concept of bio-plausibility invariably conflicts with rational and optimal system design.[14]
2. The second challenge to the development of superior performing SNNs is in hardware. Specifically, there is yet an absence of mature hardware which can perform memory and processing operations together locally to the same degree as biological neurons without introducing power and delay costs. Despite recent developments in passive local memory components such as the memristive device [137], and novel routing protocols such as AER [138], and integrated processors [139, 140], the silicon hardware available to today's SNN designer is far less mature, optimized, and usable than commercial off-the-shelf (COTS) hardware available to the ANN practitioner [141, 142]. This disparity in technology maturity means that even in ideal applications such as TOF SPAD imaging where signal modality, information redundancy, and noise characteristics significantly favor an event-based SNN system that only performs computation in response to sparse salient change, it is currently in practice more efficient to use an ANN that performs calculations on every segment of every frame regardless of information content than to design, build, and operate a theoretically more efficient experimental SNN. Finally, as with the diversion of resources in the theoretical domain of neuromorphic research, a significant subset of research on SNN hardware is dedicated to analog hardware implementations of SNNs, often not due to a clear engineering motivation for higher performance in immediate applications but based on principles of biological plausibility and non-immediate goals of building analog brain-like systems for the future. While analog implementation of SNN hardware does carry a realistic though distant promise of more power-efficient SNNs, the greater complexity in their design and their often idiosyncratic non-deterministic properties which require significant time investment to understand prevent their consolidation and incorporation into larger systems which can be used by independent research and development

[14] Human Brain Project: https://www.theatlantic.com/science/archive/2019/07/ten-years-human-brain-project-simulation-markram-ted-talk/594493/

teams in the same way as is achievable in deterministic digitally implemented SNN subsystems and components. Thus, the focus on analog SNN hardware often saps the ability of the already smaller neuromorphic field to deliver SNN hardware that can outperform current state-of-the-art ANN hardware.

Despite these challenges, however, with the end of Moore's law, with greater research and development interest, and with greater investment from the commercial sector, the "SNN handicap" is gradually shrinking, thus creating the space for SNNs to become a viable competitive solution in real-world applications. In this near-medium-term future scenario, particularly suitable applications such as TOF SPAD imaging where the application naturally favors an SNN solution will be the first to overcome the SNN handicap and form the leading edge of this shift toward true SNN computing.

4 Future Concepts and Summary

A type of SPD called SPAD and the development of an array of SPAD devices onto a microchip that can both detect and digitally time individual photons of light make this type of technology perfect for 3D flash LiDAR techniques and a potential solution for 3D imaging and tracking of moving objects in challenging environments. It provides groundbreaking detection and classification of targets at long ranges hidden by obscurants and clutter [60]. With low SWaP performance, and the ability to be integrated with intelligent (neuromorphic) circuit functionality and event-based processing, allows accurate detection and recognition very quickly with unsurpassed data compression and minimal computation overhead resulting in improved decision time in targeting and tracking.

Extrapolating from current trends that use a More-than-Moore approach to design which exploit advanced 3D stacking semiconductor assembly and integration methods and then forward-looking into the next decade, a 10 megapixel SPAD array imager is to be expected. For real-world 3D flash LiDAR tracking systems, these new high pixel count SPAD imagers will certainly address current FOV limitations, but also need the ability to dynamically adjust a wide range gate and process object detection algorithms in real time without slowing down tracking performance. As pixel counts continue to scale and larger datasets are generated, current frame-based IC design techniques as explored in Sect. 2.1 will struggle to find solutions to the ever-increasing data bottlenecks and inherent processing latencies involved when tracking fast-moving objects. The development of imaging devices inspired by biology (neuromorphic vision sensors) in which the pixels work in an asynchronous fashion responding only to events in the visual field, coupled with machine learning image processing and AI techniques, represents a new design paradigm to address these problems. The application of these new neuromorphic methods represents unparalleled performance and functionality improvements over what is possible with conventional frame-based LiDAR imaging systems.

The neuromorphic systems presented in this chapter provide a range of well-performing points in the event-based design space, which can be integrated with SPAD sensor hardware to provide event-based processing that drastically reduces not only the data rate coming off the sensor but also the quality of the output data as it relates to challenging tasks such as a view-invariant classification of large complex datasets. By using the same learning methodology and the same single-layer network structure and by testing across multiple design dimensions such as pooling and network size, event-based methods will outperform the frame-based system across all parameters while serving as a guide for the design of such networks in hardware. The scalability and speed of event-based (SNN) methods make it a promising candidate to address the high-end performance requirements expected for LiDAR imaging and tracking applications now and for the future. This certainly will create new challenges on how to train and evaluate such neuromorphic systems for real-world scenarios.

The continued application of SNN principles to SPAD-based imagers will likely see the emergence of more optimized hardware solutions and improved testing and characterization methods. Biological information processing systems are inherently physically embedded and continually responsive to a dynamic physical environment while always under severe energy, size, and speed constraints. Thus, their modes of sensing, processing, and response cannot truly be tested using passive, static datasets which define virtually all current standard LiDAR datasets. Furthermore, the physically embedded nature of biological processing and the associated physical size, weight, power, and speed limitations that come with it are fundamental aspects of their operation and cannot be treated as afterthoughts to be simulated or optimized in the final development phase. These are integral to the nature of the systems that neuromorphic computing seeks to emulate and represent new research problems to be solved.

As seen in earlier sections with tracking SPAD LiDAR systems is their complex interaction with the detection and tracking control system. LiDAR systems require rigorous physically integrated testing in a closed-loop repeatable environment where the output of the sensor-processor system affects its own future inputs. However, to date, neuromorphic datasets have almost without exception sought to emulate machine learning datasets with multiple neuromorphic versions of the Modified National Institute of Standards and Technology (MNIST) handwriting dataset and neuromorphic versions of the word recognition of Texas Instruments/Massachusetts Institute of Technology (TIMIT) audio dataset as examples. To date, no active, physically embedded neuromorphic datasets have been created which could test the active sensing capability of autonomous neuromorphic systems in general, and LiDAR-based systems motivate the development of the field toward neuromorphic LiDAR solutions that the field of neuromorphic computing has the greatest promise to deliver on. The development of repeatable physically embedded active closed-loop datasets is the first key step toward neuromorphic LiDAR solutions in this space. There is still much research effort required to explore opportunities for data generation needed for better training of AI-based systems and model validation.

In final, it is expected there will be continued growth in the application of advanced 3D technologies and architectures in developing future smart vision sensors,[15] particularly those that use neuromorphic-based SPD devices for SPL systems. These new vision sensors will be a system-on-a-chip (SoC), incorporating an SPD array image sensor and neuromorphic processing elements, including neuromorphic digital event-based circuitry and a neuromorphic processor. These sensors will allow for high dynamics and very small reaction time, with an overall low power budget, ideal for compact, mobile, and low SWaP applications. They will use novel algorithms for processing sensor data and for automatic target recognition (ATR) based on the development of new machine learning (neuromorphic) algorithms for feature extraction and classification. All this would mean a capability enhancement for a whole range of commercial applications, which includes meteorology, space, augmented reality, remote sensing, and autonomous robotics. For defense, this enabling technology will ultimately drive the development of new neuromorphic-based SPL systems that will have the potential to defeat stealth technology; improve surveillance imaging from space, over land, and underwater; and detect fast-moving targets partially concealed behind obscurants or hidden among clutter. The impact of More-than-Moore using 3D integration techniques to neuromorphic computing and LiDAR-based perception systems will provide both powerful and new sensing capabilities for the future, which otherwise would not be possible by using standard monolithic IC approaches or frame-based IC design techniques limited by Moore's law.

References

1. National Research Council, *Laser Radar: Progress and Opportunities in Active Electro-Optical Sensing* (The National Academies Press, Washington, DC, 2014)
2. M. Laurenzis, M. Rebert, S. Schertzer, F. Christnacher, Tracking and prediction of small unmanned aerial vehicles' flight behaviour and three-dimensional flight path from laser gated viewing images, in *Proceedings of SPIE, Laser Radar Technology and Applications XXIV*, vol. 110050D, (SPIE, 2019), pp. 1–12
3. V. Molebny, P. McManamon, O. Steinvall, T. Kobayashi, W. Chen, Laser radar: Historical prospective—From the East to the West. Opt. Eng. **56**(3), 1–24 (2016)
4. M. Lanzagorta, *Quantum Radar* (Morgan and Claypool, 2011)
5. F. Madonini, F. Severini, F. Zappa, F. Villa, Single photon avalanche diode arrays for quantum imaging and microscopy. Adv. Quant. Technol. **4**(7), 1–26 (2021)
6. R. Tobin, A. Halimi, A. McCarthy, X. Ren, K.J. McEwan, S. McLaughlin, G. Buller, Long-range depth profiling of camouflaged targets using single-photon detection. Opt. Eng. **57**(3), 1–10 (2018)
7. E. Lisman, *Non-acoustic Submarine Detection* (Centre for Strategic & International Studies, Washington, DC, 2019)

[15] Samsung eXtended-Cube (X-Cube) TSV Chip Stacking: https://news.samsung.com/global/samsung-announces-availability-of-its-silicon-proven-3d-ic-technology-for-high-performance-applications

8. W. Armbruster, M. Hammer, Segmentation, classification and pose estimation of maritime targets in flash-ladar imagery, in *Proceedings of SPIE 8542, Electro-Optical Remote Sensing, Photonic Technologies, and Applications VI*, vol. 85420K, (SPIE, 2012), pp. 1–11
9. C. Bruschini, H. Homulle, E. Charbon, Ten years of biophotonics single-photon SPAD imager applications – Retrospective and outlook, in *Proceedings of SPIE 10069, Multiphoton Microscopy in the Biomedical Sciences XVII*, San Francisco, California (2017)
10. I. Gyongy, N. Calder, A. Davies, N.A.W. Dutton, R.R. Duncan, C. Rickman, P. Dalgarno, R.K. Henderson, A 256 × 256, 100-kfps, 61% fill-factor SPAD image sensor for time-resolved microscopy applications. IEEE Trans. Electron Devices **65**(2), 547–554 (2018)
11. A.C. Ulku, C. Bruschini, I.M. Antolovic, Y. Kuo, R. Ankri, S. Weiss, X. Michalet, E. Charbon, A 512 × 512 SPAD image sensor with integrated gating for widefield FLIM. IEEE J. Sel. Top. Quantum Electron. **25**(1), 1–12 (2019)
12. C. Bruschini, H. Homulle, I.M. Antolovic, S. Burri, E. Charbon, Single-photon SPAD imagers in biophotonics: Review and Outlook. Light Sci. Appl. **8**(1), 1–41 (2019)
13. F. Zhao, H. Jiang, Z. Liu, Recent development of automotive LiDAR technology, industry and trends, in *Proceedings of SPIE 11179, Eleventh International Conference on Digital Image Processing (ICDIP 2019)*, vol. 11179, (SPIE, 2019), pp. 1–8
14. M. Dummer, K. Johnson, S. Rothwell, K. Tatah, M. Hibbs-Brenner, The role of VCSELs in 3D sensing, in *Proceedings of SPIE 11692, Optical Interconnects XXI*, vol. 116920C, (SPIE, 2021), pp. 1–14
15. F. Villa, F. Severini, F. Madonini, F. Zappa, SPADs and SiPMs arrays for long-range high-speed light detection and ranging (LiDAR). Sensors **21**(11), 1–23., Article no. 3839 (2021)
16. C. Zhang, N. Zhang, Z. Ma, L. Wang, Y. Qin, J. Jia, K. Zang, A 240 x 160 3D stacked SPAD dToF image sensor with rolling shutter and in pixel histogram for mobile device. IEEE Open J. Solid State Circuits Soc. **2**, 3–11 (2022)
17. R.D. Habbit Jr., R.O. Nellums, A.D. Niese, J.L. Rodriguez, Utilization of flash ladar for cooperative and uncooperative rendezvous and capture, in *Proceedings of SPIE 5088, Space Systems Technology and Operations*, Orlando, Florida, United States (2003)
18. G. Mandlburger, H. Lehner, N. Pfeifer, A comparison of single photon and full waveform LiDAR. ISPRS Ann. Photogramm. Remote Sens. Spat. Inf. Sci. **IV-2/W5**, 397–404 (2019)
19. R. Agishev, A. Comeron, J. Bach, A. Rodriguez, M. Sicard, J. Riu, S. Royo, Lidar with SiPM: Some capabilities and limitations in real environment. Opt. Laser Technol. **49**, 86–90 (2013)
20. N. Strout, United States Tasks 4 companies with tracking hypersonic weapon. Def. News **34**(21), 16 (2019)
21. P. McManamon, B. Javidi, E. Watson, M. DaneshPanah, R. Schulein, New paradigms for active and passive 3D remote object sensing, visualization, and recognition, in *Proceedings of SPIE 6967, Automatic Target Recognition XVIII*, vol. 69670E, (SPIE, 2008), pp. 1–15
22. F. Blais, Review of 20 years of range sensor development. J. Electron. Imaging **13**(1), 231–240 (2004)
23. A. Spickermann, D. Durini, S. Brocker, W. Brockherde, B.J. Hosticka, A. Grabmaier, Pulsed time-of-flight 3D-CMOS imaging using photogate-based active pixel sensors, in *Proceedings of ESSCIRC*, Athens, Greece (2009)
24. M.B. Das, S. Bose, R. Bhattacharya, Single photon response of photomultiplier tubes. Nucl. Instrum. Methods Phys. Res., Sect. A **242**(1), 156–159 (1985)
25. D. Dussault, P. Hoess, Noise performance comparison of ICCD with CCD and EMCCD cameras, in *Proceedings of SPIE 5563, Infrared Systems and Photoelectronic Technology*, vol. 5563, (SPIE, 2004), pp. 195–204
26. S. Piatek, *Silicon Photomultipliers (SiPMs): Operation, Performance, and Possible Applications* (Hamamatsu Corporation, New Jersey Institute of Technology, 2018)
27. N. Guo, K.W. Cheung, H.T. Wong, D. Ho, CMOS time-resolved, contact, and multispectral fluorescence imaging for DNA molecular diagnostics. Sensors **14**, 20602–20619 (2014)

28. T. Okina, S. Yamada, Y. Sakata, S. Kasuga, M. Takemoto, Y. Nose, H. Koshida, M. Tamaru, Y. Sugiura, S. Saito, S. Koyama, M. Mori, Y. Hirose, M. Sawada, A. Odagawa, T. Tanaka, A 1200×900 6μm 450fps Geiger-mode vertical avalanche photodiodes CMOS image sensor for a 250m time-of-flight ranging system using direct-indirect-mixed frame synthesis with configurable-depth-resolution down to 10cm, in *IEEE International Solid-State Circuits Conference – (ISSCC)*, San Francisco, CA, USA (2020)
29. S. Bellisai, F. Villa, S. Tisa, D. Bronzi, F. Zappa, Indirect time-of-flight 3D ranging, in *Proceedings of SPIE 8268, Quantum Sensing and Nanophotonic Devices IX*, vol. 82681C, (SPIE, 2012), pp. 1–8
30. I. Gyongy, G.M. Martin, A. Turpin, A. Ruget, A. Halimi, R. Henderson, J. Leach, High-speed vision with a 3D-stacked SPAD image sensor, in *Proceedings of SPIE 11721, Advanced Photon Counting Techniques XV*, vol. 1172105, (SPIE, 2021), pp. 1–7
31. J.S. Massa, G.S. Buller, A.C. Walker, S. Cova, M. Umasuthan, A.M. Wallace, Time-of-flight optical ranging system based on time-correlated single-photon counting. Appl. Opt. **37**(31), 7298–7304 (1998)
32. M.A. Albota, B.F. Aull, D.G. Fouche, R.M. Heinrichs, D.G. Kocher, R.M. Marino, J.G. Mooney, N.R. Newbury, M.E. O'Brien, B.E. Player, B.C. Willard, J.J. Zayhowski, Three-dimensional imaging laser radars with Geiger-mode avalanche photodiode arrays. Linc. Lab. J. **13**(2), 351–370 (2002)
33. Y. Kang, L. Li, D. Liu, D. Li, T. Zhang, W. Zhao, Fast long-range photon counting depth imaging with sparse single-photon data. IEEE Photonics J. **10**(3), 1–10 (2018)
34. A. McCarthy, R.J. Collins, N.J. Krichel, V. Fernández, A.M. Wallace, G.S. Buller, Long-range time-of-flight scanning sensor based on high-speed time-correlated single-photon counting. Appl. Opt. **48**(32), 6241–6251 (2009)
35. N.J. Krichel, A. McCarthy, R.J. Collins, V. Fernández, A.M. Wallace, G.S. Buller, Scanning of low-signature targets using time-correlated single-photon counting, in *Proceedings of SPIE 7482, Electro-Optical Remote Sensing, Photonic Technologies and Applications III*, vol. 748202, (SPIE, 2009), pp. 1–14
36. C. Fu, H. Zheng, G. Wang, Y. Zhou, H. Chen, Y. He, J. Liu, J. Sun, Z. Xu, Three-dimensional imaging via time-correlated single-photon counting. Appl. Sci. **10**, 1–10 (2020)
37. R. Lamb, A technology review of time-of-flight photon counting for advanced remote sensing, in *Proceedings of SPIE 7681, Advanced Photon Counting Techniques IV*, vol. 768107, (SPIE, 2010), pp. 1–12
38. R. Tobin, A. Halimi, A. McCarthy, P.J. Soan, G.S. Buller, Robust real-time 3D imaging of moving scenes through atmospheric obscurant using single-photon LiDAR. Sci. Rep. **11**(11236), 1–13 (2021)
39. Z.-P. Li, X. Huang, Y. Cao, B. Wang, Y.-H. Li, W. Jin, C. Yu, J. Zhang, Q. Zhang, C.-Z. Peng, F. Xu, J.-W. Pan, Single-photon computational 3D imaging at 45 km. arXiv:1904.10341, 1–22 (2019)
40. Z.-P. Li, J.-T. Ye, X. Huang, P.-Y. Jiang, Y. Cao, Y. Hong, C. Yu, J. Zhang, Q. Zhang, C.-Z. Peng, F. Xu, J.-W. Pan, Single-photon imaging over 200 km. Optica **8**(3), 344–349 (2021)
41. A. Migdall, S. Polyakov, J. Fan, J. Bienfang, *Single-Photon Generation and Detection: Physics and Applications* (Academic Press, 2013)
42. A. Kirmani, D. Venkatraman, D. Shin, A. Colaço, F.N.C. Wong, J.H. Shapiro, V.K. Goyal, First-photon imaging. Science **343**(6166), 58–61 (2014)
43. X. Liu, J. Shi, X. Wu, G. Zeng, Fast first-photon ghost imaging. Sci. Rep. **8**(5012), 1–8 (2018)
44. D. Shin, F. Xu, D. Venkatraman, R. Lussana, F. Villa, F. Zappa, V.K. Goyal, F.N. Wong, J.H. Shapiro, Photon-efficient imaging with a single-photon camera. Nat. Commun. **7**(12046), 1–8 (2016)
45. A. McCarthy, X. Ren, A.D. Frera, N.R. Gemmell, N.J. Krichel, C. Scarcella, A. Ruggeri, A. Tosi, G.S. Buller, Kilometer-range depth imaging at 1550 nm wavelength using an InGaAs/InP single-photon avalanche diode detector. Opt. Express **21**(19), 22098–22113 (2013)

46. A. Gupta, A. Ingle, M. Gupta, Asynchronous single-photon 3D imaging, in *IEEE/CVF International Conference on Computer Vision (ICCV)* (2019), pp. 7908–7917
47. H.J. Kong, T.H. Kim, S.E. Jo, M.S. Oh, Smart three-dimensional imaging ladar using two Geiger-mode avalanche photodiodes. Opt. Express **19**, 19323–19329 (2011)
48. P. Huang, W. He, G. Gu, Q. Chen, Depth imaging denoising of photon-counting lidar. Appl. Opt. **58**(16), 4390–4394 (2019)
49. J. Tachella, Y. Altmann, N. Mellado, A. McCarthy, R. Tobin, G.S. Buller, J.-Y. Tourneret, S. McLaughlin, Real-time 3D reconstruction from single-photon lidar data using plug-and-play point cloud denoisers. Nat. Commun. **10**(4984), 1–6 (2019)
50. C. Wu, J. Liu, X. Huang, Z.-P. Li, C. Yu, J.-T. Ye, J. Zhang, Q. Zhang, X. Dou, V.K. Goyal, F. Xu, J.-W. Pan, Non–line-of-sight imaging over 1.43 km. Proc. Natl. Acad. Sci. **118**(10), 1–7 (2021)
51. H. Tan, J. Peng, Z. Xiong, D. Liu, X. Huang, Z.-P. Li, Y. Hong, F. Xu, Deep learning based single-photon 3D imaging with multiple returns, in *International Conference on 3D Vision (3DV)*, Fukuoka, Japan (2020)
52. A. Aßmann, B. Stewart, A.M. Wallace, Deep learning for LiDAR waveforms with multiple returns, in *28th European Signal Processing Conference (EUSIPCO)*, Amsterdam, Netherlands (2021)
53. N. Radwell, S.D. Johnson, M.P. Edgar, C.F. Higham, R. Murray-Smith, M.J. Padgett, Deep learning optimized single-pixel LiDAR. Appl. Phys. Lett. **115**(231101), 1–5 (2019)
54. J. Peng, Z. Xiong, X. Huang, Z.-P. Li, D. Liu, F. Xu, Photon-efficient 3D imaging with a non-local neural network, in *European Conference on Computer Vision*, Glasgow, UK (2020)
55. D.B. Lindell, M. O'Toole, G. Wetzstein, Single-photon 3D imaging with deep sensor fusion. ACM Trans. Graph. **37**(4), 1–12 (2018)
56. P. Kirkland, V. Kapitany, A. Lyons, J. Soraghan, A. Turpin, D. Faccio, G.D. Caterina, Imaging from temporal data via spiking convolutional neural networks, in *Proceedings of SPIE – The International Society for Optical Engineering*, vol. 11540, (SPIE, 2020), pp. 1–20
57. L. Yan, Q. Hao, X. Cheng, Review of non-scanning laser 3D imaging radars: Basic principles, latest developments, and future directions, in *International Conference on Optical Instruments and Technology*, Beijing, China (2019)
58. R.D. Richmond, S.C. Cain, *Direct Detection LADAR Systems* (SPIE Press, Bellingham, 2010)
59. B.F. Aull, A.H. Loomis, D.J. Young, R.M. Heinrichs, B.J. Felton, P.J. Daniels, D.J. Landers, Geiger-mode avalanche photodiodes for three-dimensional imaging. Linc. Lab. J. **13**(2), 335–350 (2002)
60. R.M. Marino, W.R. Davis, Jigsaw: A foliage-penetrating 3D imaging laser radar system. Linc. Lab. J. **15**(1), 23–36 (2005)
61. B. Lohani, S. Chacko, S. Ghosh, S. Sasidharan, Surveillance system based on Flash LiDAR. Indian Cartogr. **XXXII**, 77–85 (2012)
62. K.J. Kasunic, *Optical Systems Engineering* (McGraw-Hill, Boulder, 2011)
63. M. O'Toole, F. Heide, D.B. Lindell, K. Zang, S. Diamond, G. Wetzstein, Reconstructing transient images from single-photon sensors, in *IEEE Conference on Computer Vision and Pattern Recognition (CVPR)*, (IEEE, 2017), pp. 2289–2297
64. S. Ma, S. Gupta, A.C. Ulku, C. Bruschini, E. Charbon, M. Gupta, Quanta burst photography. ACM Trans. Graph. **39**(4), 79:1–79:16 (2020)
65. P.F. McManamon, Review of ladar: A historic, yet emerging, sensor technology with rich phenomenology. Opt. Eng. **51**(6), 1–12 (2012)
66. G. Williams, Limitations of Geiger-mode arrays for Flash LADAR, in *Proceedings of SPIE 7684, Laser Radar Technology and Applications XV*, vol. 768414, (SPIE, 2010), pp. 1–19
67. A. Buchner, S. Hadrath, R. Burkard, F.M. Kolb, J. Ruskowski, M. Ligges, A. Grabmaier, Analytical evaluation of signal-to-noise ratios for avalanche and single-photon avalanche diodes. Sensors **21**(2887), 1–18 (2021)
68. P.F. McManamon, P. Banks, J. Beck, D.G. Fried, A.S. Huntington, E.A. Watson, Comparison of flash lidar detector options. Opt. Eng. **56**(3), 1–23 (2017)

69. K.J. Gordon, P.A. Hiskett, R.A. Lamb, Advanced 3D imaging lidar concepts for long range sensing, in *Proceedings of SPIE 9114, Advanced Photon Counting Techniques VIII*, vol. 91140G, (SPIE, 2014), pp. 1–7
70. T. Baba, Y. Suzuki, K. Makino, T. Fujita, T. Hashi, S. Adachi, S. Nakamura, K. Yamamoto, Development of an InGaAs SPAD 2D array for flash LIDAR, in *Proceedings of SPIE 10540, Quantum Sensing and Nano Electronics and Photonics XV*, vol. 105400L, (SPIE, 2018), pp. 1–14
71. A.S. Huntington, *InGaAs Avalanche Photodiodes for Ranging and LiDAR* (Woodhead Publishing, Beaverton, 2020)
72. S. Afshar, T. Hamilton, L. Davis, A.V. Schaik, D. Delic, Event-based processing of single photon avalanche diode sensors. IEEE Sens. J. **20**(14), 7677–7691 (2020)
73. J. Mau, V. Devrelis, G. Day, J. Trumpf, D. Delic, The use of statistical mixture models to reduce noise in SPAD images of fog obscured environments. SPIE Future Sens. Technol. **11525**, 1–10 (2020)
74. C. Palestini, Microelectronic 3D imaging and neuromorphic recognition for autonomous UAVs, in *Advanced Technologies for Security Applications. Proceedings of the NATO Science for Peace and Security "Cluster Workshop on Advanced Technologies"*, (Springer, Leuven, 2019), pp. 185–194
75. L. Hespel, N. Riviere, M. Fraces, P.-E. Dupouy, A. Coyac, P. Barillot, S. Fauquex, A. Plyer, M. Tauvy, M. Jacquart, I. Vin, E. Nascimben, C. Perez, J.P. Velayguet, D. Gorce, 2D and 3D flash laser imaging for long-range surveillance in maritime border security: Detection and identification for counter UAS applications, in *Proceedings of SPIE 10191, Laser Radar Technology and Applications XXII*, vol. 10191, (SPIE, 2017), pp. 1–11
76. M. Laurenzis, E. Bacher, F. Christnacher, Measuring laser reflection cross-sections of small unmanned aerial vehicles for laser detection, ranging and tracking, in *Proceedings of SPIE 10191, Laser Radar Technology and Applications XXII*, vol. 101910D, (SPIE, 2017), pp. 1–9
77. F.M.D. Rocca, H. Mai, S.W. Hutchings, T.A. Abbas, K. Buckbee, A. Tsiamis, P. Lomax, I. Gyongy, N.A.W. Dutton, R.K. Henderson, A 128 × 128 SPAD motion-triggered time-of-flight image sensor with in-pixel histogram and column-parallel vision processor. IEEE J. Solid State Circuits **55**(7), 1762–1775 (2020)
78. B. Göhler, P. Lutzmann, Range accuracy of a gated-viewing system compared to a 3D flash LADAR under different turbulence conditions, in *Proceedings of SPIE 7835, Electro-Optical Remote Sensing, Photonic Technologies and Applications IV*, vol. 783504, (SPIE, 2010), pp. 1–12
79. X. Wang, Y. Cao, W. Cui, X. Liu, S. Fan, Y. Zhou, Y. Li, Three-dimensional range-gated flash LIDAR for land surface remote sensing, in *Proceedings of SPIE 9260, Land Surface Remote Sensing II*, vol. 92604L, (SPIE, 2014), pp. 1–10
80. S.S. Jahromi, J.-P. Jansson, P. Keränen, E.A. Avrutin, B.S. Ryvkin, J.T. Kostamovaara, Solid-state block-based pulsed laser illuminator for single-photon avalanche diode detection-based time-of-flight 3D range imaging. Opt. Eng. **60**(5), 1–11 (2021)
81. S. Chan, A. Halimi, F. Zhu, I. Gyongy, R.K. Henderson, R. Bowman, S. McLaughlin, G.S. Buller, J. Leach, Long-range depth imaging using a single-photon detector array and non-local data fusion. Sci. Rep. **9**(8075), 1–10 (2019)
82. M. Henriksson, P. Jonsson, Photon-counting panoramic three-dimensional imaging using a Geiger-mode avalanche photodiode array. Opt. Eng. **57**(9), 1–12 (2018)
83. P.-y. Yi, P. Tong, Y.-j. Zhao, Application comparison of LIDAR single-pulse and multi-pulse mode in high elevation difference area, in *Proceedings of SPIE 9543, Third International Symposium on Laser Interaction with Matter*, vol. 954312, (SPIE, 2015), pp. 1–6
84. R.B. Roth, J. Thompson, Practical application of multiple pulse in air (MPIA) lidar in large-area surveys, in *The International Archives of the Photogrammetry, Remote Sensing and Spatial Information Sciences*, vol. XXXVII (2008), pp. 183–188
85. D.S. Hall, P.J. Kerstens, Multiple pulse, lidar based 3-D imaging. United States Patent US 2017/0219695 A1, 3 Aug 2017

86. A.N. Belbachir, M. Hofstatter, M. Litzenberger, P. Schon, High-precision shape representation using a neuromorphic vision sensor with synchronous address-event communication interface. Meas. Sci. Technol., IOP **20**(10), 1–9 (2009)
87. P. Lichtsteiner, C. Posch, T. Delbruck, A 128×128 120 dB 15 μs latency asynchronous temporal contrast vision sensor. IEEE J. Solid State Circuits **43**(2), 566–576 (2008)
88. T. Finateu, A. Niwa, D. Matolin, K. Tsuchimoto, A. Mascheroni, E. Reynaud, P. Mostafalu, F. Brady, L. Chotard, F. LeGoff, H. Takahashi, H. Wakabayashi, Y. Oike, C. Posch, A 1280×720 back-illuminated stacked temporal contrast event-based vision sensor with 4.86μm pixels, 1.066GEPS readout, programmable event-rate controller and compressive data-formatting pipeline, in *IEEE International Conference on Solid-State Circuits (ISSCC)*, San Francisco, CA, USA (2020)
89. S. Afshar, High speed event-based visual processing in the presence of noise. Ph.D. thesis, Western Sydney University, Sydney, 2020
90. Y. Hirose, S. Koyama, T. Okino, A. Inoue, S. Saito, Y. Nose, M. Ishii, S. Yamahira, S. Kasuga, M. Mori, T. Kabe, K. Nakanishi, M. Usuda, A. Odagawa, T. Tanaka, A 400×400-pixel 6μm-pitch vertical avalanche photodiodes CMOS image sensor based on 150ps-fast capacitive relaxation quenching in Geiger mode for synthesis of arbitrary gain images, in *IEEE International Solid State Circuits Conference*, San Francisco, California (2019)
91. K. Morimoto, A. Ardelean, M.-L. Wu, A.C. Ulku, I.M.M. Antolovic, V. Zickus, V. Kapitany, A. Fatima, A. Turpin, R. Insall, J. Whitelaw, L. Machesky, C. Bruschini, D. Faccio, E. Charbon, Megapixel time-gated SPAD image sensor for scientific imaging applications scientific imaging applications, in *Proceedings Volume 11654, High-Speed Biomedical Imaging and Spectroscopy VI*, Online Only (2021)
92. F. Piron, D. Morrison, M.R. Yuce, J.-M. Redoute, A review of single-photon avalanche diode time-of-flight imaging sensor arrays. IEEE Sensors J. **21**(11), 12654–12666 (2021)
93. K. Morimoto, A. Ardelean, M.-L. Wu, A.C. Ulku, I.M. Antolovic, C. Bruschini, E. Charbon, A megapixel time-gated SPAD image sensor for 2D and 3D imaging applications. arXiv:1912.12910, 1–11 (2019)
94. F. Zappa, S. Tisa, S. Cova, Principles and features of single-photon avalanche diode arrays. Sens. Actuators A Phys. **140**(1), 103–112 (2007)
95. M.S.A. Shawkat, N. McFarlane, A CMOS perimeter gated SPAD based mini-digital silicon photomultiplier, in *IEEE 61st International Midwest Symposium on Circuits and Systems (MWSCAS)*, (IEEE, 2018), pp. 302–305
96. D. Bronzi, F. Villa, S. Tisa, A. Tosi, F. Zappa, SPAD figures of merit for photon-counting, photon-timing, and imaging applications: A review. IEEE Sensors J. **16**(1), 3–12 (2016)
97. D. Bronzi, S. Tisa, F. Villa, S. Bellisai, A. Tosi, F. Zappa, Fast sensing and quenching of CMOS SPADs for minimal afterpulsing effects. IEEE Photon. Technol. Lett. **25**(8), 776–779 (2013)
98. Z. Zhang, C. Ma, X. Su, S. Chen, S. Wang, W. Zhu, Single photon imaging Lidar using three Geiger-mode avalanche diodes, in *Proceedings of SPIE 11338, AOPC: Optical Sensing and Imaging Technology*, Beijing, China (2019)
99. D.V. Delic, SPAD array structures and methods of operation. International Patent WO 2017/004663 A1, 12 Jan 2017
100. W. Xu, S. Zhen, H. Xiong, B. Zhao, Z. Liu, Y. Zhang, Z. Ke, B. Zhang, Design of 128×32 GM-APD array ROIC with multi-echo detection for single photon 3D LiDAR, in *Proceedings SPIE 11763, Seventh Symposium on Novel Photoelectronic Detection Technology*, Kunming, China (SPIE, 2020)
101. M. Beer, J.F. Hasse, J. Ruskowski, R. Kokozinski, Background light rejection in SPAD-based LiDAR sensors by adaptive photon coincidence detection. Sensors **18**(12), 1–16 (2018)
102. P. Padmanabhan, C. Zhang, M. Cazzaniga, B. Efe, A.R. Ximenes, M.-J. Lee, E. Charbon, A 256×128 3D-stacked (45nm) SPAD FLASH LiDAR with 7-level coincidence detection and progressive gating for 100m range and 10klux background light. IEEE ISSCC **64**, 112–114 (2021)

103. F. Severini, V. Sesta, F. Madonini, A. Incoronato, F. Villa, F. Zappa, SPAD array for LiDAR with region-of-interest selection and smart TDC routing, in *Proceedings of SPIE 11771, Quantum Optics and Photon Counting*, vol. 117710F, (SPIE, 2021), pp. 1–6
104. S.W. Hutchings, N. Johnston, I. Gyongy, T.A. Abbas, N.A.W. Dutton, M. Tyler, S. Chan, J. Leach, R.K. Henderson, A reconfigurable 3-D-stacked SPAD imager with in-pixel histogramming for flash LIDAR or high-speed time-of-flight imaging. IEEE J. Solid State Circuits **54**(11), 2947–2956 (2019)
105. C. Zhang, S. Lindner, I.M. Antolovic, J.M. Pavia, M. Wolf, E. Charbon, A 30-frames/s, 252×144 SPAD flash LiDAR with 1728 dual-clock 48.8-ps TDCs, and pixel-wise integrated histogramming. IEEE J. Solid State Circuits **54**(4), 1137–1151 (2019)
106. M. Sheehan, J. Tachella, M.E. Davies, A sketching framework for reduced data transfer in photon counting lidar. arXiv:2102.08732, 1–16 (2021)
107. A. Aßmann, B. Stewart, J.F.C. Mota, A.M. Wallace, Compressive super-pixel LiDAR for high-framerate 3D depth imaging, in *IEEE Global Conference on Signal and Information Processing (GlobalSIP)*, Ottawa, Canada (2019)
108. C. Zhang, S. Lindner, I.M. Antolovic, M. Wolf, E. Charbon, A CMOS SPAD imager with collision detection and 128 dynamically reallocating TDCs for single-photon counting and 3D time-of-flight imaging. Sensors **18**(11), 1–19 (2018)
109. D. Morrison, S. Kennedy, D. Delic, M. Yuce, J.-M. Redoute, A triple integration timing scheme for SPAD time of flight imaging sensors in 130 nm CMOS, in *25th IEEE International Conference on Electronics, Circuits and Systems (ICECS)*, Bordeaux, France (2018)
110. D. Morrison, S. Kennedy, D. Delic, M.R. Yuce, J.-M. Redoute, A 64 × 64 SPAD flash LIDAR sensor using a triple integration timing technique with 1.95 mm depth resolution. IEEE Sensors J. **21**(10), 11361–11373 (2021)
111. D. Morrison, D. Delic, M.R. Yuce, J.M. Redoute, Multistage linear feedback shift register counters with reduced decoding logic in 130-nm CMOS for large-scale array applications. IEEE Trans. VLSI Syst. **27**(1), 103–115 (2019)
112. A. Ruget, S. McLaughlin, R.K. Henderson, I. Gyongy, A. Halimi, J. Leach, Robust super-resolution depth imaging via a multi-feature fusion deep network. arXiv:2011.11444, 1–20 (2021)
113. G. Liu, J. Ke, Deep-learning for super-resolution full-waveform lidar, in *Proceedings of SPIE 11187, Optoelectronic Imaging and Multimedia Technology*, vol. 11187, (SPIE, 2019), pp. 1–9
114. I. Gyongy, S.W. Hutchings, A. Halimi, M. Tyler, S. Chan, F. Zhu, S. McLaughlin, R.K. Henderson, J. Leach, High-speed 3D sensing via hybrid-mode imaging and guided upsampling. Optica **7**(10), 1253–1260 (2020)
115. D.V. Delic, S. Afshar, Neuromorphic single photon avalanche detector (SPAD) array microchip. US Patent US 2020/0326414 A1, 15 Oct 2020
116. W. Woods, D. Delic, B. Smith, L. Świerkowski, G. Day, V. Devrelis, R. Joyce, Object detection and recognition using laser radars incorporating novel SPAD technology, in *Proceedings of SPIE, Laser Radar Technology and Applications XXIV*, vol. 11005, (SPIE, 2019), pp. 1–9
117. F. Guerrieri, S. Tisa, A. Tosi, F. Zappa, Single-photon camera for high-sensitivity high speed applications. Proc. SPIE-IS&T Electron. Imaging **7536**, 753605 (2010)
118. G.C. Boisset, B. Robertson, W.S. Hsiao, M.R. Taghizadeh, J. Simmons, K. Song, M. Matin, D.A. Thompson, D.V. Plant, On-die diffractive alignment structures for packaging of microlens arrays with 2-D optoelectronic device arrays. IEEE Photon. Technol. Lett. **8**(7), 918–920 (1996)
119. S. Burri, Y. Maruyama, X. Michalet, F. Regazzoni, C. Bruschini, E. Charbon, Architecture and applications of a high resolution gated SPAD image sensor. Opt. Express **22**(14), 17573–17589 (2014)

120. G. Intermite, A. McCarthy, R.E. Warburton, X. Ren, F. Villa, R. Lussana, A.J. Waddie, M.R. Taghizadeh, A. Tosi, F. Zappa, G.S. Buller, Fill-factor improvement of Si CMOS single-photon avalanche diode detector arrays by integration of diffractive microlens arrays. Opt. Express **23**(26), 33777–33791 (2015)
121. Y. Cheng, J. Cao, F. Zhang, Q. Hao, Design and modeling of pulsed-laser three-dimensional imaging system inspired by compound and human hybrid eye. Sci. Rep. **8**(17164), 1–13 (2018)
122. P. Vivet, G. Sicard, L. Millet, S. Chevobbe, K.B. Chehida, L.A.C. MonteAlegre, M. Bouvier, A. Valentian, M. Lepecq, T. Dombek, O. Bichler, S. Thuries, D. Lattard, C. Severine, P. Batude, F. Clermidy, Advanced 3D technologies and architectures for 3D smart image sensors, in *2019 Design, Automation & Test in Europe Conference & Exhibition (DATE)*, Florence, Italy (2019)
123. M.-J. Lee, E. Charbon, Progress in single-photon avalanche diode image sensors in standard CMOS: From two-dimensional monolithic to three-dimensional-stacked technology. Jpn. J. Appl. Phys. **57**(10), 1–7 (2018)
124. A.R. Ximenes, P. Padmanabhan, M.-J. Lee, Y. Yamashita, D.-N. Yaung, E. Charbon, A modular, direct time-of-flight depth sensor in 45/65-nm 3-D-stacked CMOS technology. IEEE J. Solid State Circuits **54**(11), 3203–3214 (2019)
125. R.K. Henderson, N. Johnston, S.W. Hutchings, I. Gyongy, T.A. Abbas, N. Dutton, M. Tyler, S. Chan, J. Leach, A 256×256 40nm/90nm CMOS 3D-stacked 120dB dynamic-range reconfigurable time-resolved SPAD imager, in *IEEE International Solid-State Circuits Conference*, San Francisco, California (2019)
126. E. Charbon, C. Bruschini, M.-J. Lee, 3D-stacked CMOS SPAD image sensors: Technology and applications, in *25th IEEE International Conference on Electronics, Circuits and Systems (ICECS)*, Bordeaux, France (2018)
127. S. Tisa, F. Zappa, A. Tosi, S. Cova, Electronics for single photon avalanche diode arrays. Sens. Actuators A Phys. **140**(1), 113–122 (2007)
128. S. Lindner, C. Zhang, I.M. Antolovic, M. Wolf, E. Charbon, A 252×144 SPAD pixel flash lidar with 1728 dual-clock 48.8 PS TDCs, integrated histogramming and 14.9-to-1 compression in 180 NM CMOS technology, in *IEEE Symposium on VLSI Circuits*, (IEEE, 2018), pp. 69–70
129. J. Mau, S. Afshar, T. Hamilton, A.V. Schaik, R. Lussana, A. Panella, J. Trumpf, D. Delic, Embedded implementation of a random feature detecting network for real-time classification of time-of-flight SPAD array recordings, in *Laser Radar Technology and Applications XXIV*, Baltimore (2019)
130. G. Mora-Martín, A. Turpin, A. Ruget, A. Halimi, R. Henderson, J. Leach, I. Gyongy, High-speed object detection with a single-photon time-of-flight image sensor. Opt. Express **29**(21), 1–13 (2021)
131. A. Berkovich, T. Datta-Chaudhuri, P. Abshire, A scalable 20 × 20 fully asynchronous SPAD-based imaging sensor with AER readout, in *IEEE International Symposium on Circuits and Systems (ISCAS)*, Lisbon, Portugal (2015)
132. M.S.A. Shawkat, N. McFarlane, A CMOS perimeter gated SPAD based digital silicon photomultiplier with asynchronous AER readout for PET applications, in *IEEE Biomedical Circuits and Systems Conference (BioCAS)*, (IEEE, 2018), pp. 1–4
133. M.S.A. Shawkat, S. Sayyarparaju, N. McFarlane, G.S. Rose, Single photon avalanche diode based vision sensor with on-chip memristive spiking neuromorphic processing, in *IEEE 63rd International Midwest Symposium on Circuits and Systems (MWSCAS)*, Springfield, MA, USA (2020)
134. T. Wunderlich, A.F. Kungl, E. Muller, A. Hartel, Y. Stradmann, S.A. Aamir, A. Grubl, A. Heimbrecht, K. Schreiber, D. Stockel, C. Pehle, S. Billaudelle, G. Kiene, C. Mauch, J. Schemmel, K. Meier, M.A. Petrovici, Demonstrating advantages of neuromorphic computation: A pilot study. Front. Neurosci. **13**(260), 1–15 (2019)

135. A. Islam, Y. Xu, T. Monk, S. Afshar, A.V. Schaik, Noise-robust text-dependent speaker identification using cochlear models. J. Acoust. Soc. Am. **151**(500), 500–516 (2022)
136. S. Afshar, L. George, J. Tapson, A.V. Schaik, T.J. Hamilton, Racing to learn: Statistical inference and learning in a single spiking neuron with adaptive kernels. Front. Neurosci. **8**(377), 1–18 (2014)
137. Z. Wang, S. Joshi, S.E. Savel'ev, H. Jiang, R. Midya, P. Lin, M. Hu, N. Ge, J.P. Strachan, Z. Li, Q. Wu, M. Barnell, G.-L. Li, H.L. Xin, R.S. Williams, Q. Xia, J.J. Yang, Memristors with diffusive dynamics as synaptic emulators for neuromorphic computing. Nat. Mater. **16**(1), 101–108 (2017)
138. K.A. Boahen, T.S. Lande, Communicating neuronal ensembles between neuromorphic chips, in *Neuromorphic Systems Engineering: Neural Networks in Silicon*, (Springer US, Boston, 1998), pp. 229–259
139. M. Davies, N. Srinivasa, T.-H. Lin, G. Chinya, Y. Cao, S.H. Choday, G. Dimou, P. Joshi, N. Imam, S. Jain, Y. Liao, C.-K. Lin, A. Lines, R. Liu, D. Mathaikutty, S. McCoy, A. Paul, J. Tse, G. Venkataramanan, Y.-H. Weng, A. Wild, Y. Yang, H. Wang, Loihi: A neuromorphic manycore processor with on-chip learning. IEEE Micro **38**(1), 82–99 (2018)
140. F. Akopyan, J. Sawada, A. Cassidy, R. Alvarez-Icaza, P. Merolla, N. Imam, Y. Nakamura, P. Datta, G.-J. Nam, B. Taba, M. Beakes, B. Brezzo, J.B. Kuang, R. Manohar, W.P. Risk, B. Jackson, D. Modha, TrueNorth: Design and tool flow of a 65 mW 1 million neuron programmable neurosynaptic chip. IEEE Trans. Comput. Aided Des. Integr. Circuits Syst. **34**(10), 1537–1557 (2015)
141. M. Abadi, P. Barham, J. Chen, Z. Chen, A. Davis, J. Dean, M. Devin, S. Ghemawat, G. Irving, M. Isard, M. Kudlur, J. Levenberg, R. Monga, S. Moore, D.G. Murray, B. Steiner, P. Tucker, V. Vasudevan, P. Warden, M. Wicke, Y. Yu, X. Zheng, TensorFlow: A system for large-scale machine learning, in *12th USENIX Symposium on Operating Systems Design and Implementation (OSDI '16)*, Savannah, GA, USA (2016)
142. E. Lindholm, J. Nickolls, S. Oberman, J. Montrym, NVIDIA Tesla: A unified graphics and computing architecture. IEEE Micro **28**(2), 39–55 (2008)

Chapter 7
Integrated Sensing Devices for Brain-Computer Interfaces

Tien-Thong Nguyen Do, Ngoc My Hanh Duong, and Chin-Teng Lin

1 Introduction

The human brain is a complex organ. All of human consciousness, our memories, our cognition, and our physical activities are handled and coordinated by the brain. Thus, understanding how the brain works and the function of each brain region can help us to explore the potential of all humankind. As a key part of this endeavor, brain-computer interfaces (BCI) allow us to decode a user's intentions through brain signals and then to convert them into machine commands [47]. Perceiving neural activity in the brain is obviously the first and most important step in the workings of a BCI system. Therefore, the quality of the sensors used to measure brain activity largely determines the reliability of the entire system.

BCI systems comprise several main components [30, 76], each with their own purpose and function. Some components are responsible for brain signal acquisition. Others handle tasks like signal processing, feature extraction, or feature decoding. Others still provide application user interfaces. Figure 7.1 illustrates the main functions of a BCI.

Brain signal acquisition involves measuring the brain's signals as input to a BCI system and then sending those signals to the components that handle signal processing. The main signal processing procedure is noise removal, which is important for ensuring a good signal-to-noise ratio. Feature extraction and feature

T.-T. N. Do · C.-T. Lin (✉)
Australian Artificial Intelligence Institute, Faculty of Engineering and IT, University of Technology Sydney, Sydney, NSW, Australia
e-mail: NguyenTienThong.Do@uts.edu.au; Chin-Teng.Lin@uts.edu.au

N. M. H. Duong
Optical Nanomaterial Group, Institute for Quantum Electronics, Department of Physics, ETH Zurich, Zurich, Switzerland
e-mail: nduoeng@phys.ethz.ch

F. Iacopi, F. Balestra (eds.), *More-than-Moore Devices and Integration for Semiconductors*, https://doi.org/10.1007/978-3-031-21610-7_7

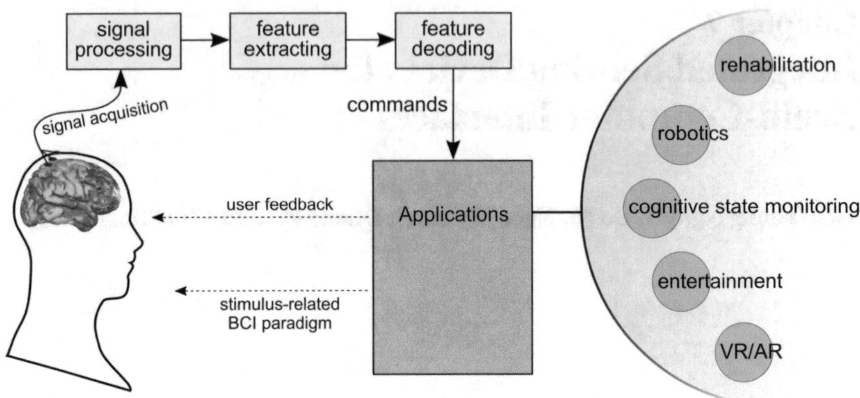

Fig. 7.1 There are five main components in a typical, basic BCI system configuration: (1) a scenario design based on a BCI paradigm; (2) modules for brain signal acquisition; (3) signal processing and feature extraction; (4) feature decoding or classifier; and (5) a user feedback system. (Reprinted from Lin and Do [47])

decoding come next. These components translate the brainwave data into the user's intentions. Lastly, the application components issue the user's intentions to various practical tools as commands. These tools can take many forms, such as a robot, a healthcare device, a virtual/augmented reality (VR/AR) system, and so on.

Brain signals can be acquired via two main types of sensors: invasive and non-invasive. The quality of the signals captured by invasive sensors is generally much higher than non-invasive sensors, and less artifact removal is typically required for a high signal-to-noise ratio. However, invasive sensors usually require surgery to implant under the scalp, and so non-invasive sensors are more commonly used. Non-invasive sensors are also easier to use and operate. They include technologies like electroencephalogram (EEG) sensors. Due to their popularity, this book chapter is mainly focused on non-invasive sensor types and their application in the field of BCI.

2 The Principles of Brain Signal Acquisition

2.1 How EEG Measurement Works

EEG is a non-invasive method of recording electrical impulse activity in the brain. Normally, conductive electrodes are placed on the scalp, which capture the small electrical potentials arising from neural activities. Among the great advantages of this method are its fine-grained time resolution, its cost effectiveness, and its portability. In fact, with EEG, one can track neuronal activity down to the millisecond range. Additionally, EEG can be used for live tracking of neuroimaging with real-world applications outside of the laboratory.

EEG measurement schemes comprise several devices, including electrodes, biopotential amplifiers, A/D converters, recording devices, and data ports [55, 78]. Among these devices, electrodes are the most crucial components as these are the devices that perceive the electrical brain activity. Again, electrodes can be invasive or non-invasive (surface devices). As with all invasive components, signal precision is better because of their direct penetration into the body, and signal-to-noise ratios tend to be much higher given the lack of interference caused by the skin or hair. The downside is that invasive sensors come at a risk to the patient and implanting them can be painful. Non-invasive sensors do not have these disadvantages, but there is a trade-off, which is increased background noise caused by the scalp, skin, and hair. Also, most electrodes also need to satisfy some specific conditions, and the materials for the EEG applications need to be biocompatible.

The same methods of recording signals are used for both types of electrodes. Here, the scalp acts as a bridge providing the potential differences between the signal an electrode captures, either active or passive, and a reference to another surface electrode. In other words, the electrodes form a basic electrical circuit. The role of an active electrode is to (i) compensate for the wide impedance range of the sensor; (ii) eliminate coupling between the cable and sources of interference; and (iii) limit the movement of cables and connectors. Conversely, passive electrodes can record EEG signals without the need to modify them. Ionic currents are converted directly into electric currents via signal and reference sensors. Further, other external electromagnetic signals beyond brain signals can be recorded in the 50–60 Hz range depending on the country. Notably, the noise measured by two electrodes is likely to be the same, but the signals captured will differ depending upon where the electrode is positioned on the head.

2.2 *Typical Methodologies for Non-invasive EEG Setup and Analysis*

The setup of non-invasive EEG systems is quite straightforward. The participant stays still, while the experimenter applies conductive gel (for typical gel-based sensor) to each electrode to ensure that impedance remains lower than a certain threshold (see Fig. 7.2). This gel generally improves the quality of signal data, raising the signal-to-noise ratio and making the system less sensitive to external noise. As a rule of thumb, best practice dictates that the impedance of all passive electrodes is kept at less than 5 or 10 kΩ, while the threshold for active electrodes is generally a bit higher at up to 25 kΩ (e.g., LiveAmps System, Brain Products, Gilching, Germany).

After placing all the electrodes, the experimenter normally carries out three basic checks before starting their experiments:

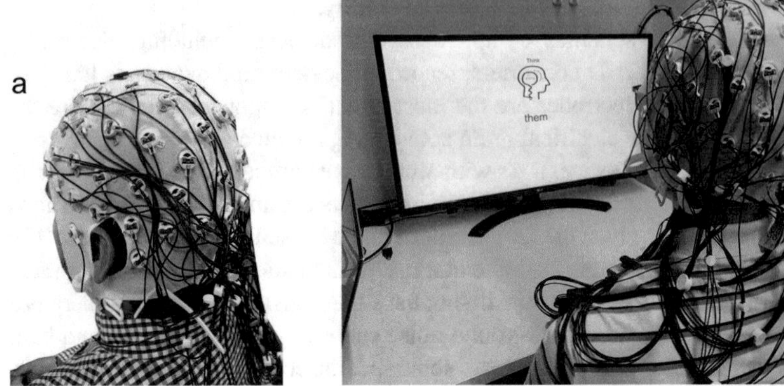

Fig. 7.2 The conventional EEG experimental setup with active EEG sensor in LiveAmps System – Brain Products (**a**) Participant with the active sensor cap. (**b**) One example of the stationary experiment

- The data passing through all channels should be in the normal range.
- There should be a strong peak in or nearby the frontal area when the participant blinks their eyes.
- An alpha oscillation should appear in or nearby the occipital area when the participant closes their eyes, and this oscillation should disappear when the eye is opened again.

A typical brain signal obtained through the scalp can have an amplitude ranging from 0.5 to 100 μV and a frequency in the range of 0.5–100 Hz. Moreover, brain signals also form five different patterns following the wave shapes of sinusoidal curves. Namely, these patterns are delta, theta, alpha, beta, and gamma waves [13]. Figure 7.3 shows the five different patterns. We already know that particular frequency bands appearing in specific brain regions strongly relate to the cognitive status of the participant. For example, theta waves in the frontal cortex indicate mental workload status [31, 60], while high gamma bands in the same region relate to a person's emotional state [79]. Thus, the feature extraction will be replied and depended on the experiment design.

As usual routine, the obtained signal data will be firstly applied band-pass filtered (normally 1–100 Hz). Then epochs are extracted around the onset of activity pertaining to the event of interest. With this information, the data can then be analyzed. There are two main forms of analysis: event-related potential (ERPs), which looks at brain signals captured with respect to the time domain, and event-related spectral perturbations (ERSPs), where signals are analyzed with respect to the time-frequency domain. Figure 7.4 illustrates this latter kind of analysis.

Several frameworks and toolboxes are available to support non-invasive EEG data analysis, such as the EEGLAB toolbox [14], FieldTrip [62], MNE [28], and Brainstorm [71]. Using these toolboxes can save researchers time with pre-processing and, overall, make analysis easier.

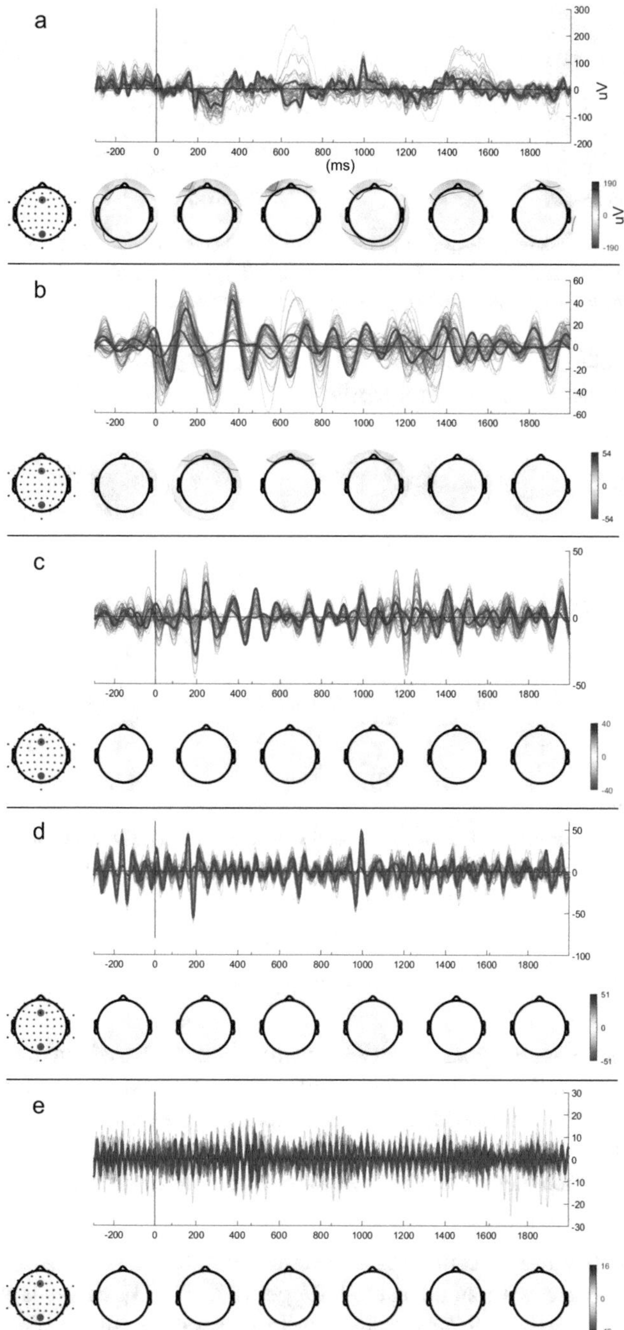

Fig. 7.3 The EEG data in different bands. (**a**) 1–40 Hz. (**b**) Theta (4–8 Hz). (**c**) Alpha (8–12 Hz). (**d**) Beta (12–30 Hz). (**e**) Low gamma (30–40 Hz). (This figure was created with the EEGVIS toolbox [22])

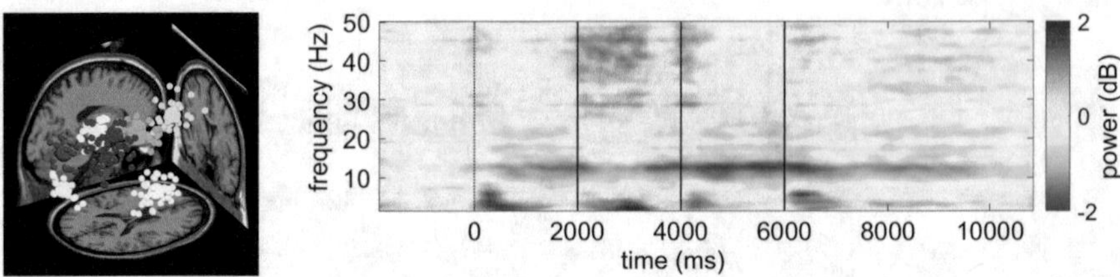

Fig. 7.4 Time-frequency analysis of EEG data. The left side shows the results of clustering from a dipole fitting routine [61] with the EEG channel data. The right side shows the ERSP results for one specific cluster being the parietal cortex (colored in blue on the left)

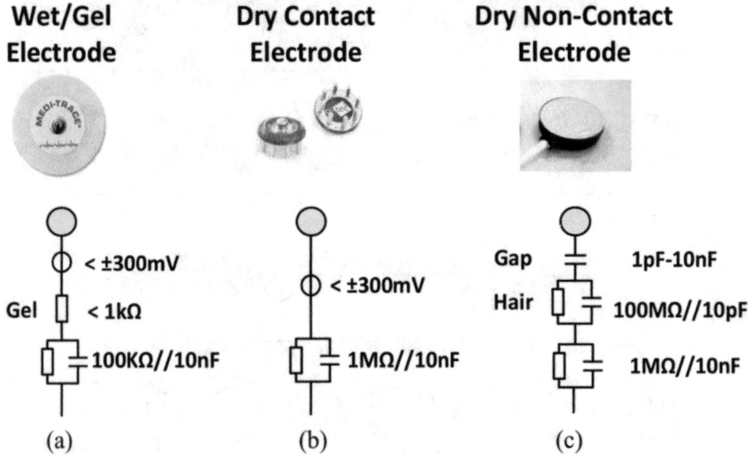

Fig. 7.5 Common types of EEG electrodes and equivalent electrical models of electrode-tissue interfaces. (**a**) Ag/AgCl electrode, (**b**) g.tec's dry electrode, and (**c**) QUASAR's capacitive electrode. (Reprinted from Xu et al. [78])

2.3 Non-invasive EEG Electrode Types

There are three basic types of EEG electrodes, each based on the amount of electrolyte used at the electrode-skin interface. The three kinds are wet electrodes, semi-dry/quasi-dry electrodes, and dry electrodes (see Fig. 7.5). The data received from the sensor is then converted into different stages of the EEG circuit board [55, 78] (Fig. 7.6). Each electrode type is explained in more detail next.

Wet Electrodes

The most widely used electrode is a silver/silver chloride (Ag/AgCl) disk that is affixed to the head with an adhesive EC2 gel [27, 68, 73]. Silver is widely used in metallic skin-surface electrodes [6] due to its solubility in salt. Silver chloride quickly saturates and reaches equilibrium state. Alternative metals that can be used for EEG electrodes are tin (Sn), gold (Au), and platinum (Pt). Generally, the impedance for most wet electrodes is below 5–10 kΩ and should be measured prior

a

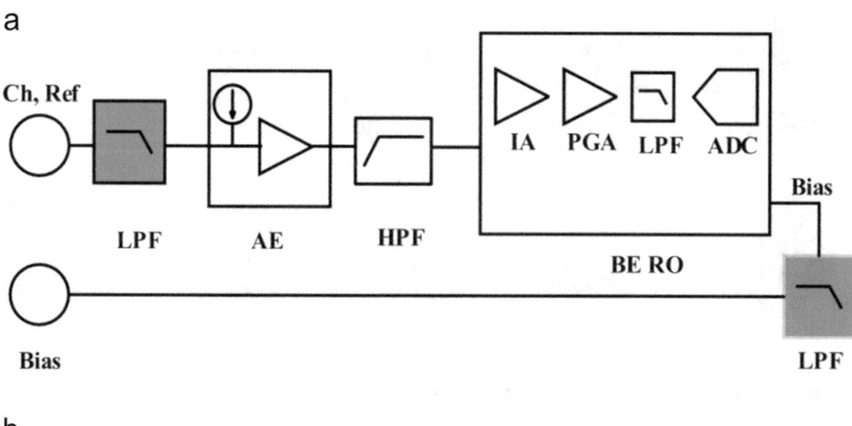

b

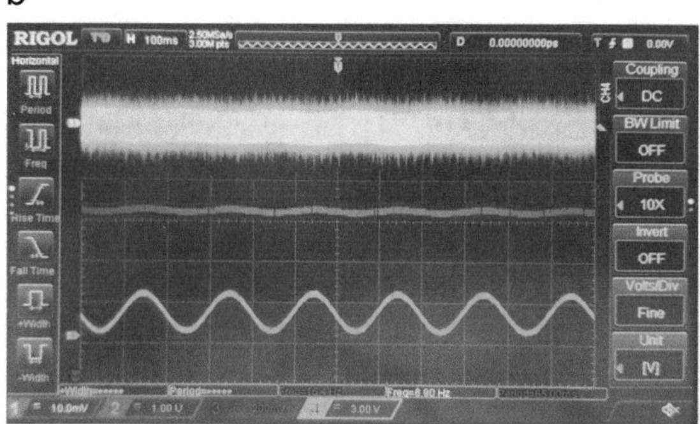

Fig. 7.6 High-level block diagram of an EEG circuit. (**a**) The introduction of RC low-pass filters (LPF) between the channel (Ch), the reference (Ref) dry electrodes, and the active electrode (AE) integrated circuits (ICs). RC is also placed between the bias output from back-end readout (BE RO) IC and the bias electrode. The high-pass filter (HPF) shown in the diagram ensures an AC coupling between AE and BE. (**b**) The output of a low-pass filter (purple) and a high-pass filter after amplification (blue) from functional simulated data (green). (Reprinted from Mihajlović et al. [55])

to acquiring the signals. Fig. 7.5 depicts an equivalent circuit model of electrode-tissue contact for different types of electrodes.

A primary requirement for the use of these types of sensors is that the electrode needs to make a low impedance connection with the scalp [39, 44]. This can be achieved by parting the hair with a Q-tip and using rubbing alcohol rub to clean the scalp area. Conductive gel is then added to act as a bridge between the electrode and the scalp. Performing this procedure, however, is time-consuming as it takes time to apply the gel. Further, in the end, even though the electrolyte gel is considered minimally invasive, it is still sticky and can leave patients' hair and scalp dirty [51]. It can also cause an allergic reaction in some rare cases. Plus, the gel can dry out over time, and, as it does, its transductive properties degrade. Thus, this entire approach requires a trained specialist to set up the measurement system and acquire the data.

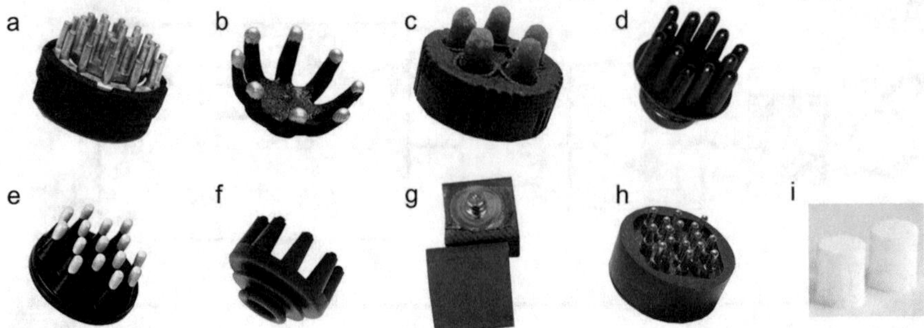

Fig. 7.7 Sensor types – several types of dry electrodes in the market. (**a**) Wearable sensing. (**b**, **c**) Cognionics. (**d**) OpenBCI sensor. (**e**) IMEC. (**f**) g.tec g.SAHARA. (**g**, **h**) MINDO. (**i**) mBrainTrain sponge sensor (semi-dry)

Dry EEG Electrodes

An alternative platform is dry EEG electrodes [15, 74]. Dry electrodes can record high-quality signals without the need for conductive gel, which reduces setup time. There are a wide range of different electrode designs [21, 26, 72], as illustrated in Fig. 7.7. Dry electrodes can also be made out of a variety of materials [77], in order to improve the quality of collected signals in different situations. For example, fingered electrodes [36] can be used to enhance the quality of the contact in the hair regions, hence achieving a higher signal-to-noise ratio. The rationale of using fingers or prongs is that these electrodes can push the hair apart and make close contact with the scalp. These sensors are also beneficial as they are commercially available and can be connected to EEG amplifiers, which can help to minimize the travel distance of the EEG signal before it is buffered.

Semi-Dry Electrodes

Another promising sensor is based on semi-dry electrodes, which stand as a compromise between wet and dry interfaces [41, 42, 56]. This system uses a minimal amount of electrolyte gel to improve interfacial impedance. This lesser volume of gel acts as a small connection between the scalp and the electrode but decreases the domains that can be degraded due to the gel solvent evaporating. Further, semi-dry electrodes offer a more comfortable experience for the users. That said, semi-dry electrode technology still needs work. The stability of these systems over time is a problem, and more work needs to be done on new design strategies to control the release of the electrolyte solution into the interface for long-term use. For example, mBrainTrain (https://mbraintrain.com/) and OpenBCI (https://shop.openbci.com/products/gelfree-bci-cap-kit?variant=40785117249694) are both working on saline solutions delivered by a sponge-like system to allow semi-dry electrodes to feasibly work with some current real-world applications.

3 Applications in BCI

Conventionally, BCI applications are performed in the stationary setup, where participants remain in a fixed position with limited movement. The stationary setup is beneficial for getting better signal quality and reducing the impact of noisy artifacts associated with human movement and muscle signals. However, the stationary setup pays off with the limited flexibility in the experience design, where participants can perform limited actions within a narrow experimental design scenario. This section of the chapter discusses some typical BCI applications that have been successfully reported in the literature.

3.1 Conventional BCI Applications

BCI applications can translate user intentions from brain data into digital commands; however, due to the sensitivity of the brain sensor, most BCI applications require the subject to remain stationary when they interact with the system (see Fig. 7.2). Such strict constraints limit the practicability of BCI, rendering them less suitable for applications like rehabilitation, prosthetics, or restoring the motor functions of patients with disabilities.

Several BCI paradigms have been demonstrated to successfully decode neural signals into action. These paradigms rely on specific neural properties at specific brain regions. For example, visual evoked potential (VEP) uses information extracted from the visual cortex, while the motor imagery paradigm uses information from the motor sensory region to translate the actions into intentions.

3.2 Mobile BCI Applications

In contrast to conventional setups, mobile and semi-mobile setups allow users to operate the BCI system in a more natural way [20]. Applications here tend toward the passive paradigms, such as cognitive driving studies [16, 18] or estimating the mental workload or fatigue status of subjects and adapting the information overhead to boost overall system performance [17, 19]. Furthermore, another example is the error-related potentials (ErrP), when the BCI system identifies the potential error operation of the system and AI algorithms could further correct it. However, at the current stage, the stability and robustness of mobile applications need to be increased, and this requires overcoming several challenges associated with the quality of mobile sensors and the robustness of noise removal algorithms.

3.3 Other Non-EEG BCI Applications

Beyond EEG technology, there are other brain imaging modalities that can be used to capture user intentions from brain activity.

MEG MEG is another imaging modality with high temporal resolution that has been used in various BCI applications and especially with motor imagery [25, 54, 67]. It is, however, a bulky and expensive technology, which has to date limited its use to laboratory conditions. Only very recently has a mobile MEG system been developed that can capture the cognitive status of a subject while the participant is engaged in natural activity, such as spatial navigation [10]. However, the mobile MEG is still bulky compared to current mobile EEG systems. Further, Shah and Wakai [44] are looking at integrating this mobile MEG with quantum sensors for even greater functionality. These breakthroughs are sure to bring great opportunities to the research community in not only neural science but also neural engineering. One important aspect of MEG systems to be aware of is that they tend to weigh a great deal. Hence, researchers need to be cognizant of the cognitive and physical toll wearing the mobile MEG system has on their participants during long experiments.

fMRI fMRI brain imaging is another common method of capturing neuronal activity that has been pair with BCI. It is particularly common in the area of neuroprosthetics and rehabilitation. Unlike EEG technology, fMRI provides higher spatial resolution that can reveal which brain regions strongly relate to a performed task. More importantly, fMRI can reduce the experiment setup and training time for participants. For example, with fMRI, scholars have been able to decode coordinated hand movements [53], such as Bleichner et al. [9], who were able to decode four different gestures from the American Sign Language alphabet, "L," "F," "W," and "Y," with up to 63% accuracy. However, due to its low temporal resolution, fMRI imaging is hard to extend to online closed-loop BCI systems, where participants perform tasks and receive feedback in real time.

fNIR Where EEG systems measure neural activity via the postsynaptic potential of ensembles of neurons, fNIR technology gauges hemodynamic responses using optical sensors. fNIR systems deliver higher spatial resolution than EEGs but have lower temporal resolution [50]. fNIR has been used in various BCI applications. For example, Batula et al. [7] used fNIR to measure the differences between motor preparation and motor movement, while Benitez-Andonegui et al. [8] demonstrated motor imagery using a combination of a BCI and an AR scenario.

Hybrid BCI Systems Each neural monitoring technology has its own strengths and limitations. For example, EEG applications are limited to measuring the brain's cortical activity, while fMRI is generally limited to offline data analysis. Integrating multiple brain imaging modalities can therefore play to the strengths of one technology while overcoming the limitations of an individual modality. Such integration is called hybrid BCI. For example, a hybrid EEG/fNIRS system could benefit from the high temporal resolution of an EEG system as well as the high spatial resolution of fNIRS [49]. It is worth noting, however, that hybrid systems tend to increase the cost of running experiments and software development and maintenance can be far more complex.

4 Future Electrode Directions

There are three key factors that need to be considered for the future direction of BCIs. These are users, industry, and research. The vision for advanced BCIs is user-friendly interfaces for applications that are able to acquire and convert a range of different brain signals. Further, bridging BCIs with applications in games, health, and education is a long-term goal in the field. For instance, the ability to integrate BCIs into plug-and-play devices is expected to be available in the near future as is using BCIs to treat diseases like epilepsy, depression, Parkinson's disease, and schizophrenia.

Although there has been much progress in sensor technology, most commercial sensors in common use are based on wet electrodes. Developing flexible sensors that are wearable, stable, and comfortable for users remains a significant challenge. Indeed, there are many distinct factors ruling the performance of a wearable EEG electrode. The interplay between these different factors needs to be explored more. In making practical dry electrodes, one needs to consider not only materials and making novel shapes (Fig. 7.7) but also the electrode design, which stands as a multidisciplinary and complex problem. In this chapter, some of promising alternative platforms for future sensors are reviewed.

Smart Electrodes

Currently, we have seen the advent of wireless sensors, which allow for mobility when collecting EEG measurements. With rapid developments in sensing technology, it is expected that there will be smart electrodes in the near future. These sensors will automate the process of adapting to user behaviors and movements, which relies on computational intelligence. The multi-channel and real-time acquisition are also expected advancements, which will likely require the assistance of different kinds of filters. Advances in nanofabrication technologies are another development set to affect sensor technologies. Electrodes at the nano-scale with high detection limits could do much to help capture sensitive neural signals. Moreover, combining artificial intelligence and sensing technologies is also likely to lead to advanced sensors with high sensing capabilities.

Wearable Pressure Sensors

Wearable pressure sensors that imitate the human skin have already been used in a broad range of applications. These work by transducing pressure and converting that pressure into electronic signals. They are attached to the human body, acquiring signals that provide information about the health status of the user. In this way, these sensors are likely to play a significant role in the next generation of monitoring systems [4, 11, 12, 24, 48] and personal healthcare [2, 63, 64].

More specifically, the force or pressure from external mechanical stimuli or deformations is converted into readable electronic signals by these sensors, and these signals can be measured using common electronic equipment. Piezoresistive, capacitive, piezoelectric, and triboelectric effects are the most common sensing mechanisms [29].

Wireless Transfer Implantable Sensors
In clinical procedures, monitoring the physiological reaction of patients and simultaneously providing proper therapeutic responses are important [45, 52, 75, 80]. Nevertheless, it remains technically challenging to provide medical treatment to the patients while maintaining ongoing monitoring at the point of care on a real-time basis [37, 38, 46]. In this context, implantable bioelectronic devices are promising solution. Energy storage is perhaps the most significant hurdle for miniaturizing these devices [58]. At present, either these devices do not last as long as they need to, or additional and repetitive surgeries are required to replace batteries when they are exhausted. Additionally, long-term use can result in mechanical stress and immunological reactions to the surrounding tissues [5]. So far, these issues have substantially limited the practicality of these devices, and, for this reason, they are not used more often in the areas where they could help most, such as with cardiovascular and neurological disease [34, 65, 70]. One future proposal for solving the energy problem is to use biocompatible energy storage, such as batteries or supercapacitors [40, 43, 69]. Alternatively, biodegradable materials can be used to create temporary implantable devices that disappear after a predetermined period. This approach eliminates the need for conducting removal/replacement surgery and widens the range of applications open to bioelectronic devices. Although not a permanent solution, such an approach might be a good stopgap until wireless power transfer becomes a feasible reality [1, 3, 57].

Photonic Sensors
Photonic sensors are lightweight, consume less energy, and provide low latency while opening the opportunity for integrated EEG devices, which could lead to a wider choice of wearable EEG devices.

In conventional sensor platforms and measuring instruments, the elements of an electronic circuit can malfunction. The result is often interference in the measuring probe caused by metallic cables or shorted currents. In this context, photonic-based E-field sensors, which are mostly composed of non-metallic materials, are a highly suitable alternative. In fact, photonic sensors have several distinct advantages compared to their traditional metallic-based counterparts. These include:

- No metallic elements. Thus, the sensors are explosion-proof and non-sensitive to radiofrequencies and other external electromagnetic fields.
- A small lightweight footprint with great flexibility that allows access to restricted sensing areas.

- Chemically stable and environmentally inert. Thus, maintenance is easier and so are storage and transportation.
- A user-friendly interface with optical data communication systems and secure data transmission.

The real strength of photonic E-field sensors lies in the use of dielectric materials. This is because there is no potential interference source with weak electrical signals derived from mapping the brain. Therefore, photonic-based EEG probes can be highly sensitive. An array of high light confinement photonic crystals can be promising for miniaturized devices.

One study [59], for example, proposes a photonic sensor model that makes use of a dome-shaped micro-scale laser as a sensing element. The device operates on the principle of morphology-dependent resonance (MDR), where the resonant wavelength is actively controlled by changes in the morphologies of a micro-scale laser that is affected by the external environment. The authors modeled performance on the Simulink (R) platform (MathWorks Inc., Natick, Massachusetts, USA), which provided similar results to actual photonic sensors. The sensors are designed as a ring resonator with the main design parameters to consider being the radius of the sphere and its refractive index. The physical environment that causes the phase shift in the sensors is certain sources of strains, such as pressure or an electric field. These deform the physical properties of the resonator, which lead to a response. Notably, the photonic sensor being modeled in this work was sensitive to 4×10^{-4} nm/Pa given a 4:1 polymeric ratio.

Fiber-Based and Integrated E-Field Sensors

Typically, there are two types of photonic E-field sensors, i.e., fiber-based and integrated. These sensors rely upon interferometric architectures, such as the Mach-Zehnder (M-Z) interferometer, including the four-port coupler and three-port coupler interferometric configuration [33, 66]. However, there is requirement for metal electrodes to be placed near the sensing waveguides of the M-Z arms, which serves as an antenna to collect the field from an electro-optical crystal. Even though it is easy to design and implement, these sensors have limited spatial resolution because of the long interaction length needed to obtain optical signals. Moreover, these sensors require auxiliary electrical connections, making them less attractive than a remote sensing method that only uses optical fiber. Lastly, biasing point drift is another issue that needs to be resolved for highly sensitive measurements.

An alternative approach is to rely on the use of liquid crystals embedded in a polymer matrix. By applying an electric field, the liquid crystals can be oriented in a certain direction, which changes their refractive index. However, this kind of sensor has low sensitivity due to a weak coupling between the evanescent field and the liquid crystal droplets. There are also other kinds of sensors based on E-field fibers that rely on piezoelectric or electrostrictive transducers to generate phase shifts in the optical signal that propagates along the optical fiber. The best electro-absorbent sensor proposed to date showed a minimum detectable E-field of 0.1 V/m and a bandwidth of 6 GHz. It is also possible to introduce cavities into E-field sensors to generate resonance frequency shifts, such as the Fabry-Pérot cavity and the disk resonator.

5 Challenges and Outlooks

Breakthroughs in brain sensor development will play an important role in future generations of BCIs. The current non-invasive BCI systems can quickly decode user intentions, but they are limited to just a few commands and so are not yet developed to a level where they are useful for daily activities. For truly high information transfer rates along with high spatial and temporal resolution, BCI systems still rely on invasive sensors. Even so, new and innovative brain imaging systems that can provide both high temporal and spatial resolution and less sensitivity to noise are needed.

In addition, high signal-to-noise ratio, low power consumption, and miniaturized BCI system also play an important role. These features allow the BCI system to be used effectively and easily for normal users in other applications, e.g., entertainment and ambulatory. The low noise and low power consumption could help provide reliable and stable signals while ensuring the prolonged usage of the system. More importantly, miniaturization plays an essential role in realizing these above features in daily applications.

The future should see more wearable and portable devices along with novel and flexible electrode materials that bring more comfort to users. To make dry electrodes a practical reality, we need to start considering new materials and novel shapes for both the sensors and the headsets. Novel sensor types are also likely to bring great advancements to the field, such as those based on ultrasound [81], photons, epitaxial graphene [23], and tattoos [32, 35]. Few things are certain, but one fact that will definitely take priority in the development of the field is that a wide variety of factors rule the performance of wearable EEG electrodes, and the interplay between these numerous factors needs to be comprehensively explored if BCIs are to deliver fully robust performances in the future.

References

1. K. Agarwal, R. Jegadeesan, Y.-X. Guo, N.V. Thakor, Wireless power transfer strategies for implantable bioelectronics. IEEE Rev. Biomed. Eng. **10**, 136–161 (2017)
2. Y. AL-Handarish, O.M. Omisore, T. Igbe, S. Han, H. Li, W. DU, J. Zhang, L. Wang, A survey of tactile-sensing systems and their applications in biomedical engineering. Adv. Mater. Sci. Eng. **2020**, 1 (2020)
3. A.B. Amar, A.B. Kouki, H. Cao, Power approaches for implantable medical devices. Sensors **15**, 28889–28914 (2015)
4. M. Amit, L. Chukoskie, A.J. Skalsky, H. Garudadri, T.N. Ng, Flexible pressure sensors for objective assessment of motor disorders. Adv. Funct. Mater. **30**, 1905241 (2020)

5. T. Araki, F. Yoshida, T. Uemura, Y. Noda, S. Yoshimoto, T. Kaiju, T. Suzuki, H. Hamanaka, K. Baba, H. Hayakawa, Long-term implantable, flexible, and transparent neural interface based on Ag/Au core–shell nanowires. Adv. Healthc. Mater. **8**, 1900130 (2019)
6. S.I. Arman, A. Ahmed, A. Syed, Cost-effective EEG signal acquisition and recording system. Int. J. Biosci. Biochem. Bioinforma. **2**, 301 (2012)
7. A.M. Batula, J.A. Mark, Y.E. Kim, H. Ayaz, Comparison of brain activation during motor imagery and motor movement using fNIRS. Comput. Intell. Neurosci. **2017**, 1 (2017)
8. A. Benitez-Andonegui, R. Burden, R. Benning, R. Möckel, M. Lührs, B. Sorger, An augmented-reality fNIRS-based brain-computer interface: A proof-of-concept study. Front. Neurosci. **14**, 346 (2020)
9. M.G. Bleichner, J.M. Jansma, J. Sellmeijer, M. Raemaekers, N.F. Ramsey, Give me a sign: Decoding complex coordinated hand movements using high-field fMRI. Brain Topogr. **27**, 248–257 (2014)
10. E. Boto, N. Holmes, J. Leggett, G. Roberts, V. Shah, S.S. Meyer, L.D. Muñoz, K.J. Mullinger, T.M. Tierney, S. Bestmann, Moving magnetoencephalography towards real-world applications with a wearable system. Nature **555**, 657–661 (2018)
11. C.M. Boutry, A. Nguyen, Q.O. Lawal, A. Chortos, S. Rondeau-Gagné, Z. Bao, A sensitive and biodegradable pressure sensor array for cardiovascular monitoring. Adv. Mater. **27**, 6954–6961 (2015)
12. C.M. Boutry, L. Beker, Y. Kaizawa, C. Vassos, H. Tran, A.C. Hinckley, R. Pfattner, S. Niu, J. Li, J. Claverie, Biodegradable and flexible arterial-pulse sensor for the wireless monitoring of blood flow. Nat. Biomed. Eng. **3**, 47–57 (2019)
13. M.X. Cohen, *Analyzing Neural Time Series Data: Theory and Practice* (MIT press, Cambridge, 2014)
14. A. Delorme, T. Mullen, C. Kothe, Z.A. Acar, N. Bigdely-Shamlo, A. Vankov, S. Makeig, EEGLAB, SIFT, NFT, BCILAB, and ERICA: New tools for advanced EEG processing. Comput. Intell. Neurosci. **2011**, 10 (2011)
15. G. Di Flumeri, P. Aricò, G. Borghini, N. Sciaraffa, A. Di Florio, F. Babiloni, The dry revolution: Evaluation of three different EEG dry electrode types in terms of signal spectral features, mental states classification and usability. Sensors **19**, 1365 (2019)
16. T.-T.N. Do, C.-H. Chuang, S.-J. Hsiao, C.-T. Lin, Y.-K. Wang, Neural comodulation of independent brain processes related to multitasking. IEEE Trans. Neural Syst. Rehabil. Eng. **27**, 1160–1169 (2019)
17. T.-T.N. Do, A.K. Singh, C.A.T. Cortes, C.-T. Lin, Estimating the cognitive load in physical spatial navigation, in *2020 IEEE Symposium Series on Computational Intelligence (SSCI)*, (IEEE, 2020), pp. 568–575
18. T.-T.N. Do, Y.-K. Wang, C.-T. Lin, Increase in brain effective connectivity in multitasking but not in a high-fatigue state. IEEE Trans. Cogn. Dev. Syst. **13**(3), 566–574 (2020)
19. T.-T.N. Do, T.-P. Jung, C.-T. Lin, Retrosplenial segregation reflects the navigation load during ambulatory movement. IEEE Trans. Neural Syst. Rehabil. Eng. **29**, 488–496 (2021)
20. T.-T.N. Do, C.-T. Lin, K. Gramann, Human brain dynamics in active spatial navigation. Sci. Rep. **11**, 1–12 (2021)
21. G. Edlinger, G. Krausz, C. Guger, A dry electrode concept for SMR, P300 and SSVEP based BCIs, in *2012 ICME International Conference on Complex Medical Engineering (CME)*, (IEEE, 2012), pp. 186–190
22. B. Ehinger, EEGVIS toolbox. *Osnabrück. [Google Scholar]*, (2018)
23. S.N. Faisal, M. Amjadipour, K. Izzo, J.A. Singer, A. Bendavid, C.-T. Lin, F. Iacopi, Non-invasive on-skin sensors for brain machine interfaces with epitaxial graphene. J. Neural Eng. **18**, 066035 (2021)
24. W. Fan, Q. He, K. Meng, X. Tan, Z. Zhou, G. Zhang, J. Yang, Z.L. Wang, Machine-knitted washable sensor array textile for precise epidermal physiological signal monitoring. Sci. Adv. **6**, eaay2840 (2020)
25. S.T. Foldes, D.J. Weber, J.L. Collinger, MEG-based neurofeedback for hand rehabilitation. J. Neuroeng. Rehabil. **12**, 1–9 (2015)

26. C. Fonseca, J.S. Cunha, R.E. Martins, V. Ferreira, J.M. De Sa, M. Barbosa, A.M. Da Silva, A novel dry active electrode for EEG recording. IEEE Trans. Biomed. Eng. **54**, 162–165 (2006)
27. L. Geddes, L. Baker, A. Moore, Optimum electrolytic chloriding of silver electrodes. Med. Biol. Eng. **7**, 49–56 (1969)
28. A. Gramfort, M. Luessi, E. Larson, D.A. Engemann, D. Strohmeier, C. Brodbeck, R. Goj, M. Jas, T. Brooks, L. Parkkonen, MEG and EEG data analysis with MNE-Python. Front. Neurosci. **7**, 267 (2013)
29. M.L. Hammock, A. Chortos, B.C.K. Tee, J.B.H. Tok, Z. Bao, 25th anniversary article: The evolution of electronic skin (e-skin): A brief history, design considerations, and recent progress. Adv. Mater. **25**, 5997–6038 (2013)
30. B. He, H. Yuan, J. Meng, S. Gao, Brain–computer interfaces, in *Neural Engineering*, ed. by B. He, (Springer, Cham, 2020)
31. J. Jacobs, G. Hwang, T. Curran, M.J. Kahana, EEG oscillations and recognition memory: Theta correlates of memory retrieval and decision making. NeuroImage **32**, 978–987 (2006)
32. S. Kabiri Ameri, R. Ho, H. Jang, L. Tao, Y. Wang, L. Wang, D.M. Schnyer, D. Akinwande, N. Lu, Graphene electronic tattoo sensors. ACS Nano **11**, 7634–7641 (2017)
33. M. Kanda, K.D. Masterson, Optically sensed EM-field probes for pulsed fields. Proc. IEEE **80**, 209–215 (1992)
34. D.-H. Kim, R. Ghaffari, N. Lu, S. Wang, S.P. Lee, H. Keum, R. D'angelo, L. Klinker, Y. Su, C. Lu, Electronic sensor and actuator webs for large-area complex geometry cardiac mapping and therapy. Proc. Natl. Acad. Sci. **109**, 19910–19915 (2012)
35. J. Kim, W.R. De Araujo, I.A. Samek, A.J. Bandodkar, W. Jia, B. Brunetti, T.R. Paixao, J. Wang, Wearable temporary tattoo sensor for real-time trace metal monitoring in human sweat. Electrochem. Commun. **51**, 41–45 (2015)
36. S. Krachunov, A.J. Casson, 3D printed dry EEG electrodes. Sensors **16**, 1635 (2016)
37. S.R. Krishnan, H.M. Arafa, K. Kwon, Y. Deng, C.-J. Su, J.T. Reeder, J. Freudman, I. Stankiewicz, H.-M. Chen, R. Loza, Continuous, noninvasive wireless monitoring of flow of cerebrospinal fluid through shunts in patients with hydrocephalus. NPJ Digit. Med. **3**, 1–11 (2020)
38. J.W. Kwak, M. Han, Z. Xie, H.U. Chung, J.Y. Lee, R. Avila, J. Yohay, X. Chen, C. Liang, M. Patel, Wireless sensors for continuous, multimodal measurements at the skin interface with lower limb prostheses. Sci. Transl. Med. **12**, eabc4327 (2020)
39. S. Leach, K.-Y. Chung, L. Tüshaus, R. Huber, W. Karlen, A protocol for comparing dry and wet EEG electrodes during sleep. Front. Neurosci. **14**, 586 (2020)
40. G. Lee, S.K. Kang, S.M. Won, P. Gutruf, Y.R. Jeong, J. Koo, S.S. Lee, J.A. Rogers, J.S. Ha, Fully biodegradable microsupercapacitor for power storage in transient electronics. Adv. Energy. Mater. **7**, 1700157 (2017)
41. G. Li, D. Zhang, S. Wang, Y.Y. Duan, Novel passive ceramic based semi-dry electrodes for recording electroencephalography signals from the hairy scalp. Sensors Actuators B Chem. **237**, 167–178 (2016)
42. G. Li, S. Wang, Y.Y. Duan, Towards conductive-gel-free electrodes: Understanding the wet electrode, semi-dry electrode and dry electrode-skin interface impedance using electrochemical impedance spectroscopy fitting. Sensors Actuators B Chem. **277**, 250–260 (2018)
43. H. Li, C. Zhao, X. Wang, J. Meng, Y. Zou, S. Noreen, L. Zhao, Z. Liu, H. Ouyang, P. Tan, Fully bioabsorbable capacitor as an energy storage unit for implantable medical electronics. Adv. Sci. **6**, 1801625 (2019)
44. G. Li, J. Wu, Y. Xia, Q. He, H. Jin, Review of semi-dry electrodes for EEG recording. J. Neural Eng. **17**, 051004 (2020)
45. R. Li, H. Qi, Y. Ma, Y. Deng, S. Liu, Y. Jie, J. Jing, J. He, X. Zhang, L. Wheatley, A flexible and physically transient electrochemical sensor for real-time wireless nitric oxide monitoring. Nat. Commun. **11**, 1–11 (2020)
46. C. Lim, Y. Shin, J. Jung, J.H. Kim, S. Lee, D.-H. Kim, Stretchable conductive nanocomposite based on alginate hydrogel and silver nanowires for wearable electronics. APL Mater. **7**, 031502 (2018)

47. C.-T. Lin, T.-T.N. Do, Direct-sense brain-computer interfaces and wearable computers. IEEE Trans. Syst. Man Cybern. Syst. **51**, 298–312 (2021)
48. Z. Liu, Y. Ma, H. Ouyang, B. Shi, N. Li, D. Jiang, F. Xie, D. Qu, Y. Zou, Y. Huang, Transcatheter self-powered ultrasensitive endocardial pressure sensor. Adv. Funct. Mater. **29**, 1807560 (2019)
49. Z. Liu, J. Shore, M. Wang, F. Yuan, A. Buss, X. Zhao, A systematic review on hybrid EEG/fNIRS in brain-computer interface. Biomed. Signal Process. Control **68**, 102595 (2021)
50. S. Lloyd-Fox, A. Blasi, C. Elwell, Illuminating the developing brain: The past, present and future of functional near infrared spectroscopy. Neurosci. Biobehav. Rev. **34**, 269–284 (2010)
51. M.A. Lopez-Gordo, D. Sanchez-Morillo, F.P. Valle, Dry EEG electrodes. Sensors **14**, 12847–12870 (2014)
52. D. Lu, Y. Yan, R. Avila, I. Kandela, I. Stepien, M.H. Seo, W. Bai, Q. Yang, C. Li, C.R. Haney, Bioresorbable, wireless, passive sensors as temporary implants for monitoring regional body temperature. Adv. Healthc. Mater. **9**, 2000942 (2020)
53. S. Martin, P. Brunner, C. Holdgraf, H.-J. Heinze, N.E. Crone, J. Rieger, G. Schalk, R.T. Knight, B.N. Pasley, Decoding spectrotemporal features of overt and covert speech from the human cortex. Front. Neuroeng. **7**, 14 (2014)
54. J. Mellinger, G. Schalk, C. Braun, H. Preissl, W. Rosenstiel, N. Birbaumer, A. Kübler, An MEG-based brain–computer interface (BCI). NeuroImage **36**, 581–593 (2007)
55. V. Mihajlović, S. Patki, J. Xu, Noninvasive wearable brain sensing. IEEE Sensors J. **18**, 7860–7867 (2018)
56. A.R. Mota, L. Duarte, D. Rodrigues, A. Martins, A. Machado, F. Vaz, P. Fiedler, J. Haueisen, J. Nóbrega, C. Fonseca, Development of a quasi-dry electrode for EEG recording. Sensors Actuators A Phys. **199**, 310–317 (2013)
57. B.D. Nelson, S.S. Karipott, Y. Wang, K.G. Ong, Wireless technologies for implantable devices. Sensors **20**, 4604 (2020)
58. K.A. Ng, A. Rusly, G.G.L. Gammad, N. Le, S.-C. Liu, K.-W. Leong, M. Zhang, J.S. Ho, J. Yoo, S.-C. Yen, A 3-mbps, 802.11 g-based EMG recording system with fully implantable 5-electrode EMGxbrk acquisition device. IEEE Trans. Biomed. Circuits Syst. **14**, 889–902 (2020)
59. I.L. Olokodana, S.P. Mohanty, E. Kougianos, M. Manzo, Towards photonic sensor based Brain-Computer Interface (BCI), in *2018 IEEE International Smart Cities Conference (ISC2)*, (IEEE, 2018), pp. 1–5
60. J. Onton, A. Delorme, S. Makeig, Frontal midline EEG dynamics during working memory. NeuroImage **27**, 341–356 (2005)
61. R. Oostenveld, T.F. Oostendorp, Validating the boundary element method for forward and inverse EEG computations in the presence of a hole in the skull. Hum. Brain Mapp. **17**, 179–192 (2002)
62. R. Oostenveld, P. Fries, E. Maris, J.-M. Schoffelen, FieldTrip: Open source software for advanced analysis of MEG, EEG, and invasive electrophysiological data. Comput. Intell. Neurosci. **2011**, 156869 (2011)
63. H. Ouyang, J. Tian, G. Sun, Y. Zou, Z. Liu, H. Li, L. Zhao, B. Shi, Y. Fan, Y. Fan, Self-powered pulse sensor for antidiastole of cardiovascular disease. Adv. Mater. **29**, 1703456 (2017)
64. H. Pan, G. Xie, W. Pang, S. Wang, Y. Wang, Z. Jiang, X. Du, H. Tai, Surface engineering of a 3D topological network for ultrasensitive piezoresistive pressure sensors. ACS Appl. Mater. Interfaces **12**, 38805–38812 (2020)
65. J. Park, S. Choi, A.H. Janardhan, S.-Y. Lee, S. Raut, J. Soares, K. Shin, S. Yang, C. Lee, K.-W. Kang, Electromechanical cardioplasty using a wrapped elasto-conductive epicardial mesh. Sci. Transl. Med. **8**, 344ra86-344ra86 (2016)
66. V. Passaro, F. Dell'olio, F. De LEONARDIS, Electromagnetic field photonic sensors. Prog. Quantum Electron. **30**, 45–73 (2006)
67. S. Roy, D. Rathee, A. Chowdhury, K. Mccreadie, G. Prasad, Assessing impact of channel selection on decoding of motor and cognitive imagery from MEG data. J. Neural Eng. **17**, 056037 (2020)

68. M. Schwartz, Electrodes and the measurement of bioelectric events. J. Clin. Eng. **2**, 169 (1977)
69. H. Sheng, J. Zhou, B. Li, Y. He, X. Zhang, J. Liang, J. Zhou, Q. Su, E. Xie, W. Lan, A thin, deformable, high-performance supercapacitor implant that can be biodegraded and bioabsorbed within an animal body. Sci. Adv. **7**, eabe3097 (2021)
70. S.H. Sunwoo, S.I. Han, H. Kang, Y.S. Cho, D. Jung, C. Lim, C. Lim, M.J. Cha, S.P. Lee, T. Hyeon, Stretchable low-impedance nanocomposite comprised of Ag–Au core–shell nanowires and Pt black for epicardial recording and stimulation. Adv. Mater. Technol. **5**, 1900768 (2020)
71. F. Tadel, S. Baillet, J.C. Mosher, D. Pantazis, R.M. Leahy, Brainstorm: A user-friendly application for MEG/EEG analysis. Comput. Intell. Neurosci. **2011**, 1 (2011)
72. B.A. Taheri, R.T. Knight, R.L. Smith, A dry electrode for EEG recording. Electroencephalogr. Clin. Neurophysiol. **90**, 376–383 (1994)
73. P. Tallgren, S. Vanhatalo, K. Kaila, J. Voipio, Evaluation of commercially available electrodes and gels for recording of slow EEG potentials. Clin. Neurophysiol. **116**, 799–806 (2005)
74. B. Vasconcelos, P. Fiedler, R. Machts, J. Haueisen, C. Fonseca, The arch electrode: A novel dry electrode concept for improved wearing comfort. Front. Neurosci. **15**, 1378 (2021)
75. L. Veeramuthu, M. Venkatesan, J.-S. Benas, C.-J. Cho, C.-C. Lee, F.-K. Lieu, J.-H. Lin, R.-H. Lee, C.-C. Kuo, Recent progress in conducting polymer composite/nanofiber-based strain and pressure sensors. Polymers **13**, 4281 (2021)
76. J.R. Wolpaw, N. Birbaumer, D.J. Mcfarland, G. Pfurtscheller, T.M. Vaughan, Brain–computer interfaces for communication and control. Clin. Neurophysiol. **113**, 767–791 (2002)
77. H. Wu, G. Yang, K. Zhu, S. Liu, W. Guo, Z. Jiang, Z. Li, Materials, devices, and systems of on-skin electrodes for electrophysiological monitoring and human–machine interfaces. Adv. Sci. **8**, 2001938 (2021)
78. J. Xu, S. Mitra, C. Van Hoof, R.F. Yazicioglu, K.A. Makinwa, Active electrodes for wearable EEG acquisition: Review and electronics design methodology. IEEE Rev. Biomed. Eng. **10**, 187–198 (2017)
79. K. Yang, L. Tong, J. Shu, N. Zhuang, B. Yan, Y. Zeng, High gamma band EEG closely related to emotion: Evidence from functional network. Front. Hum. Neurosci. **14**, 89 (2020)
80. S. Yoo, J. Lee, H. Joo, S.H. Sunwoo, S. Kim, D.H. Kim, Wireless power transfer and telemetry for implantable bioelectronics. Adv. Healthc. Mater. **10**, 2100614 (2021)
81. M. Zhang, Z. Tang, X. Liu, J. Van der Spiegel, Electronic neural interfaces. Nat. Electron. **3**, 191–200 (2020)

Index

B
Bipolar junction transistors (BJTs), 49, 57–59, 63, 91
Brain-computer interfaces (BCIs), ix, x, 241–254

C
Copper interconnect, 174–176

D
Deep learning, 199, 207, 222, 223
Device modelling, 117–121
Diodes, 30, 49, 111, 192

E
Electroencephalogram (EEG), 242–248, 250–254
Electromagnetic, ix, 3, 7–9, 11, 12, 21–30, 32, 36–38, 127–129, 131, 134–136, 139, 141, 143, 148, 191, 243, 252
Electron design automation (EDA), 108, 111, 120, 121
Electrostatic, 3–7, 11, 12, 21–23, 59, 108, 116
Energy autonomy, 21, 38, 39
Energy harvesting, 1–39
Energy release rate, 160, 162, 164–167, 171–176, 178–180
Event-based vision, 208–211

F
Flexibe electronics, vi, viii, 105–122
Fracture, ix, 159–161, 163–168, 170–176, 178–182

G
Gallium nitride (GaN), 47–91

H
High-electron mobility transistors (HEMTs), 49, 50, 76–83, 87–91
High-power-devices, 83

I
Inductive, 8, 21–23, 25, 27, 32–34, 38, 39
Insulated gate bipolar transistors (IGBTs), 48, 49, 51, 58, 63, 64, 72–75, 91

J
Junction gate field-effect transistors (JFETs), 49, 50, 59–64, 68, 69, 75, 82, 88, 91

L
Low-k materials, 169, 170

M
Mechanical properties, 10, 162, 163, 180–182
Metal oxide devices, 112
Metamaterial, ix, 128–130, 134, 141, 143–148, 151
Metasurface, ix, 127–151
Metawaveguides, 127–151
Microcrack steering, 161
Microsystems, 8, 12, 13, 20–22, 25, 27–29, 31, 35–39
MOSFET, 32, 33, 35, 36, 48–50, 58, 59, 61, 63–74, 78, 82–88, 91

F. Iacopi, F. Balestra (eds.), *More-than-Moore Devices and Integration for Semiconductors*, https://doi.org/10.1007/978-3-031-21610-7

N

Non-invasive, x, 21, 22, 143, 242–248, 254

O

Organic devices, 20
Organic light-emitting diodes (OLEDs), 111–112, 114, 121
Organic photovoltaics (OPV), 111–114, 121
OTFTs, 118, 119

P

Piezoelectric, 3, 9–12, 21, 22, 27, 29, 32, 34, 37, 89, 252, 253
Power management, 1–39
Printed electronics, viii, 105–122

R

Reliability, ix, 21, 27, 39, 47–49, 55, 57–59, 61, 68, 69, 75, 81–82, 85, 89, 91, 109, 113, 117, 121, 157–161, 170, 179–181, 241

S

Sensor, 3, 106, 191, 241
Silicon carbide (SiC), 47–91
Single-photon avalanche diode (SPAD) imagers, 192, 195, 202, 206, 215, 220, 221, 223–225, 228–230
Single-photon sensors, 192

T

Terahertz, 127–151
Thermal, ix, 13–15, 17, 18, 20, 21, 28, 37, 48, 50, 55, 57, 59, 63, 66, 67, 78, 80, 81, 83, 89–91, 111, 112, 117, 121, 159, 179, 192, 196, 200
3D flash time-of-flight (TOF) LiDAR, 194, 199, 200
3D imaging, 160, 168, 175–177, 181, 193, 195–205, 223, 230

W

Wavefront, 129, 133, 134, 137
Wearable devices, 252, 254

X

X-ray microscopy, 168–171, 174, 175, 178, 179

Printed in the United States
by Baker & Taylor Publisher Services